W9-AGV-442

DIFFERENTIAL AND INTEGRAL

Differential and Integral

A Constructive Introduction to Classical Analysis

PAUL LORENZEN

translated by

JOHN BACON

UNIVERSITY OF TEXAS PRESS

AUSTIN AND LONDON

International Standard Book Number 0-292-70114-4

Library of Congress Catalog Card Number 76-155895

AMS 1970 Subject Classifications 02E05, 02H13, 26-00, 26-02, 40A05, 53A15, 53A45, 58A05

Manufactured in the United States of America

DEDICATED

TO THE MEMORY OF

HERMANN WEYL

TRANSLATOR'S FOREWORD

I have tried to make the English edition of this book a straightforward counterpart of Professor Lorenzen's original. The translation is thus neither literal nor literary, though it is nearer to being the former. Since no two connoisseurs agree on what constitutes a good translation, it will not please everyone. My purpose will be fulfilled if I have made Professor Lorenzen's book readable in English.

To facilitate cross-reference, the page numbers of the original German edition have been printed in the margin. But for translation, the only changes in this edition are necessary corrections of the arguments on pages 78 and 154 and minor additions to pages 173 and 282.

For invaluable help with the translation, especially with the technical terms, my warm thanks go to Norman M. Martin. I am equally indebted to Christian Thiel for his thorough perusal and criticism of the original draft. Finally, Professor Lorenzen himself obligingly went over the entire translation with Dr. Thiel and me.

This edition would not have been possible but for a grant from the National Translation Center of Austin, Texas, funded by the Ford Foundation. I thank Keith Botsford and Fred Wieck, successive directors of the Center, for their cooperation and encouragement. In the preparation of copies of the final manuscript, Irwin C. Lieb came to my assistance in securing a subvention from The University of Texas Graduate School, for which I am grateful. For help in reading proofs, I thank Barbara Stocking.

John R. Silber originally suggested the translation to me and was instrumental in arranging for me to do it. This edition owes its existence to him as much as to anyone else.

John Bacon

Fordham University

PREFACE

Since the appearance of my *Einführung in die operative Logik und Mathematik* (1955), I have given a simplified account of the grounding of logic in *Metamathematik* (1962). I now put forth a grounding of analysis which is likewise considerably simplified: as was the case in H. Weyl's *Das Kontinuum* (1918), no construction of "higher" language levels is undertaken. For this reason the book is dedicated to the memory of Hermann Weyl.

For the opportunity of clarifying all of the more difficult points in the manuscript in conversations with him, I should again like to thank Mr. Otto Haupt.

For many good suggestions concerning the proofs I am grateful to Messrs. G. Pickert, K. Schütte, K. Lorenz, and E. Wette.

My thanks are also due to the publishers for their obliging cooperation.

Paul Lorenzen

University of Erlangen-Nuremberg
17 November 1964

CONTENTS

Translator's Foreword vii

Preface ix

Introduction 1

I. Foundations

§1. The Natural Numbers and the Logical Particles . . 6
§2. The Rational Numbers 15
§3. Sets and Functions 20
§4. Inductive Definitions. 29

II. Real Numbers

§5. Rational Sequences 42
§6. Definition of the Real Numbers 54
§7. The Completeness of the Field of Real Numbers . . 60

III. One-Place Functions

§8. Continuous Functions 71
§9. The Elementary Functions. 83
§10. Sequences of Functions 98
§11. Differentiation. 114
§12. Taylor Series 131
§13. Integration 142

IV. Many-Place Functions

§14. The N-Dimensional Number Space 155
§15. Continuous and Differentiable Functions 163
§16. Mappings 172
§17. Multiple Differentiation and Integration 183

V. Differential Geometry

§18. Analysis and Geometry 199
§19. Curves, Surfaces, and Figures 203
§20. Tensors and Differential Forms 212
§21. Alternating Multiplication and Differentiation . . . 232
§22. Multidimensional Integrals 243
§23. Metric Differential Geometry. 273

Bibliography 293

Index of Symbols 295

Index of Subjects 297

Index of Names 303

DIFFERENTIAL AND INTEGRAL

INTRODUCTION

This introduction to classical analysis aims to present the differential and integral calculus in accordance with the demands which must currently be placed on a rigorous mathematical treatment. The essential content of classical analysis, as it developed from the seventeenth through the nineteenth centuries, comprises in my view the following:

1. The representation of all elementary functions with the help of Taylor series.
2. The reduction of integration to the inversion of differentiation by means of the Fundamental Theorem.
3. The reduction of multidimensional to one-dimensional integrals by means of the differential calculus and the Stokes formula.

In order to secure this classical content, it has been the custom for some decades to use an "axiomatic foundation." The construction of real numbers and real functions (which are the objects of classical analysis) has given way to an axiom system. The question whether there is a model for the axiom system is banished from mathematics. Moreover, since the axiom system is formulated with the help of the concept of set, either the "axiomatic foundation" of analysis stays at a naïve level, or an axiom system for set theory has to be added. This second alternative has the advantage of giving a single axiom system, from which inferences can be drawn solely according to the rules of the elementary logical calculus. Every assertion of classical analysis is thereby transformed into an assertion of the deducibility of a particular formula in a particular calculus.

That is an advantage insofar as machines are capable of checking the proofs of such assertions (viz., deductions in some calculus). However, the transformation of classical analysis into a calculus has the disadvantage that the choice of the calculus can only be justified pragmatically. It is claimed that this is the way modern science has arisen—with the help of the chosen calculus—and that this is why we want to work with it.

2 It is further claimed, however, if often only tacitly, that this pragmatic justification is the *sole* possibility of justification, unless the mathematician should wish to venture upon the slick ice of metaphysics. Such an assertion is itself nothing but unfounded metaphysics—or at any rate an unfounded claim.

The present account aims to show that neither the metaphysics of pragmatism nor any sort of vulgarized-Platonic speculations about the existence of mathematical objects "in themselves" is needed for classical analysis.

The point of departure is not therefore an axiom system for real numbers (or for sets); rather, a step-by-step construction of the natural numbers, the real numbers, and sets of these is carried out. This use of constructions for the grounding of analysis is quite compatible with the *axiomatic method* as currently practiced by Bourbaki. Once the objects of analysis have been constructed, the deductive connections among sentences concerning them are best clarified by taking the objects together with some of the relations defined among them as models of suitable structures (fields, lattices, topological spaces, etc.).

Thus the axiomatic method—unless it is misconceived as mere formalism—always presupposes the construction of models. The present constructive introduction lays stress entirely on the construction of the objects and on a first proof of the theorems which make the "discovery" of structures in analysis possible.

For elementary arithmetic the controversy over a constructive versus an axiomatic foundation (by means of the Peano axiom system) is in essence only a verbal controversy. Since no formalist denies being able to understand statements about the deducibility of formulas in a calculus, he is still less in a position to deny understanding statements like $2 \cdot 2 = 4$. For this involves nothing else than the deducibility of a formula in a very simple calculus, that of the multiplication table. A difference between constructivism and formalism arises only in the case of logically complex statements, say "$m \cdot n = n \cdot m$ for all m, n." The formalist interprets this statement as the assertion of the deducibility of a formula, namely $\bigwedge_{m,n} m \cdot n = n \cdot m$ in the Peano formalism. As to the source of the Peano axioms he keeps quiet. The constructivist on the other hand continues to interpret the statement as a statement about multiplication. He proves it by giving a procedure for getting a deduction of the formula $n \cdot m = p$ out of a deduction of the formula $m \cdot n = p$.

3 Since, after all, any mathematician can see that a deduction of the formula $\bigwedge_{m,n} m \cdot n = n \cdot m$ in the Peano formalism can always be read as "giving a procedure" (for getting from $m \cdot n = p$ to

$n \cdot m = p$), there remains for elementary arithmetic a difference only in the way of talking about arithmetic, not in the arithmetic itself.

By way of modern foundational research, anyone who will can inform himself sufficiently about these matters. However, we know from experience that meticulously making every step explicit in the proof of logically compound statements is unnecessary for an initial comprehension. (The "natural" understanding of logic which one has learned with one's "natural" language is enough.) Therefore this book presupposes a knowledge of elementary logic (but no acquaintance at all with the concept of set).

Thus §1 essentially just recapitulates the construction of the natural numbers and their equality and introduces a symbolic notation for the logical particles.

The definition of the rational numbers can be carried out on this basis in the usual way by construction and abstraction. But in order to be able to define the real numbers, it is necessary first to clarify the notions of "set" and "function." Sets are obtained (in §3) by abstraction from formulas, and functions by abstraction from terms. These notions thereby become dependent upon the "indefinite" possibilities for the construction of formulas and terms (cf. §4).

The occurrence of this indefiniteness is the crucial difference of analysis from arithmetic. That is why the formalist introduces "set" as a basic undefined concept and works with axiom systems that are taken to be tried-and-tested crystallizations of naïve ways of operating with sets. Thus he works with formalized naïvetés, which at best are interpreted, as in so-called semantics, by metalinguistic naïvetés.

In place of naïve operation with sets or one of the formalizations thereof, this book offers a construction or, more precisely, a consideration of the possibilities for the construction of formulas and terms.

For those familiar with the various attempts at axiomatizing set theory, it may be remarked that the constructions used here could of course be axiomatized. The result is a ramified theory of types, but one so simplified that only two types of variables are left, definite and indefinite variables.

Since, however, there is a constructive model for the ramified theory of types—just as for the Peano axiom system—and since statements are only to be made about this model, no axiomatization has been attempted here. It was more important first of all to show how indefiniteness may be appropriately handled in the

4 proofs of the classical results concerning the completeness of the real numbers and then the continuity, differentiability, etc. of the real functions.

Not all currently conventional proofs can be taken over, but in each case certain classical methods of proof can be applied so as to uphold the results. The preservation of the classical stock of theorems (without significant modification of the proofs) by distinguishing definite from indefinite concepts—this is the essential content of the book.

Beyond that, importance is attached to a consistent symbolism, especially for differentials and differential quotients. In addition, the presentation departs from that of previous textbooks in several points, which are briefly listed here for Chapters II–IV:

1. Instead of local compactness, the analytic induction principle is everywhere employed for the real numbers (§7).
2. The elementary functions are introduced "genetically" in §9 by roughly following their historical development, without anticipating either differentiation or integration (or for that matter arc length). In place of sin and cos, two functions Sin and Cos are first defined by

$$i^{\xi} = \mathrm{Cos}\ \xi + i\ \mathrm{Sin}\ \xi,$$

   and the connection with the ancient problem of computing the length of chords is explained.
3. The mean-value theorem is everywhere replaced—in accordance with Dieudonné's example—by the *bound-theorem*: Every bound of differential quotients is also a bound of the difference quotients (§11).
4. Instead of being proved for continuous functions, integrability (the existence of a primitive function) is proved directly for jump-continuous functions (these are everywhere continuous except for denumerably many jumps: §13).
5. The N-dimensional number-space is treated as an N-fold product space. Thus norms are used only insofar as they are replaceable by topologically equivalent norms (§14).
6. As a generalization of the definition of integration $\mathcal{J}$ by

$$\mathcal{J}\mathcal{D} = \Delta$$

   ($\mathcal{D}\phi$ is the derivative of ϕ, $\Delta\phi$ the difference), the total mixed difference $\Delta^N = \Delta_1 \ldots \Delta_N$ is formed for N-place functions as an N-fold difference (with respect to all arguments, independently of the order), and similarly the total mixed derivative $\mathscr{D}^N = \mathscr{D}_1 \ldots \mathscr{D}_N$. The N-fold integral $\mathscr{J}^N$ (over intervals) can then be defined (§17) by 5

$$\mathscr{J}^N \mathscr{D}^N = \Delta^N.$$

This definition of the integral leads to the interchangeability of differentiation and integration (i.e., the Leibniz formula for differentiation with respect to one parameter), where the integrands occurring need only be integrable (not necessarily continuous: §17).

A crucial difficulty in any introduction to analysis—as every textbook shows—is a proof of the transformation formula for the N-dimensional integral. In Chapter V of this book, a new route is taken. Multidimensional integration is defined from the beginning as a geometric operation, and thus as invariant with respect to locally affine coordinate transformations. This requires first of all the introduction of the Leibniz-Cartan differential calculus, including Poincaré's theorem (§21). Differential forms are defined as alternating tensors in the tangent space (§20). $\mathscr{J}\mathscr{D} = \Delta$ is replaced by the geometric

$$\int d = \Delta$$

for curve integrals.

The generalization to multidimensional integrals is the Stokes formula for differential forms,

$$\int d = \oint$$

(where $\oint$ denotes the boundary integral).

Just as the curve integral can be defined by $\int d = \Delta$, the multidimensional integral is defined inductively according to dimension by means of the Stokes formula (§22). All recourse to the Riemann integral is thus avoided. Instead, full additivity is demonstrated for the integral employed. The appropriate domains of integration for that purpose prove to be almost open domains (which are distinguished from open sets only by a null set).

No metric is used until §23, when a metric for Riemannian spaces in general is introduced. The content tensor (ε-tensor) leads to the orthocomplement * for alternating tensors (i.e., for differential forms) and thus to codifferentiation $\delta = *d*$. These in turn yield the differential operators of vector analysis and the special metric cases of Stokes' formula for arbitrary coordinate systems.

6

I. FOUNDATIONS

§1. The Natural Numbers and the Logical Particles

We learn how to use the *natural numbers* 1, 2, 3, . . . in elementary school, employing from the outset this very decimal notation. However important this notation is in practice, it is theoretically inessential. All the theorems of arithmetic would still be formulable and provable if we had only the primitive notation of the Stone Age,

$$|,\ ||,\ |||,\ \ldots,$$

at our disposal.

The *construction* of these primitive *numerals* is easily described: Begin with |, and from a numeral n construct the successor $n|$ (i.e., by adding a |).

We symbolize these *rules of construction* by means of the notation . . . ⇒ . . ., to be read as "from . . . construct . . .":

$$\begin{array}{c} \Rightarrow | \\ n \Rightarrow n|\,. \end{array} \qquad (1.1)$$

The letter "n" here is a *variable* for numerals, i.e., a symbol for which any numeral whatsoever may be substituted. Thus it is not the numerals that are "variable," but only the various substitutions we can make. In contrast to the variables, the numerals are called "*constants*." The numerals are customarily understood as representations of *numbers*. We need not reflect here on the distinction thus drawn between numerals and numbers, since in our numerals |, ||, |||, . . . we have chosen a *faithful* representation of the numbers: different numerals represent different numbers. (The distinction is based on an abstraction, as will be described in more detail in §2.)

Numerals are defined as those *figures* which can be constructed according to the rules (1.1). Thus the following *statements* about numerals are *true* or, as we also say, "*valid*":

| is a numeral.

7 If n is a numeral, then $n|$ is a numeral.

We write these in a logically standardized formulation as

$|$ is numeral
n is numeral $\rightarrow n|$ is numeral.

And in general we use the notation $A \rightarrow B$ for "if A, then B." Statements of the form $A \rightarrow B$ (where A and B are already statements) are called *subjunctions*. A is the *antecedent* and B the *consequent* of the subjunction.

Moving now to numbers, we obtain from the above statements

$|$ is number
n is number $\rightarrow n|$ is number. (1.2)

Such statements as (1.2) involving $\rightarrow$ are something quite different from the rules (1.1) using $\Rightarrow$. Only statements can be true or false, not rules.

Understanding statements of the form "if A, then B" presupposes an understanding of the *logical particle* "if . . . , then . . ." For this purpose, the beginner can rest assured that he has already learned sufficient logic, since he has after all learned to speak.

On the basis of this "natural" logic, the following is also evident. The numerals are defined as those figures which can be constructed according to the rules (1.1). Now let $A(|), A(||), \ldots$ be any statements about numbers such that

(1) $A(|)$
(2) for all m, $A(m) \rightarrow A(m|)$.

Then it will also be the case that

(3) for all n, $A(n)$.

A new logical particle appears here: the *quantifier* "for all." We write "for all x" symbolically as $\bigwedge_x$. Thus in place of (2) and (3) we get

(2′) $\bigwedge_m .A(m) \rightarrow A(m|)$.
(3′) $\bigwedge_n A(n)$.

The assertion (3′), or more precisely the assertion of the statement (3′), can be defended as follows. If an n is put forth of which $A(n)$ is to be asserted, then there is a construction for this n in accordance with the rules (1.1):

$$|, \quad ||, \ldots, \quad m, \quad m|, \ldots, \quad n.$$

If we now write out—in a sense parallel to this—the statements

$$A(|),\quad A(||),\ldots,\quad A(m),\quad A(m|),\ldots,\quad A(n)$$

8 then the truth of each of these statements follows successively by (1) and (2), exactly as did the constructibility of n by (1.1).

We can summarize this as follows. If the statement $A(|)$ holds; and if it is true for all m that if $A(m)$ holds, $A(m|)$ does too; then $A(n)$ holds for all n.

In order to formulate this principle in standardized logical notation, we have only to introduce a symbol $\wedge$ for the *connective* "and," whereupon we get the following *principle of induction*:

$$A(|) \wedge \bigwedge_m .A(m) \to A(m|). \dot{\to} \bigwedge_n A(n). \tag{1.3}$$

A dot has been placed over the second arrow $\to$ here to indicate that the statement as a whole is to be grouped at that point into two parts, and not, say, at the $\wedge$. The statement $\bigwedge_m .A(m) \to A(m|).$ contains two dots on the line in place of the usual parentheses.

The numerals are constructed so that, like the number words, they can be used for counting. The particular way in which the strokes are drawn—whether thickly or thinly, for example—makes no difference. We regard numerals as *identical* just in case they result from identical constructions. What does that mean? It means first of all that every numeral $|$ is to count as identical or equal to every other numeral $|$. Next, it means that a one-step construction applied to equal numerals m and n will yield equal numerals in turn (viz., $m|$ and $n|$). If we use the form $m = n$ for *identity* (equality) statements, then what we are saying is that the *true* identity statements are those and only those figures (formulas) which can be constructed according to the following rules:

$$\begin{gathered} \Rightarrow | = | \\ m = n \Rightarrow m| = n|. \end{gathered} \tag{1.4}$$

On the basis of this *construction* the following statements hold:

$$\begin{gathered} | = | \\ m = n \to m| = n|. \end{gathered} \tag{1.5}$$

But it is also true that

$$m| = n| \to m = n, \tag{1.6}$$

for if $m| = n|$ is constructible, the last construction step will have to have proceeded from $m = n$. Hence $m = n$ must be constructible.

(1.5) and (1.6) together yield the result that $m = n$ holds if—and only if—$m| = n|$ holds. The connective "*if and only if*," which occurs here for the first time and for which we have no single word in English, will be symbolized from now on by $\leftrightarrow$.

9

$A \leftrightarrow B$ is thus defined as $A \rightarrow B \land B \rightarrow A$. We write such definitions in the following form (with $\leftrightharpoons$ as the sign of definition):

$$A \leftrightarrow B \leftrightharpoons A \rightarrow B \dot{\land} B \rightarrow A.$$

With the new symbol $\leftrightarrow$ we can write

$$m = n \leftrightarrow m| = n|.$$

An identity statement $m = n$ which cannot be constructed according to (1.4) is false. We then write $\neg\, m = n$ (with $\neg$ for the logical particle "*not*") or, more briefly, $m \neq n$. On this definition of *nonidentity* (inequality) $\neq$, it is, e.g., true that

$$m| \neq | \quad \text{(for all } m). \tag{1.7}$$

In order to dispute this, one would have to assert the constructibility of $m| = |$. But none of the rules of (1.4) could yield this as a final step.

Likewise we get

$$| \neq n|. \tag{1.8}$$

Although we have not *proved* the above statements by logically deducing them from others (how, after all, would that have been possible?), we must keep in mind that they are not arbitrary stipulations or so-called axioms either. The numbers and numerical equality are already defined by the rules (1.1) and (1.4). Even so, (1.2)–(1.3), (1.6)–(1.7) are customarily called the "Peano axioms."

Regarding the definition of the true identity statements as those constructible by (1.4), one further explanation is in order. Why have we called the relation thus defined between natural numbers "identity," denoting it with the usual identity sign =? Identity is a logical notion: it is defined in logic so that $x = y$ for any two objects x and y is to mean that, for any arbitrary statement A, $A(x)$ holds if and only if $A(y)$ does.

Again, we use the sign $\leftrightharpoons$ to express this definition of identity:

$$x = y \leftrightharpoons \text{for all } A,\ A(x) \leftrightarrow A(y).$$

The generalization "for all statements A" cannot be rendered by the usual quantifier $\bigwedge$, for we have to do here with something new, something that our ordinary language fails to bring out. When we use the quantifier $\bigwedge_n$ with a variable n for numbers, the *range of values* of n is circumscribed by a construction [only those numerals constructible according to (1.1) can be substituted

10 for the variable]. For the *statement variable A*, on the other hand, no "range of values" is marked out. It is left open what could count as a statement, e.g., statements from other natural languages, statements from the language of mathematicians in the year 3000, or what not.

The possibility of meaningfully employing such unrestricted or "*indefinite* variables," as we shall say technically, rests on the fact that the specification of A as a statement already suffices to enable us to assert the truth of certain statements about A. For example, $A \to A$ can be asserted. For the logical truths, we need not know what language (past, present, or future) the component statements come from. It is enough to presuppose that *the component statements are signs* (*configurations*) *that can be asserted.* What it means to "assert" a *sign*, e.g., an English sentence, is something that everyone knows who has learned to speak at all. It is an activity that is learned by practice. The distinction of this activity from others, such as mere prattling, likewise has to be learned practically.

Elementary arithmetic has no need of indefinite variables and thus no need of indefinite quantifiers (it gets by with the ordinary ones, which from now on we shall call "*definite*"), whereas a characteristic of infinitesimal mathematics, or so-called analysis, is that the use of indefinite quantifiers is required too (cf. §4). We shall write the indefinite universal quantifier as a boldface **⋀**.

We can then symbolize the definition of identity as

$$x = y \leftrightharpoons \boldsymbol{\bigwedge}_A . A(x) \leftrightarrow A(y). \tag{1.9}$$

The justification for calling the relation defined by means of (1.4) "identity" stems from the fact that we can still exercise control over what is to count as an *arithmetical* statement about the numerals constructed in accordance with (1.1). We want to admit only those statements A for which the following "principle of identity" holds:

$$m = n \mathrel{\dot{\to}} A(m) \leftrightarrow A(n).$$

That means that the statements A are to be *invariant* under the replacement of a numeral m by another numeral n which is "identical" with m in the sense of (1.4). Thus, e.g., statements about the thickness or the color of a stroke | are not arithmetical statements about the numeral |. This limitation on arithmetical statements serves to restrict our attention to those aspects of numerals which are essential for counting. Thus we speak of the numerals "as numbers." To speak of a numeral "as a number" means precisely to restrict oneself to invariant statements about numerals. Under the restriction of A to arithmetical statements the following now holds:

$$m = n \leftrightarrow \bigwedge_A . A(m) \leftrightarrow A(n). \qquad (1.10)$$

Only (1.10) together with (1.9) justifies using the identity sign in (1.4). *11*

What remains, then, is to prove (1.10). As a preliminary, we note the following so-called *laws of identity,* which hold on the basis of definition (1.4):

$$\begin{aligned} n &= n && \textit{(reflexivity)} \\ m = p \wedge n = p &\dot{\rightarrow} m = n. && \textit{(comparativity)} \end{aligned} \qquad (1.11)$$

We give the proof here in full, since it is the first occasion we have had to acquaint ourselves with a proof by the principle of induction (1.3).

In order to prove $n = n$, all we need, according to the principle of induction, are

(1) $| = |$
(2) $m = m \rightarrow m| = m|$.

Both of these premises have already been established (1.5).

The proof of comparativity is lengthier, because the statement contains three variables. Let us abbreviate it as $B(m, n, p)$, and the statement "for all p, $B(m, n, p)$" as $A(m, n)$. We will prove the following assertions:

(1) $A(|, |)$

(2) $A(m|, |)$

(3) $A(|, n|)$

(4) $A(m, n) \rightarrow A(m|, n|)$.

It is easy to see that by the principle of induction, $A(m, n)$ will thereby have been proved for all m, n.

Assertion (1) is trivial:

$$\bigwedge_p . | = p \wedge | = p \dot{\rightarrow} | = |.$$

Assertions (2)–(4) are proved by induction on the variable p. The premises for the induction of (2) follow from

(5) $m| = | \wedge | = | \dot{\rightarrow} m| = |$
(6) $m| = p| \wedge | = p| \dot{\rightarrow} m| = |$.

In both cases, in view of (1.7)–(1.8), the antecedents contain false statements ($m| = |$ and $| = p|$), so the subjunctions are true. (3) is handled similarly to (2).

(4) reads

$$A(m, n) \to \bigwedge_p B(m|, n|, p).$$

By the principle of induction, we could prove this if we had

$$\begin{array}{ll} (7) & B(m|, n|, |) \\ (8) & A(m, n) \to B(m|, n|, p|). \end{array}$$

12 (7), viz., $m| = | \wedge n| = | \stackrel{.}{\to} m| = n|$, again follows from (1.7). In place of (8), it will suffice to prove $B(m, n, p) \to B(m|, n|, p|)$, i.e.,

$$m = p \wedge n = p \stackrel{.}{\to} m = n \stackrel{..}{\to} m| = p| \wedge n| = p| \stackrel{.}{\to} m| = n|.$$

Proof. From $m| = p|$ and $n| = p|$ we get $m = p$ and $n = p$ by (1.6). These together with the antecedent yield $m = n$, from which $m| = n|$ follows by (1.5).

With that the laws of identity (1.11) are proved. From them we get the *symmetry* of identity (substituting m for p in comparativity):

$$m = n \leftrightarrow n = m. \tag{1.12}$$

As a further consequence we conclude that the statement $p = q$ is an *arithmetical* statement $A(q)$. For $A(q)$ is invariant; that is to say,

$$m = n \stackrel{.}{\to} p = m \leftrightarrow p = n,$$

and this follows directly from the laws of identity.

With that we have just about established (1.10). Let us now assume for all arithmetical statements A that

$$A(m) \leftrightarrow A(n).$$

Then, in particular, for all p,

$$p = m \leftrightarrow p = n,$$

whence also

$$m = m \leftrightarrow m = n.$$

Reflexivity then gives us the desired $m = n$.

With this proof of (1.10) we have justified the use of the identity sign in (1.4).

Addition and multiplication are operations that construct a third number out of two given numbers. For the moment let us write $\frac{m+n}{p}$ or $\frac{m \cdot n}{p}$ when p is the result respectively of *adding* or *multiplying* m and n. Then the construction rules for such statements will be

$$\left.\begin{array}{rl} & \Rightarrow \frac{m + |}{m|} \\ \frac{m+n}{p} & \Rightarrow \frac{m+n|}{p|} \\ & \Rightarrow \frac{| \cdot n}{n} \\ \frac{m \cdot n}{p} ,, \frac{p+n}{q} & \Rightarrow \frac{m| \cdot n}{q} \end{array}\right\} \quad (1.13)$$

The last rule contains two premises. We separate the premises of rules by a double comma ,, . 13

It can be proved (by simple induction, which we omit here) that these rules uniquely determine the number p in $\frac{m+n}{p}$ and $\frac{m \cdot n}{p}$. Thus we have

$$\left.\begin{array}{l} \frac{m+n}{p_1} \wedge \frac{m+n}{p_2} \mathrel{\dot{\to}} p_1 = p_2 \\ \frac{m \cdot n}{p_1} \wedge \frac{m \cdot n}{p_2} \mathrel{\dot{\to}} p_1 = p_2 \end{array}\right\} \quad (1.14)$$

In light of (1.14), we write $m + n = p$ and $m \cdot n = p$ in place of $\frac{m+n}{p}$ and $\frac{m \cdot n}{p}$.

An equation such as

$$(m_1 + n_1) \cdot (m_2 + n_2) = q$$

is then to be understood as saying that, for appropriate numbers p_1 and p_2,

$$m_1 + n_1 = p_1$$
$$m_2 + n_2 = p_2$$
$$p_1 \cdot p_2 = q.$$

Expressions that can be constructed out of the constant | and the variables with the help of + and · are called "*terms*." If we now return to the notation 1 for |, then $1 + (m + 1) \cdot (1 + 1)$ and $(m \cdot (n + 1)) \cdot p$ are examples of terms. Terms without variables are called *constant terms*.

From now on we shall use 2 as an abbreviation for 1 + 1, 3 for 2 + 1, etc.; in other words, the customary decimal notation.

We introduce the new variables $S, T, \ldots$ for terms as just defined so as to be able to formulate explicitly what an equation between arbitrary terms means. If the right-hand term is a numeral, then the following two rules suffice:

$$\begin{aligned} S = m,, T = n,, m + n = p &\Rightarrow S + T = p \\ S = m,, T = n,, m \cdot n = p &\Rightarrow S \cdot T = p. \end{aligned} \tag{1.15}$$

For arbitrary equations between terms we must add

$$S = p,, T = p \Rightarrow S = T. \tag{1.16}$$

Only now are we in a position to write down the familiar *"laws of arithmetic"* (which again we shall not prove here—for details, see E. Landau: *Foundations of Analysis*):

14

commutativity:	$m + n = n + m$	$m \cdot n = n \cdot m$
associativity:	$m + (n + p) = (m + n) + p$	
	$m \cdot (n \cdot p) = (m \cdot n) \cdot p$	
distributivity:	$(m + n) \cdot p = m \cdot p + n \cdot p.$	

Subtraction and *division*, unlike addition and multiplication, are not always applicable to the natural numbers. These operations are defined as follows:

$$\left.\begin{aligned} m - n = p &\leftrightharpoons m = p + n \\ \frac{m}{n} = p &\leftrightharpoons m = p \cdot n. \end{aligned}\right\} \tag{1.17}$$

The use of equations here is justified by the validity of

$$\left.\begin{aligned} m = p_1 + n \wedge m = p_2 + n &\dot{\rightarrow} p_1 = p_2 \\ m = p_1 \cdot n \wedge m = p_2 \cdot n &\dot{\rightarrow} p_1 = p_2. \end{aligned}\right\} \tag{1.18}$$

With that, we have mentioned all the properties of the natural numbers that will be needed in the next section for the construction of the rational numbers.

In our discussion so far, we have used several logical particles, viz., $\neg$, the *connectives* $\rightarrow$ and $\wedge$, and the *universal quantifier* $\wedge$ ($\bigwedge$ for indefinite variables). In terms of these particles, we were able to define the additional connective $\leftrightarrow$.

In the sequel we shall use two more logical particles. From the standpoint of classical logic, they may be defined as follows:

$$A \vee B \leftrightharpoons \neg . \neg A \wedge \neg B.$$

$$\bigvee_x A(x) \leftrightharpoons \neg \bigwedge_x \neg A(x).$$

$\vee$ is a connective which may be rendered as "or." $\bigvee_x A(x)$ is read in classical logic as "for some x, $A(x)$" or, especially in mathematical contexts, as "there is an x such that $A(x)$" or "there exists an x such that $A(x)$." $\bigvee$ is therefore called the *existential quantifier*. With indefinite variables we write **V**.

As convenient abbreviations we introduce the restricted quantifiers. In place of $\bigwedge_x . C(x) \rightarrow A(x)$. we write $\bigwedge_{x \atop C(x)} A(x)$, read as "for all x such that $C(x)$, $A(x)$." For $\bigvee_x . C(x) \wedge A(x)$. we write $\bigvee_{x \atop C(x)} A(x)$, read as "for some x such that $C(x)$, $A(x)$."

The classical negation rules

$$\neg \bigwedge_x A(x) \leftrightarrow \bigvee_x \neg A(x)$$

$$\neg \bigvee_x A(x) \leftrightarrow \bigwedge_x \neg A(x)$$

hold for the restricted quantifiers too:

$$\neg \bigwedge_{x \atop C(x)} A(x) \leftrightarrow \bigvee_{x \atop C(x)} \neg A(x)$$

$$\neg \bigvee_{x \atop C(x)} A(x) \leftrightarrow \bigwedge_{x \atop C(x)} \neg A(x).$$

In this book we shall use the logical particles according to the rules of so-called *classical logic*. For the beginner this means that he can stick with the uncritical use of the logical particles that is familiar to him from ordinary language. To be sure, that means not taking into account the fact that in mathematics we are no longer dealing with "real" objects, but with infinitely many merely constructible objects.

Our treatment can be called "constructive" even so, since classical logic can itself be justified from the standpoint of a *constructive logic* by way of so-called consistency proofs (cf. Lorenzen 1962). Furthermore, for indefinite quantifiers we shall use only constructively valid inferences.

§2. The Rational Numbers

The general applicability of division is achieved by introducing a new kind of number, the positive rational number. For this purpose we use a method that is fundamental to all our theoretical thinking, the method of *abstraction*. Since the real numbers to be introduced in Chapter II likewise rest on abstractions, let us see briefly what the essentials of the method are.

By constructing expressions of the form $\frac{m}{n}$, i.e., the so-called *fractions* (with numerator m and denominator n), we gain nothing, for the fractions $\frac{1}{2}$ and $\frac{2}{4}$ are evidently different. The fractions $\frac{2}{1}$ and $\frac{4}{2}$, on the other hand, are terms representing the same number (i.e., $\frac{2}{1} = p$ and $\frac{4}{2} = p$ hold for some p, namely 2), whence by (1.6) the equation $\frac{2}{1} = \frac{4}{2}$ holds. But how can we get to the equation $\frac{1}{2} = \frac{2}{4}$, when these fractions are not terms for numbers at all? We observe that, if two fractions $\frac{m_1}{n_1}$ and $\frac{m_2}{n_2}$ *are* numerical terms, then

$$\frac{m_1}{n_1} = \frac{m_2}{n_2} \leftrightarrow m_1 \cdot n_2 = m_2 \cdot n_1. \tag{2.1}$$

16 To prevent misunderstandings, let us for the time being use $//$ instead of the fraction bar. We now "construe" or *construct* figures of the form $m//n$ as fractions. For these we define a relation $\sim$ ("*equivalent*") as follows:

$$m_1//n_1 \sim m_2//n_2 \leftrightharpoons m_1 \cdot n_2 = m_2 \cdot n_1. \tag{2.2}$$

Equivalent fractions can be different. We now proceed to *abstract* from such differences. We do so by restricting ourselves to those statements about fractions whose validity remains unchanged when a fraction is replaced by an equivalent one. Such statements we shall call "*invariant*" (with respect to $\sim$).

An example of invariant statements are those of the form "$m//n$ is whole," given the definition

$$m//n \text{ is whole} \leftrightharpoons \bigvee_p m = p \cdot n.$$

The invariance in this case consists in the (easily proved) validity of

$$m_1 \cdot n_2 = m_2 \cdot n_1 \dot{\rightarrow} m_1 = p \cdot n_1 \leftrightarrow m_2 = p \cdot n_2.$$

In general, a statement $A(m//n)$ is invariant with respect to $\sim$ if

$$m_1 \cdot n_2 = m_2 \cdot n_1 \dot{\rightarrow} A(m_1//n_1) \leftrightarrow A(m_2//n_2).$$

To bring out the invariance explicitly, we use the new notation $A(m/n)$ when $A(m//n)$ is invariant. Now we can assert, e.g., "1/2 is not whole."

We write m/n for $m//n$ only when we are dealing with invariant statements (with respect to $\sim$). This *restriction* of statements to invariant ones is what we call "*abstraction.*" Since all statements admissible under this restriction which hold for m_1/n_1 also hold for m_2/n_2, provided only that $m_1//n_1 \sim m_2//n_2$, we are now justified in calling m_1/n_1 and m_2/n_2 "identical" or "equal." By "*equality*" we mean, after all, interchangeability in (admissible) statements. Writing again "=" for this kind of equality (for the natural numbers, too, equality means that what is true of a number is true of all equal ones), we get

$$m_1//n_1 \sim m_2//n_2 \rightarrow m_1/n_1 = m_2/n_2 . \tag{2.3}$$

To show that the converse of (2.3) holds, we first establish that the statement $m_1//n_1 \sim m_2//n_2$ is itself invariant with respect to $\sim$. The statement contains two arguments, so invariance must hold for each argument; i.e.,

$$m_1//n_1 \sim m_1'//n_1' \dot{\rightarrow} m_1//n_1 \sim m_2//n_2 \leftrightarrow m_1'//n_1' \sim m_2//n_2$$

$$m_2//n_2 \sim m_2'//n_2' \dot{\rightarrow} m_1//n_1 \sim m_2//n_2 \leftrightarrow m_1//n_1 \sim m_2'//n_2'.$$

It is easy to see that these invariances are equivalent to the symmetry and transitivity of $\sim$; i.e.,

$$\left.\begin{array}{r} m_1//n_1 \sim m_2//n_2 \rightarrow m_2//n_2 \sim m_1//n_1 \\ m_1//n_1 \sim m_2//n_2 \wedge m_2//n_2 \sim m_3//n_3 \dot{\rightarrow} m_1//n_1 \sim m_3//n_3. \end{array}\right\} \tag{2.4}$$

17

The statements (2.4) in turn are easily verified in view of (2.2).

The relation $\sim$ is also reflexive, i.e.,

$$m//n \sim m//n. \tag{2.5}$$

Abstraction of the sort used here to move from the fractions $m//n$ to the terms m/n (which are now called "terms for *positive rational numbers*") is similarly applicable wherever we have a reflexive and comparative relation. Such relations are called "*equivalence relations*." Every equivalence relation permits a transition from the objects for which it is defined to new "*abstract objects*." This transition or *abstraction* is carried out as above by *restricting* statements about the original objects to those statements which are *invariant* with respect to the equivalence relation.

Abstraction is not a mental but rather a logical process, i.e., an operation on statements.

For the positive rational numbers m/n we have obtained

$$m_1/n_1 = m_2/n_2 \leftrightarrow m_1 \cdot n_2 = m_2 \cdot n_1. \tag{2.6}$$

From now on we shall use the new variables $\delta, \varepsilon, \ldots$ for the positive rational numbers. Thus only terms of the form m/n may be substituted for these variables.

We assume familiarity with the definitions of *addition* and *multiplication* (and the invariance thereof) for positive rational numbers; viz.,

$$m_1/n_1 + m_2/n_2 \leftrightharpoons (m_1n_2 + m_2n_1)/n_1n_2$$

$$m_1/n_1 \cdot m_2/n_2 \leftrightharpoons m_1m_2/n_1n_2.$$

The same *laws of arithmetic* hold as for the natural numbers. If we then reintroduce *division* as the inverse of multiplication,

$$\frac{\delta_1}{\delta_2} = \varepsilon \leftrightharpoons \delta_1 = \varepsilon \cdot \delta_2, \tag{2.7}$$

the operation is now *unlimitedly applicable*:

$$\bigwedge_{\delta_1,\delta_2} \bigvee_{\varepsilon} \ \delta_1 = \varepsilon \cdot \delta_2.$$

Addition and multiplication of positive rational numbers are so defined that

$$m/1 + n/1 = (m + n)/1$$

$$m/1 \cdot n/1 = mn/1.$$

18 Since

$$m/1 = n/1 \leftrightarrow m = n$$

also holds, we can—without confusion—write $m/1$ more briefly as m.

Furthermore, we get $m/n = \frac{m}{n}$ $\left[\text{since by (2.7) } m/n = \frac{m/1}{n/1}\right]$, so that the distinctive fraction bars become superfluous. We therefore write *positive rational numbers* from now on in the form $\frac{m}{n}$.

Subtraction is not unlimitedly applicable to the positive rational numbers. If $\varepsilon + \delta_1 = \delta_2$ holds for some ε (in which case $\varepsilon = \delta_2 - \delta_1$ is uniquely determined), then we write $\delta_1 < \delta_2$; i.e.,

$$\delta_1 < \delta_2 \leftrightharpoons \bigvee_{\varepsilon} \ \varepsilon + \delta_1 = \delta_2. \tag{2.8}$$

The relation $<$ is called the *natural ordering relation* on the positive rational numbers. For $\delta_2 < \delta_1$ we also use $\delta_1 > \delta_2$:

$$\delta_1 > \delta_2 \leftrightharpoons \delta_2 < \delta_1. \tag{2.9}$$

The transition from the positive rational numbers to the rational numbers in general serves to make subtraction unlimitedly applicable. This transition is accomplished without abstraction. It suffices to adopt 0 and $-\delta$, $-\varepsilon$, . . . as "new" numbers in addition to the positive rationals δ, ε, . . . This presents no problem.

All it means is that along with constant terms for positive rational numbers we will now use these terms with a prefixed minus sign -, and also 0, as additional constants. The appellation "constants for rational numbers" for these terms is justified by laying down an appropriate equality relation and arithmetical operations for such constants; viz.,

$$\left.\begin{array}{l} 0 = 0 \\ 0 \neq \varepsilon \\ 0 \neq -\varepsilon \\ \delta \neq -\varepsilon \\ -\delta = -\varepsilon \leftrightarrow \delta = \varepsilon \\ (-\delta) + \varepsilon = \varepsilon + (-\delta) = \begin{cases} \varepsilon - \delta & \text{if} \quad \delta < \varepsilon \\ 0 & \text{if} \quad \delta = \varepsilon \\ -(\delta - \varepsilon) & \text{if} \quad \delta > \varepsilon \end{cases} \\ (-\delta) + (-\varepsilon) = -(\delta + \varepsilon) \\ (-\delta) \cdot \varepsilon = \delta \cdot (-\varepsilon) = -(\delta \cdot \varepsilon) \\ (-\delta) \cdot (-\varepsilon) = \delta \cdot \varepsilon . \end{array}\right\} \tag{2.10}$$

If we use $r, s, \ldots$ for the rational numbers (i.e., for 0, δ, or $-\delta$), then calculations involving 0 are determined by *19*

$$\begin{aligned} 0 + r &= r + 0 = r \\ 0 \cdot r &= r \cdot 0 = 0. \end{aligned} \tag{2.11}$$

The laws of identity (equality), addition, and multiplication all hold for the rational numbers, as well as the *unlimited applicability* of subtraction and the applicability of division $\frac{r}{s}$ with the sole limitation that $s \neq 0$.

The *ordering relation* $<$ is defined by

$$r < s \leftrightharpoons \bigvee_{\delta} r + \delta = s. \tag{2.12}$$

This ordering of the rational numbers is subject to the following *order laws*:

totality	$r \neq s \dot{\rightarrow} r < s \vee s < r$
transitivity	$r < s \wedge s < t \dot{\rightarrow} r < t$
monotony	$\begin{cases} r < s \rightarrow r + t < s + t \\ r < s \wedge 0 < t \dot{\rightarrow} r \cdot t < s \cdot t. \end{cases}$

Again we write $r > s$ if and only if $s < r$, and also

$$r \leqslant s \leftrightharpoons r < s \vee r = s$$

$$r \geqslant s \leftrightharpoons s \leqslant r.$$

For the *ordering relation* $\leqslant$ the following principles clearly hold:

$$r \leqslant s \vee s \leqslant r$$

$$r \leqslant s \wedge s \leqslant r \dot{\to} r = s$$

$$r \leqslant s \wedge s \leqslant t \dot{\to} r \leqslant t$$

$$r \leqslant s \to r + t \leqslant s + t$$

$$r \leqslant s \wedge 0 \leqslant t \dot{\to} r \cdot t \leqslant s \cdot t.$$

The absolute value $|r|$ of r is defined as follows:

$$|r| \leftrightharpoons \begin{cases} \delta & \text{if} \quad r = \delta \\ 0 & \text{if} \quad r = 0 \\ \delta & \text{if} \quad r = -\delta. \end{cases}$$

We then have

$$|r + s| \leqslant |r| + |s|$$

and

$$|r \cdot s| = |r| \cdot |s|.$$

Looking back, we see that by construction the rational numbers $r, s, \ldots$ include

1. the *natural* numbers $m, n, \ldots$
2. the *whole* numbers or *integers* (i.e., the natural numbers together with 0 and the *negative integers* $-m, -n, \ldots$)
3. the positive rational numbers $\delta, \varepsilon, \ldots$

§3. Sets and Functions

In the preceding sections, we have obtained the natural numbers and the rational numbers by an interplay of construction and abstraction. To arrive at the real numbers, which are required for differentiation and integration, we shall have to apply both methods several times more. For the time being, though, we shall stay with the rational numbers and use both methods to introduce sets and functions. These two concepts are of crucial importance for analysis. As we shall see, they are mutually definable in terms of one another, although the concept of set is in a certain sense superior to that of function, in that *for constructing sets we have the logical particles at our disposal.*

The concept of function, on the other hand, is more immediately accessible: it does not presuppose any logic. For this reason we shall begin by introducing functions.

If r, s, . . . are variables for rational numbers, then we can form "*rational terms*" as follows:

1. the variables r, s, . . . are rational terms;
2. if S, T are rational terms, so are $S + T$, $S \cdot T$, $\frac{S}{T}$;
3. if $S(r)$ is a rational term containing the variable r, then $S(c)$ is a rational term, where c is a rational constant.

As abbreviations we write r^2 for $r \cdot r$, r^3 for $r \cdot r \cdot r$, etc. Thus $\frac{r^2 + 1}{r^2 - 1}$ and $\frac{1}{r - 1}\left(\frac{1}{r + 1} + \frac{r^2}{r + 1}\right)$ are examples of rational terms.

These two terms are of course different, yet when "r" is replaced by a rational-number constant (other than ± 1), both terms become constants for the same rational number.

Thus for

$$S(r) \leftrightharpoons \frac{r^2 + 1}{r^2 - 1} \quad \text{and} \quad T(r) \leftrightharpoons \frac{1}{r - 1}\left(\frac{1}{r + 1} + \frac{r^2}{r + 1}\right)$$

we have $S(c) = T(c)$ for all rational numbers $c \neq \pm 1$.

21

In general, let us call two terms $S(r)$ and $T(r)$ containing exactly one variable r "*equivalent*" if for all numbers c (with restrictions when necessary to rule out division by 0),

$$S(c) = T(c).$$

Then this relation between terms is clearly an equivalence relation, i.e., reflexive and comparative.

The abstract objects which arise by abstraction from the difference between equivalent terms are *functions* (in this case, rational functions of rational numbers). *Functions are abstracted from terms.*

Thus we say that two terms *represent* the same function if they are equivalent. In denoting the function represented by the term $T(r)$, we should make reference to the variable r for which constants are to be substituted. Following Peano, who used the symbol ℩ in a similar connection, let us write ℩$_r T(r)$ for the function represented [by the term $T(r)$ with respect to the variable r]. We can read this as "*the function of r abstracted from $T(r)$*" or simply "the function from r to $T(r)$."

℩$_r(r + 1)$ and ℩$_r(1 + r)$ thus denote the same function. If by way of abbreviation we refer to the function ℩$_r T(r)$ represented by the term $T(r)$ as "f", we are now in a position to say what we mean by the "*value*" of a function f for the "*argument*" c: the value of the function for c is none other than the number $T(c)$. Thus, e.g., the value of the function $\daleth_r \frac{r^2+1}{r^2-1}$ for the argument 3 is the number $\frac{3^2+1}{3^2-1}$, i.e., $\frac{10}{8}$ or $\frac{5}{4}$. The value of a function f for an argument c does not depend on which particular term is used to represent the function: if $f = \daleth_r T(r)$ and $f = \daleth_r S(r)$, then by our definition $S(c) = T(c)$ must always be the case. Thus, take any term representing f and substitute c for the variable in the term. The resulting constant term denotes the desired value of the function f for the argument c.

The value of the function is usually written as "$f(c)$." This notation goes back to the beginnings of the infinitesimal calculus with Leibniz and Bernoulli. However, it has the disadvantage of obscuring the distinction between the term $T(c)$, the function f, and the value of the function f for c. Similarly, one frequently speaks of the function $f(x)$, using an appropriate variable x, rather than of the function f.

In order to make quite clear that for us it is a matter of a function f and a number c (as argument) determining a new num-

22 ber, the value, let us use "$f \daleth c$" to denote the value of the function, or simply "fc" when no misunderstanding is possible. Thus the inverted iota—this time without an attached variable—is used just as is, say, the multiplication sign. "$a \cdot c$" denotes a number, and "$f \daleth c$" likewise. The only difference is that "a" denotes a number and "f" a function.

These definitions lead to the following result:

$$f = \daleth_r T(r) \leftrightarrow \bigwedge_c fc = T(c). \tag{3.1}$$

For two functions f and g we define their *composition* $f \daleth g$ by

$$f \daleth g \,\dot{\daleth}\, c \leftrightharpoons f \,\dot{\daleth}\, g \daleth c.$$

With rational terms we can represent the *rational functions*. In §4 it will be our task to investigate how to construct other kinds of terms, which in turn will permit the representation of new kinds of functions.

In defining the general concept of function, it is usual to leave open what sort of terms may be taken to represent a function. One says something to the effect that any "rule" which correlates certain numbers x with unique numbers y represents a function, and one adds that this rule need not be a "mathematical formula."

Such (regrettably widespread) formulations as "If to each number x there corresponds a (definite) number y, then y is called a function of x and we write $y = f(x)$" indeed avoid any mention of a "rule." On the other hand, such a formulation also says something wholly unintended: that y (i.e., a *number*, though one knows not which) rather than f is the function.

Still other attempts to define the idea of a function without recourse to terms (or "rules") run as follows: "If each number x is correlated with exactly one number y, then the correlation is called a function."

The defect of this would-be definition is that it presupposes an understanding of the concept of correlation.

If, however, we wish to define the concept of correlation, we are led back to the set concept, so that ultimately the concept of a function is based upon that of a set.

To explain this connection, we must next define what is to be understood by "set" (without, of course, using the concept of function). The definition of sets proceeds analogously to the definition of functions. In the latter case we started with terms; now we start instead with *formulas*. As far as we have developed the theory of the rational numbers, we have as formulas primarily equations and inequalities between (rational) terms; e.g., *23*

$$r + 1 < r^2 + s, \quad r^3 - 1 = 0, \quad 2 \cdot 2 = 4.$$

The essential feature of such formulas is not that they consist of mathematical symbols, but that they are statements. Whether we write "$2 \cdot 2 = 4$" or "two times two is four," both are statements. By a statement we mean any sign that can meaningfully be asserted (cf. §1). For present purposes, equations and inequalities of the sort already introduced will do as elementary statements. Let us furthermore agree that all logical compounds of elementary statements, i.e., combinations using the logical particles, as for example

$$2 \cdot 2 = 4 \wedge 2 \cdot 3 = 5, \qquad \bigwedge_r 1 \cdot r = r,$$

shall likewise be admitted as statements.

Not just true statements are to be admitted, of course, but also false ones (as well as statements that are neither true nor false, should there be such a thing). We use as logical particles the connectives already introduced,

$$\wedge, \quad \vee, \quad \rightarrow, \quad \leftrightarrow,$$

as well as the quantifiers

$$\bigwedge, \quad \bigvee$$

and the negation sign $\neg$.

Statements are distinguished from other formulas by the fact that they contain no *free variables*, i.e., variables which could be replaced by constants. In this respect statements resemble constant terms.

If any constants in a statement are replaced by variables, it becomes a *statement form*; e.g.,

$$r^2 < 1, \quad r + s = s + r, \quad \bigvee_r r < s.$$

These are to be distinguished from statements (from here on).

Statements and statement forms are together called "*formulas*." Since the only elementary formulas we have so far are the equations and inequalities between terms, all other formulas are logically compounded from these elementary formulas.

Only after these preliminaries are we in a position to introduce the set concept. It arises through a process of abstraction corresponding to that which took us from terms to functions.

Let $x, y, \ldots$ be variables for objects of some sort (e.g., natural numbers or rational numbers).

24 We consider two formulas $A(x)$ and $B(x)$ containing just one *free* variable (i.e., one replaceable by constants, and thus not "bound" by a quantifier).

Two such formulas are called equivalent if all objects c in the range of x satisfy

$$A(c) \leftrightarrow B(c).$$

For example, $r > 1$ and $1 < r$ are trivially equivalent. This relation between formulas is reflexive and comparative, and hence an equivalence relation. The abstract objects obtained by abstracting from the difference between equivalent formulas are called "*sets*."

Thus we say that two formulas represent the same set if and only if they are equivalent. Our notation for the set represented by the formula $A(x)$ should again make reference to the variable x. For this purpose we shall use the Greek letter ϵ (again following Peano). We denote the set represented by $A(x)$ with respect to x as $\epsilon_x A(x)$, read as "the set of xs abstracted from $A(x)$" or simply "the set of xs such that $A(x)$."

Just as we defined the number $f \,\iota\, c$ as $T(c)$ for $f = \iota_x T(x)$, so we now define the statement $c \in K$ for a set $K = \epsilon_x A(x)$ by means of $A(c)$:

$$K = \epsilon_x A(x) \leftrightarrow \bigwedge_c . c \in K \leftrightarrow A(c). \tag{3.2}$$

The objects c of which $c \in K$ is true are called the *members* of K.

With n as a variable for natural numbers, e.g., $K = \epsilon_n(n$ is prime) is the set of prime numbers. It will be the case that

$$n \in K \leftrightarrow n \text{ is prime}.$$

The prime numbers are the members of this set K.

A set like $\epsilon_n(n = 1 \vee n = 2 \vee n = 3)$ is called finite. Its members are 1, 2, and 3.

A *finite set* with members $c_1, c_2, \ldots, c_n$ is denoted briefly by $\{c_1, c_2, \ldots, c_n\}$. Thus we have

$$c \in \{c_1, c_2, \ldots, c_n\} \leftrightarrow c = c_1 \vee c = c_2 \vee \ldots \vee c = c_n.$$

Every finite set can be represented simply as a system of objects $c_1, \ldots, c_n$. For two systems $a_1, \ldots, a_m$ and $b_1, \ldots, b_n$,

$$\{a_1, \ldots, a_m\} = \{b_1, \ldots, b_n\}$$

holds just in case the two systems arise from one another by permutation and gemination or contraction (i.e., the replacement of $\ldots, c, \ldots$ by $\ldots, c, c, \ldots$ or vice versa).

This equivalence relation between systems of objects gives rise to the "natural" concept of the finite set referred to in ordinary language—though usually not by the term "set." We say, e.g., "The number of the Apostles was 12" rather than "The set of Apostles was 12-membered." 25

Only for infinite sets is abstraction from formulas necessary.

Since sets are abstracted from formulas, and formulas are built up by means of logical particles, corresponding operations arise for sets. We write $x \notin K$ for $\neg\, x \in K$ and define first of all for every set K a *complement* $\llcorner K$:

$$\llcorner K \rightleftharpoons \epsilon_x x \notin K.$$

For two sets K_1, K_2 we define

$$K_1 \cap K_2 \rightleftharpoons \epsilon_x . x \in K_1 \wedge x \in K_2. \quad (\textit{intersection})$$
$$K_1 \cup K_2 \rightleftharpoons \epsilon_x . x \in K_1 \vee x \in K_2. \quad (\textit{union})$$
$$K_1 \llcorner K_2 \rightleftharpoons \epsilon_x . x \in K_1 \wedge x \notin K_2. \quad (\textit{difference})$$
$$K_1 \sqcup K_2 \rightleftharpoons K_1 \llcorner K_2 \cup K_2 \llcorner K_1. (\textit{Boolean sum})$$

For a sequence $K_* = K_1, K_2, K_3, \ldots$ [a sequence arising from the formula $A(x, n)$ by $K_n = \epsilon_x A(x, n)$] intersection and union can be generalized:

$$\bigcap K_* = \bigcap_n K_n = \epsilon_x \bigwedge_n x \in K_n$$
$$\bigcup K_* = \bigcup_n K_n = \epsilon_x \bigvee_n x \in K_n.$$

$\bigcap_{i\,1}^{\;n} K_i$ and $\bigcup_{i\,1}^{\;n} K_i$ are similarly defined for finitely many sets $K_1, \ldots, K_n$.

If $\curlywedge$ is a false statement, then $\cap \leftrightharpoons \in_x \curlywedge$ is called the *empty set*: for no x is $x \in \cap$.

Two sets K_1, K_2 are called *disjoint* if $K_1 \cap K_2 = \cap$.

K_1 is a *subset* of K_2 (or *included* in K_2) if $K_1 \llcorner K_2 = \cap$; i.e., if for no x is

$$x \in K_1 \wedge x \notin K_2.$$

We then write $K_1 \subseteq K_2$. It clearly holds that

$$K_1 \subseteq K_2 \longleftrightarrow \bigwedge_x . x \in K_1 \rightarrow x \in K_2.$$

and

$$K_1 = K_2 \longleftrightarrow K_1 \subseteq K_2 \wedge K_2 \subseteq K_1.$$

We define

$$K_1 \subset K_2 \leftrightharpoons K_1 \subseteq K_2 \wedge K_1 \neq K_2.$$

If $K_1 \subset K_2$, we say that K_1 is a *proper subset* of K_2.

The elementary part of set theory, which only contains theorems about these operations, is nothing but elementary logic in another notation.

26 A *generalization of the set concept* is achieved by also admitting formulas for abstraction which contain more than one free variable. A simple example is $r < s$. We form $\in_{r,s}(r < s)$, "the set of rs and ss such that $r < s$." Here we have two variables, whence the set is called a 2-place set. *Many-place sets* are customarily called "*relations*"—thus in §2 we referred to $\in_{r,s}(r < s)$ as an "ordering relation." An example of a 3-place relation is $P = \in_{l,m,n}(l^2 + m^2 = n^2)$, the set of Pythagorean triples l, m, n. Thus we can write $l, m, n \in P$ instead of $l^2 + m^2 = n^2$. An n-place relation is a set with n-place *member systems* (so-called $n$tuples) as members:

$$x_1, \ldots, x_n \in K \longleftrightarrow A(x_1, \ldots, x_n)$$

$$\text{for } K = \in_{x_1, \ldots, x_n} A(x_1, \ldots, x_n).$$

For 2-place relations $R = \in_{x,y} A(x, y)$ in particular we can abbreviate $x, y \in R$ as xRy ($r < s$ can be construed as such an abbreviation).

With many-place sets or relations in hand, we have also reached "correlations." For "*correlation*" is nothing but another term for a 2-place relation. The above definition of functions in terms of correlations can now be formulated as follows:

A 2-place relation R for which

(1) $\bigwedge_x \bigvee_y yRx$

(2) $$y_1Rx \wedge y_2Rx \mathrel{\dot{\rightarrow}} y_1 = y_2 \tag{3.3}$$

is called a function.

Since by assumption here a y is uniquely determined by yRx for every x, y can indeed be construed as the value of the function for the argument x. The term from which we abstract the function to be defined is formed colloquially with the *definite article*: "the y such that yRx."

We write this symbolically (again following Peano) as

$$\iota_y yRx. \tag{3.4}$$

Such terms are called *description terms* or ι-terms.

We insert here the observation that the ι-terms include in particular μ-*terms*. These are used as follows. If $A(n)$ is a statement form with one natural-number variable n, and if there is any n at all such that $A(n)$, then the expression "the smallest n such that $A(n)$" denotes a unique natural number. We can denote that natural number formally by the following ι-term:

$$\iota_n . A(n) \wedge \bigwedge_m . m < n \rightarrow \neg A(m). .$$

We write this term for short as $\mu_n A(n)$.

An expression of the form $\iota_y A(y)$ is a description term if and only if

(1) $\bigvee_y A(y)$

27

(2) $A(y_1) \wedge A(y_2) \mathrel{\dot{\rightarrow}} y_1 = y_2$;

otherwise, let us call it a *pseudodescription.*

That a μ-term is always a genuine description under the condition that $\bigvee_n A(n)$ follows [since (2) is trivially satisfied] from

$$\bigvee_n A(n) \rightarrow \bigvee_n . A(n) \wedge \bigwedge_m . m < n \rightarrow \neg A(m). .$$

But this statement is logically equivalent to

$$\bigwedge_n . \bigwedge_m . m < n \rightarrow \neg A(m). \rightarrow \neg A(n). \rightarrow \bigwedge_n \neg A(n),$$

which in turn easily arises from the principle of induction [with $B(n)$ for $\neg A(n)$] by way of

$$\bigwedge_n . \bigwedge_m . m < n \rightarrow B(m). \rightarrow B(n). \rightarrow \bigwedge_{\substack{n \\ n \leq 1}} B(n)$$

$$\bigwedge_n . \bigwedge_m . m < n \rightarrow B(m). \rightarrow B(n). \mathrel{\dot{\rightarrow}} \bigwedge_{\substack{n \\ n \leq p}} B(n) \rightarrow \bigwedge_{\substack{n \\ n \leq p+1}} B(n).$$

The following properties of the ordering relation $<$ for the natural numbers are used here:

$$\neg\, m < 1 \quad (\text{since } m + n \neq 1)$$

and $\quad m < n + 1 \dot{\rightarrow} m < n \vee m = n$

(since $\quad m + p = n + 1 \wedge p = 1 \dot{\rightarrow} m = n$

and $\quad m + p = n + 1 \wedge p = q + 1 \dot{\rightarrow} m + q = n$).

The indeterminate expressions, such as $\frac{0}{0}$, can be construed as pseudodescriptions:

$$\frac{0}{0} = \iota_x(0 \cdot x = 0).$$

We need not "forbid" pseudodescriptions, but false assertions ensue from treating pseudodescriptions as genuine descriptions.

By means of description terms we can form a function for every correlation R that fulfills the conditions (3.3); viz.,

$$f = \rceil_x \iota_y yRx,$$

or the function f such that

$$y = f \rceil x \leftrightarrow yRx.$$

Only if we refrain from distinguishing the function f from the 2-place relation R do we get the above definition of the function as a correlation. For the sake of conceptual clarity we shall retain the distinction in this book. In practice, though, it makes little difference. Just as every correlation R which fulfills (3.3) yields a function via the term (3.4), so also every function f yields a correlation satisfying (3.3), viz.,

28

$$\Gamma_f = \epsilon_{x,y}(y = f \rceil x). \tag{3.5}$$

We call this relation Γ_f the "*graph*" of the function f. This name is suggested by the familiar graphic representation of functions as curves. Such a curve can, after all, be construed as a set of geometric points. The set Γ_f, to be sure, contains the coordinate values x, y rather than the points, but Γ_f can still be called a "graph" too. We then have

$$y = f \rceil x \leftrightarrow y, x \in \Gamma_f.$$

Corresponding to many-place sets we can also form *many-place functions*. We do so by starting with terms of more than one free variable, e.g., $r^2 + s^2$.

For the term $T(x_1, \ldots, x_n)$ let

$$f = \rceil_{x_1, \ldots, x_n} T(x_1, \ldots, x_n)$$

be the function of $x_1, \ldots, x_n$ which is abstracted from $T(x_1, \ldots, x_n)$.

We then set

$$f\urcorner(c_1, \ldots, c_n) = T(c_1, \ldots, c_n).$$

$f\urcorner(c_1, \ldots, c_n)$ is the value of the function f for the argument *system* $c_1, \ldots, c_n$.

The graph of an n-place function is an $(n + 1)$-place relation. If we supply parentheses around the argument system—e.g., to distinguish $f\urcorner(x, y)$ from $(f\urcorner x), y$—then we can omit the $\urcorner$. This gives the usual notation $f(c_1, \ldots, c_n)$.

We shall not go into possible conventions for the omission of parentheses here.

For the special case $n = 2$, however, we can write $f(x, y)$ as xfy. 2-place functions are also called "*compositions.*"

In concluding our general discussion of sets and functions, there is one more point to which we call attention. Just as every function f can be correlated with a set, its graph Γ_f, which uniquely determines it, so conversely every set K can be correlated with a determining function, its "*characteristic*" χ_K.

To this end we define

29

$$\chi_K\urcorner x = \begin{cases} 1 & \text{if } x \in K \\ 0 & \text{if } x \notin K. \end{cases}$$

This definition can be written with the aid of a description term as follows:

$$\chi_K\urcorner x = \iota_y . x \in K \rightarrow y = 1 \,\dot\wedge\, x \notin K \rightarrow y = 0.$$

We then get

$$x \in K \leftrightarrow \chi_K\urcorner x = 1.$$

In an analogous manner we get characteristics of many-place functions. If K has n places, so does χ_K. Occasionally one comes across a conception of sets that identifies them with their characteristics. However, if we are to avoid losing sight of formulas and terms as two different components of the mathematician's language, we shall be well advised to introduce sets and functions independently of one another by abstraction:

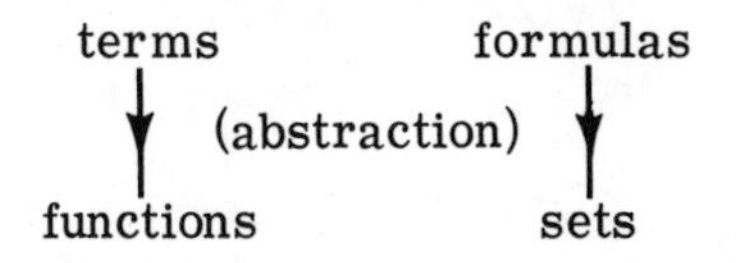

§4. Inductive Definitions

In §3 we introduced sets and functions by abstraction on formulas and terms already available. Now we shall take up the task of constructing further terms and formulas that will be needed later to define the sets and functions requisite for analysis, especially the real numbers themselves. Thus far terms and formulas arose by virtue of the availability of certain elementary terms for the rational numbers, viz., $r + s$, $r \cdot s$, $\frac{r}{s}$. From these all other rational terms were to be constructed by iterated substitution (by constants as well as variables). The terms then gave rise to elementary formulas (viz., equations and inequalities between rational terms), and all other formulas were logically compounded from the elementary formulas.

The task of getting new terms and formulas thus entails the introduction of new elementary terms and possibly even new elementary formulas (besides $r = s$, $r < s$). Suppose that we somehow or other had a new 2-place elementary term $E(r, s)$. By abstraction it would give rise to a 2-place function $f = \iota_{r,s} E(r, s)$, and in terms of this function the elementary term $E(r, s)$ could then be rewritten as $f(r, s)$.

30

Thus we can also say that we want to introduce new functions (rather than new elementary terms). As an example, consider the multiplication of natural numbers, a 2-place function introduced in §1 by certain construction rules. It is immediately seen to be characterized by

$$1 \cdot n = n$$
$$(m + 1) \cdot n = m \cdot n + n. \tag{4.1}$$

What does this "characterization" mean? Well, addition being presupposed as familiar, every function f which satisfies the equations

$$f(1, n) = n$$
$$f(m + 1, n) = f(m, n) + n \tag{4.2}$$

is the same as multiplication; i.e.,

$$f(m, n) = m \cdot n. \tag{4.3}$$

This last equation is a consequence of (4.1) and (4.2) by induction on m.

The pair of equations (4.2) is therefore called a definitional schema for multiplication or, more precisely, an *inductive definitional schema*, because the function to be defined is specified first for 1 (in the first argument place) and then for $m + 1$ by reference to the value of the function for m.

The definitional schema (4.2) is a special kind of schema. It seems natural to try to define further functions by means of similar schemata. For example, we claim that every schema of the form

$$\begin{aligned} f\urcorner 1 &= c \\ f\urcorner(n+1) &= T(f\urcorner n, n+1) \end{aligned} \tag{4.4}$$

"inductively defines" a function f, where c is a constant and $T(x, n+1)$ an already available term (containing a variable x). At any rate, it is clear that if there is a function f which satisfies (4.4), then the latter uniquely determines f.

What are we to do, however, if there is no such function f—among those so far available? It is by no means self-evident that these equations can be satisfied at all. Indeed, it is certain that the equations

31

$$\begin{aligned} f1 &= f2 \\ f(n+1) &= fn + 1, \end{aligned}$$

for example [which are, to be sure, of a different form from (4.4)], are unsatisfiable.

In §1 we assured ourselves of the "existence" of multiplication by giving rules of construction for the formulas $m \cdot n = p$. By abstraction $m \cdot n = p$ gives rise to the graph of multiplication $\Gamma_\times$, i.e.,

$$p, m, n \in \Gamma_\times \leftrightarrow p = m \cdot n.$$

Thus we can say that what we did in §1 was to give a constructive definition of the graph of multiplication. For $p, m, n \in \Gamma_\times$ is meant to hold just in case the formula $\frac{m \cdot n}{p}$ is constructible by the rules provided.

The "existence" of this graph is no problem, for the construction rules establish what sense a statement of the form $\frac{m \cdot n}{p}$ is to have. Thus $\frac{m \cdot n}{p}$ is thereby introduced as a new kind of elementary formula. Abstraction from such an elementary formula then yields the 3-place relation $\Gamma_\times$.

All that remains to be proved is that the relation $\Gamma_\times$ is indeed the graph of a function; i.e., that for any two numbers m and n there is exactly one number p such that $p, m, n \in \Gamma_\times$. This comes immediately by induction.

It turns out that, in certain respects, it is simpler to introduce sets than it is to introduce functions. For that reason, we shall also answer the question whether the schema (4.4) defines a function by first of all looking for a definition of the graph of f.

The graph Γ of f must obviously satisfy

$$\begin{aligned} y, 1 \in \Gamma &\leftrightarrow y = c \\ y, n+1 \in \Gamma &\leftrightarrow \bigvee_z . z, n \in \Gamma \wedge y = T(z, n+1). \end{aligned} \tag{4.5}$$

Once we have satisfied ourselves that there is exactly one 2-place relation Γ that fulfills (4.5), it will again be trivial to prove by induction that Γ is a graph.

What remains to be asked, then, is in what sense a schema like (4.5) inductively defines a set (or a relation).

The advantage of moving from the inductive definition of functions to that of sets is now seen to lie in the easy generalizability of the definitional schema (4.5), provided that we no longer care whether it is a graph or simply some set or other that is being defined.

32 We claim that the following equivalences

$$\begin{aligned} 1, x_1, \ldots, x_m \in K &\leftrightarrow A(x_1, \ldots, x_m) \\ n+1, x_1, \ldots, x_m \in K &\leftrightarrow B_K(n+1, x_1, \ldots, x_m) \end{aligned} \tag{4.6}$$

inductively define a set K, provided that the formula $A(x_1, \ldots, x_m)$ does *not* contain the symbol K, and that the formula $B_K(n+1, x_1, \ldots, x_m)$ may contain parts of the form $s, t_1, \ldots, t_m \in K$ (with terms $s, t_1, \ldots, t_m$) so long as $s < n+1$.

The two equivalences (4.6) can be combined into one:

$$n, x_1, \ldots, x_m \in K \stackrel{..}{\leftrightarrow} n = 1 \wedge A(x_1, \ldots, x_m) \dot{\vee} \bigvee_k . n = k+1 \wedge B_K(n, x_1, \ldots, x_m). \tag{4.7}$$

The possibility of thus combining the parts of definitional schemata for sets (as opposed to functions) is a result of the utility of the logical particles.

We can therefore base our inquiry into inductive definitions on the following *schema*:

$$n, x_1, \ldots, x_m \in K \leftrightarrow A_K(n, x_1, \ldots, x_m), \tag{4.8}$$

where the formula $A_K(n, x_1, \ldots, x_m)$ contains the symbol K only in contexts of the form $s, t_1, \ldots, t_m \in K$ such that $s < n$.

As (4.7) shows, s may contain a bound variable k. The stipulation $s(k) < n$ is then to be understood as requiring that the quantifier binding k ("for all k" or "for some k") be a *restricted quantifier*, i.e., "for all k such that $C(k)$" or "for some k such that $C(k)$," where

$$C(k) \rightarrow s(k) < n.$$

A formula $A_K(n, x_1, \ldots, x_m)$ satisfying these conditions may be referred to briefly as an *induction formula* (with respect to K).

Every induction formula (with respect to K) does indeed "*define*" an $(m + 1)$-place set K by virtue of (4.8). That is to say, the meaning of the assertion $n, x_1, \ldots, x_m \in K$ is established for every n (and for every system $x_1, \ldots, x_m$). This follows trivially by induction on n. For $n = 1-1, x_1, \ldots, x_m \in K$ is equivalent to $A_K(1, x_1, \ldots, x_m)$ by (4.8). To assert $1, x_1, \ldots, x_m \in K$ thus has the same meaning as asserting $A_K(1, x_1, \ldots, x_m)$, in which K only seems to occur, since $s < 1$ is not satisfiable.

If now the assertion of $k, y_1, \ldots, y_m \in K$ makes sense for all $k < n + 1$ and for all of $y_1, \ldots, y_m$, then by (4.8) $n + 1, x_1, \ldots, x_m \in K$ is equivalent to $A_K(n + 1, x_1, \ldots, x_m)$, of which the parts $s, t_1, \ldots, t_m \in K$ are already "defined." In this way the meaning of asserting $n, x_1, \ldots, x_m \in K$ is furthermore established only once, i.e., uniquely, for each n, for never is $n < n$.

33

These exhaustive considerations of the possibility of inductive definitions by the schema (4.8) may perhaps seem pedantic. They are nevertheless necessary to make it clear that sets and functions are objects obtained by introducing new elementary formulas. In other words, assertions in which a new symbol—e.g., K—appears make no sense until we give them a meaning. The sense or meaning need not consist in the stipulation of a procedure for deciding the truth or falsity of the assertions. Such a statement as, e.g., "All perfect numbers are even" is meaningfully assertible even though no decision procedure is established. The sense of that statement is determined by the fact that it begins with a universal quantifier:

$$\bigwedge_n . n \text{ is perfect} \rightarrow n \text{ is even.}$$

We thus know that the assertion could be refuted by giving a number n which was perfect but not even. Since there are infinitely many numbers, it is a stroke of luck when we can decide whether or not a given universal statement is refutable by a number n. The same is true of existential statements (for some $n, \ldots$).

The undecided character of many statements formed with quantifiers led the intuitionists (Brouwer 1907) to forbid use of the logical principle according to which every statement is true or false, the so-called law of excluded middle. However, the formalists (Hilbert, Gentzen) succeeded in showing that recourse to this law is "innocuous" in the following sense. By the use of excluded middle one never gets statements contradictory to those provable without excluded middle. In particular, all statements built up out of decidable elementary statements without quantifiers that are provable by excluded middle are also provable without it.

For the purposes of constructive mathematics we shall therefore use the logical particles in the manner customary outside mathematics (where we never have to do with more than an indeterminate finitude of objects). In other words, we shall use excluded middle as if it were a logical truth. For indefinite quantifiers, however, the intuitionistic usage suffices (cf. pp. 77, 114).

The definitional schema (4.8) admits of far-reaching applications. Simpler schemata, such as (4.4), are adequate for the introduction of elementary functions.

34 The definitions of the 2-place power function

$$x^1 = x$$
$$x^{n+1} = x^n \cdot x$$

and of the factorial function

$$1! = 1$$
$$(n+1)! = n! \cdot (n+1)$$

are examples of such schemata.

It should be noted, however, that the variables $x_1, \ldots, x_m$ in (4.8) need not be variables for numbers, but may be variables for any other objects whatsoever (provided they are available). If f is a function, then

$$f^{(1)} = f$$
$$f^{(n+1)} = f \urcorner f^{(n)}$$

defines the iteration of f. If we set $\Phi(n, f) = f^{(n)}$, then Φ is a function that has functions and numbers as arguments. The values of this function are again functions.

Sets of functions, sets of sets, and the like can be defined analogously. One trivial example is the set of all rational functions (functions from rational numbers to rational numbers).

The various ways just discussed of defining sets and functions do not necessarily permit us to speak of, say, the set of all possible functions, even if the arguments and values are restricted to natural numbers. Our definitional schemata only present possibilities for augmenting the stock of functions already available; no limit to this process of augmentation has been established.

The definition of functions and sets is an unlimited, open process. Statements about all sets of natural numbers, all functions of rational numbers, etc., are therefore called "*indefinite*" (cf. §1).

That the process of definition can always be continued can be proved by the following line of reasoning, which follows Cantor. Suppose we have set a limit to the use of new definitional schemata. Then up to that point we shall have introduced only finitely many different symbols into our arithmetic. As an example, the *language* of any textbook fulfills this condition: it can use only a finite alphabet of individual symbols. Every term and every formula is a finite combination of those individual symbols. Now, it is possible to *enumerate* all possible combinations of finitely many symbols. First we line the symbols up individually (in some order or other), then the finite number of combinations of 2 symbols (again in some particular order), etc. This sequence S_1, S_2, S_3, ... of all finite symbol combinations contains among other things the terms and induction formulas for the functions. *35* We now proceed to delete all other symbol combinations, and we replace the remaining terms and induction formulas by the functions they represent or define. In this way we obtain a sequence f_1, f_2, f_3, ... containing all functions definable by means so far allowed.

Then all we have to do is to stipulate that

$$F \iota n = f_n \iota n + 1,$$

thereby obtaining a function F that is not identical with any of the functions $f_1, f_2, \ldots\ldots$; for $F = f_m$ would lead to $f_m \iota m = f_m \iota m + 1$.

The function F is thus not definable by permissible means; by nevertheless defining F, we have succeeded in extending the process of definition beyond its supposed limit. Since F was explicitly defined in terms of f_n, the function f_n must itself already have been indefinable by permissible means. Talk about all symbol combinations of a given language is possible only in an extension of that language, not in the language itself.

The procedure of constructing from a given sequence of functions f_1, f_2, f_3, ... (with natural numbers as arguments and values) a further function

$$F = \iota_n . f_n \iota n + 1.$$

is called the *Cantor diagonal argument*, since the values $f_n \iota n$ lie along the diagonal of the following table of the values of the functions:

$$f_1 \text{℩} 1, \quad f_1 \text{℩} 2, \quad f_1 \text{℩} 3, \ldots\ldots$$
$$f_2 \text{℩} 1, \quad f_2 \text{℩} 2, \quad f_2 \text{℩} 3, \ldots\ldots$$
$$f_3 \text{℩} 1, \quad f_3 \text{℩} 2, \quad f_3 \text{℩} 3, \ldots\ldots$$
$$\vdots$$

It was first used by Cantor—in a somewhat different form—to prove the existence of nonalgebraic numbers. A trivial example of how it can be used is the following proof of the existence of nonconstant functions:

$$f_1 = 1, \quad f_2 = 2, \quad f_3 = 3, \ldots\ldots$$

is an enumeration of *all* constant functions. Thus

$$F \text{℩} n = f_n \text{℩} n + 1 = n + 1$$

produces a nonconstant function.

However, Cantor also "proved" by the diagonal argument the "superdenumerability" or nondenumerability of the set of all possible functions. Suppose, he argued, that the set of all possible functions were denumerable and that f_1, f_2, f_3, ... were an enumeration of that set. Then $f_n \text{℩} n + 1$ would produce a function distinct from all possible functions. Since that is contradictory, the supposition must be false. Hence the set of all possible functions is not denumerable. In this way Cantor became convinced that the infinity of natural numbers was only the lowest infinity, besides which there exist other (indeed infinitely many other) "higher infinities."

36

As the above argument shows, this conviction rests on the assumption of the existence of the set of all functions. One assumes that there is such a thing as that set. Only then, after all, can one assume it to be denumerable and reduce this assumption to absurdity. What is problematic about Cantor's proof is only the first assumption, the existence of the set of all possible functions. If we define functions—as we did in §3—by abstraction from terms, then we can construct no set of terms adequate for the representation of all possible functions. Any rule for constructing terms must work with a finite number of symbols, and hence can produce at most denumerably many terms. Thus it is precisely Cantor's diagonal argument that showed us that no construction rule can produce all functions.

At this point we have two alternatives. Either we restrict ourselves to constructible sets of functions (in which case there is no set of all possible functions and the first assumption of Cantor's "proof" is not fulfilled, so that all inferences from it are nonbinding fantasies), or we do without a constructive fulfillment of Cantor's assumptions. On the second alternative, the assumptions become the axioms of an axiomatic theory in which "function" (or "set") is a primitive undefined concept. Logically, there can be no objection to the erection of axiomatic theories (in this case axiomatic set theory). Even if the axiom system of the theory is not known to be consistent, the deducing of theorems from the axioms is an unobjectionable preoccupation. (On axiomatic set theory, cf. Quine 1963.)

The only objection to axiomatic set theories (only those, of course, for which we have no constructive model) lies in the question, What for?

In the spirit of Hilbert, theoreticians frequently argue as follows. If an axiomatic theory is consistent, then "existence" is attributed to the primitive notions of the theory; and what exists, so the argument goes, must (or in any event may) be investigated. If the concept of existence is eliminated from this argument, then it comes down to the claim that one may investigate consistent theories. Since, however, it is nobody's business here to permit or to forbid, the claim makes no sense whatsoever. All anyone can do is to decide for himself what he wants to do.

In any case, we will show in this book (following especially H. Weyl 1918) that the differential and integral calculus can be developed to its full classical extent without axiomatic set theory, i.e., strictly within the bounds of the constructible.

With inductive definitional schemata we have sufficient means in hand to define all the requisite functions and classes. Furthermore, we are quite capable of making certain statements about *all* sets or functions. This possibility deserves further explanation.

Suppose it is claimed, let us say, that for all functions f (from numbers to numbers)

$$f1 + f1 = 2 \cdot f1. \tag{4.9}$$

This evidently presents no difficulties, since after all $x + x = 2x$ is true of all numbers x. For x any term may be substituted (provided one has the term at all!). The above claim concerning all "possible" functions can thus be meaningfully interpreted as a claim about all functions so far introduced—and about those which have yet to be introduced. For we shall acknowledge a new expression as a "term" only if it can be substituted for the numerical variables.

To be able to claim that (4.9) holds we need not even know what functions are available at the time. The claim is rather compatible with every possible extension of the stock of functions by way of new terms.

Cantor's assumption of the denumerability of the set of all functions (in his proof of nondenumerability), on the other hand, is incompatible with indefiniteness. After any extension, the assumed enumeration would no longer enumerate all the functions. One may if one wishes attribute "existence" to the class of all functions, but one has no right to claim that this in any way affects the question of indefiniteness.

In the same way we can talk about "all" sets. Sets arise by abstraction from formulas, i.e. (after replacement of possible variables by constants), statements.

For example, all statements A and B satisfy

$$A \vee B \,\dot{\wedge}\, \neg A \,\ddot{\to}\, B. \tag{4.10}$$

To appreciate the validity of (4.10) we need not know all statements that are possible. All that is required is that A and B in fact be "statements" in each particular case. That is to say, their use must be the same as that established for other statements: they must be meaningfully assertible. This stipulation that a statement is something that can be meaningfully asserted also suffices to render the combination of statements by means of logical particles meaningful (cf. Lorenzen 1962).

(4.10) is also compatible with any extension of our fund of statements to include new ones.

A less trivial example is the following assertion about all functions from natural numbers to natural numbers:

$$\bigvee_n fn \leqslant f(n+1). \tag{4.11}$$

In other words, $f1 > f2 > f3 > \ldots\ldots$ cannot hold *ad infinitum*. To prove this, we let $m = f1 + 1$. By induction we can then deduce from the assumption $f1 > f2 > f3 > \ldots\ldots$ that $fn \leqslant m - n$ holds for all n. It holds for 1 by the definition of m. For $n+1$, $f(n+1) < fn \leqslant m - n$ or $f(n+1) \leqslant m - n - 1$ follows. This leads to the contradiction $fm \leqslant 0$. Thus the assumption that for all n it is not the case that $fn \leqslant f(n+1)$ must be false; i.e., it does not hold of all n that not $fn \leqslant f(n+1)$. The move from this to (4.11) is a logical inference. That is the way the quantifiers are used "classically."

The principle of induction used here can itself be formulated as a sentence about all sets:

Every set which contains 1, and which contains $n + 1$ if it contains n, contains all natural numbers.

Symbolically,

$$1 \in K \wedge \bigwedge_n . n \in K \to n+1 \in K . \dot{\to} \bigwedge_m m \in K.$$

Writing it now with an indefinite quantifier, we get

$$\bigwedge_K . 1 \in K \wedge \bigwedge_n . n \in K \rightarrow n+1 \in K . \dot{\rightarrow} \bigwedge_m m \in K.$$

Any statement containing an indefinite quantifier will be called an *indefinite* statement for short.

The logical truths belong to the indefinite statements but do not exhaust them. The usual arithmetical statements without indefinite quantifiers will occasionally be distinguished as *definite* statements.

Both the indefiniteness of the "concepts" of set and function (i.e., the indefiniteness of the construction process for formulas and terms) and the possibility of indefinite statements must always be kept in mind in a constructive treatment of analysis. An indefinite statement can be true only if it is compatible with indefiniteness: it must remain true under all extensions of the means of language construction, i.e., all extensions to include new sets or functions.

Since we defined sets by abstraction from formulas, a set would, according to the foregoing, always be definite. $K = \epsilon_x A(x)$ is the set of those xs which satisfy $A(x)$. The definite objects for which the variable x stands must already be available *before* the set is formed. Only when we reflect on the indefinite possibilities of language construction does anything indefinite crop up in addition to the definite objects that arise by construction and abstraction from the finite combinations of the individual symbols of a given language. (Every language must have a finite alphabet; otherwise it could never be learned.) Only then can we use indefinite variables, e.g., for the formulas and terms of an "arbitrary" language. By abstraction we then attain the capacity to speak of arbitrary sets and functions. This indefiniteness even comes into play if we limit ourselves—as in logic—to arbitrary statements about determinate objects.

39

However, once we have taken the step of introducing indefinite variables (and once we have realized the possibility of proving certain statements containing indefinite variables—despite their indefiniteness), it is natural to extend to formulas and terms with free indefinite variables the abstraction operations leading from formulas and terms to sets and functions. The indefinite ϵ- and ι-operators introduced in this manner (e.g., $\epsilon_f f0 = 0$, the set of functions f such that $f0 = 0$; or $\iota_f f0$, the function from f to $f0$) yield indefinite sets and functions.

The most important example of these is the indefinite set of all definite sets of natural numbers. This indefinite set is called (after Cantor) the *power set* of the set of natural numbers.

Not every set of sets is indefinite. For example, there is nothing indefinite about the infinite set whose members are the following sets:

the set of all natural numbers,
the set of all squares of natural numbers,
the set of all cubes of natural numbers,
⋮

But the power set is supposed to contain as members "all" sets of natural numbers. So as to attach some sense to this definition, we interpret the word "set"—contrary to Cantor—as "definite set."

The power set thereby becomes an indefinite set. We might consider admitting indefinite sets of natural numbers into the power set too. But in order to define indefinite sets, we must first of all introduce indefinite variables. This is done precisely by reflecting on the indefinite possibilities of language. Thus we can at least define a power set that contains all definite sets of *40* natural numbers. This book will show that for the purposes of analysis we can get by with this power set (and the corresponding indefinite set of all definite real numbers).

So far as terminology is concerned, we need not continue calling indefinite sets "sets"; we could use a new word, say "class." But to avoid deviation from the accustomed terminology, we shall retain "set" as the common term for definite and indefinite sets alike. For indefinite sets, however, we shall occasionally use the designation "class."

Let us briefly recapitulate what we have done so far. We began with the construction of the natural numbers. Other numbers, such as the rationals, resulted from finite combinations of natural numbers (e.g., by the formation of fractions) together with some abstractions. We then obtained further definite objects by constructing statements (formulas) about objects already available. This "construction" of formulas involved the formation of certain definitional schemata and subsequent combination by means of logical particles. The construction process is, however, open or indefinite. Definite sets and functions arise by abstraction from constructed formulas and terms.

The construction of statements about available objects is called a construction of a "language" treating of those objects. The construction of a language can be iterated: one can construct a language treating of the statements of a language already constructed.

The total enterprise of constructing "languages" in this way is by no means unambiguously determined. It cannot be established once and for all what all could be taken as a definitional schema. By the same token, it is left open how far the iteration of language construction can go. (There are even constructive possibilities for a transfinite iteration of language construction; cf. Lorenzen 1955.)

It will turn out that we can erect classical infinitesimal mathematics without committing ourselves to a particular *language construction*.

For the sequel we must keep in mind that all definite objects are determinate combinations of finitely many particular symbols. The only added device will be occasional abstractions with respect to certain equivalence relations.

At each stage in the construction of definite objects, therefore, all objects constructed up to that point will be denumerable. The enumeration, as a correlation between natural numbers and the objects, need not of course be representable by one of the formulas already constructed—but it will always admit of representation (by way of an inductive definitional schema) in the language treating of all objects thus far constructed.

II. REAL NUMBERS

§5. Rational Sequences

The various principles established for the rational numbers in §2—the laws of identity, the order laws, the arithmetical laws, and the general applicability of subtraction and division (excepting division by 0)—are summarized by saying that the rational numbers constitute an ordered field.

The rational numbers turn out to be inadequate at the level of elementary algebra, i.e., the theory of equations. If we have an algebraic equation to solve of the form

$$x^n + a_1x^{n-1} + \ldots + a_n = 0 \tag{5.1}$$

and if there is no whole-number solution, then there is no rational solution. The simplest examples are given by the equations

$$x^2 + 1 = 0$$

$$x^2 - 2 = 0 .$$

Let us sketch briefly a proof of the general claim.

If $x = \frac{p}{q}$ (where p and q share no common divisor) is a rational number satisfying (5.1), then

$$x^n = -a_1x^{n-1} - \ldots - a_n .$$

Therefore $\frac{p^n}{q} = -a_1p^{n-1} - \ldots - a_nq^{n-1}$ is a whole number; i.e., p^n is divisible by q, p is divisible by q, and $q = 1$. Thus x is whole.

There is the following difference between the insolubility of $x^2 + 1 = 0$ and $x^2 - 2 = 0$. The value of $x^2 - 2$ never reaches 0 but can be brought "arbitrarily close" to 0; the value of $x^2 + 1$, on the other hand, is always $\geqslant 1$.

That the value of $x^2 - 2$ approaches arbitrarily close to 0 means the following. For every (positive rational) ε there is a rational solution of the inequality

$$-\varepsilon < x^2 - 2 < \varepsilon .$$

In terms of absolute difference, this can be written as 42

$$|x^2 - 2| < \varepsilon. \tag{5.2}$$

The solutions of (5.2) are called "*approximate solutions*" of the equation $x^2 - 2 = 0$. In contrast to the equation $x^2 + 1 = 0$, the equation $x^2 - 2 = 0$ has approximate solutions for every ε. This is what is meant by saying that the equation $x^2 - 2 = 0$ has "arbitrarily precise" approximate solutions.

Procedures for computing ever closer approximate solutions were already known to the Babylonians (ca. 2000 B.C.). Those procedures are used today to obtain the decimal expansion of "the" solution (as we say). For $x^2 - 2 = 0$ they produce the following rational numbers as approximate solutions:

$$1, 1.4, 1.41, 1.414, \ldots\ldots$$

Such a procedure does not just give finitely many approximations. If the rule for the procedure is formulated mathematically, it yields the inductive definition (cf. §4) of a function f such that

$$|f(n)^2 - 2| < \frac{1}{10^{n-1}}.$$

It is this definite function f that solves the problem of finding arbitrarily precise approximations for the equation. The functions arising in this connection have natural numbers as arguments. These functions constitute such an important subclass of the class of all functions that they have been given a special name: they are called *sequences*. Occasionally one reads something like the following "definition" of sequences: a sequence (of numbers) consists of infinitely many numbers written down in a particular order.

Aside from the fact that no one will ever find infinitely many numbers or infinitely many numerals written down, we want here to understand sequences not as certain ordered sets, but as certain functions, viz., those with natural numbers as arguments. We presuppose nothing about the values of the functions. For the time being, however, we shall concern ourselves with the functions of natural numbers that have rational numbers as values. Such sequences will be called *rational sequences* for short. In a sequence, the values of the function are called the "*members*" of the sequence.

A sequence need not have infinitely many members; e.g., the members can all be identical.

As a notation for sequences we use variables for the members with one of the following symbols as a subscript: *, †, ⸸, ‡; e.g.,

r_*, $s_\dagger$ for rational sequences. This notation has the advantage that we can designate the members of the sequence s_* as s_1, s_2, ... The nth member of s_* is s_n. The term s_n is not a symbol for the sequence; rather, $s_* = \imath_n s_n$. The symbols $*$, $\dagger$, $\ddagger$, . . . are not regular variables (whether free or bound) but rather so-called *place markers*. They permit a designation of sequences without recourse to the $\imath$-operator—while at the same time not obscuring the distinction between the sequence s_* and the term s_n. Also, we permit ourselves to write $s_* = s_1, s_2, \ldots$ . .

43

Since certain equations are without (rational) solutions, it is natural to try taking sequences of approximate solutions as a substitute. The breakthrough that distinguishes analysis as it arose in the seventeenth century from ancient infinitesimal mathematics consists in regarding the sequences of approximate solutions as new mathematical objects, as new numbers. It was possible to show that the new "numbers" again form an ordered field, just as do the rational numbers.

That program will be carried out presently. It should be pointed out right now, however, that the sequences of rational numbers to be considered do not form a definite set but only a class. Greek geometry, especially Eudoxos' theory of proportions (Euclid, Book V) and Archimedes' applications of that theory (e.g., in measuring the circle, squaring the parabola, determining the tangent of the spiral) were of crucial importance for the historical development of analysis. This connection with geometry still plays an important role today, of course, but for a constructive development of the theory of real numbers the connection tends to be misleading. The existence of geometric quantities, e.g., the diagonal of the square, must always be proved from the geometric axioms, while the arithmetical construction of the real numbers is independent of the geometric axioms. Even if there were no geometry, i.e., no exact theory of spatial figures, the task would remain of finding a substitute for the missing rational solutions of algebraic equations.

By the way, the needs of arithmetic in fact quickly take us beyond the "algebraic" numbers. If we have defined the rational powers ε^r for positive rational $\varepsilon \neq 1$, i.e., $\varepsilon^{\frac{m}{n}}$ as the solution of the equation

$$x^n = \varepsilon^m, \quad \varepsilon^0 = 1, \quad \varepsilon^{-\frac{m}{n}} = \left(\frac{1}{\varepsilon}\right)^{\frac{m}{n}},$$

then such equations as

$$\varepsilon^x = \delta \tag{5.3}$$

lead to still further "numbers," the logarithms. There is a proof, which at this point we cannot even sketch, to the effect that the equation (5.3) has no algebraic solution unless it has a rational one. (This holds even if ε, δ are algebraic; cf. T. Schneider 1957.)

The historical development of analysis is for us unthinkable without its applications to geometry and physics. (Later we shall use them to provide a motivation for the definition of the integral.) The systematic construction of analysis is independent of all applications, however. In particular, the existence of the real numbers does not depend on geometric theorems, let alone geometric intuition. *44*

In the systematic introduction of real numbers we take as our guide the familiar decimal fractions. An infinite decimal expansion

$$0.a_1a_2a_3\ .\ .\ .\ .\ .$$

can be construed as a representation (albeit imperfect) of a sequence r_* of rational numbers; viz.,

$$r_1 = 0.a_1, \quad r_2 = 0.a_1a_2, \quad r_3 = 0.a_1a_2a_3, \ .\ .\ .\ .\ .$$

The salient property of these rational sequences is that their members approach each other arbitrarily closely if one just goes far enough, i.e., if the subscripts (arguments) become sufficiently large. Indeed, it is the case for $m < n$ that

$$0 \leqslant r_n - r_m < \frac{1}{10^m}.$$

Let us formulate in mathematical language this feature of arbitrarily close approximation for sufficiently large "subscripts." That the members approach arbitrarily close to each other means that the differences $r_n - r_m$ become arbitrarily small. So as to be able to ignore order, we write

$$|r_m - r_n| < \varepsilon.$$

These inequalities are to hold of every (positive rational) ε for "sufficiently large" subscripts. That is to say, they are to hold of all m, n when $m > N$ and $n > N$ is satisfied by an appropriate N.

Without using "arbitrarily" or "sufficiently" or similar words, we can express this as follows:

For every ε there is an N such that for all $m > N$ and $n > N$

$$|r_m - r_n| < \varepsilon.$$

In symbols, finally, this property reads

$$\bigwedge_{\varepsilon}\bigvee_{N}\bigwedge_{m,n}.m > N \wedge n > N \dot{\to} |r_m - r_n| < \varepsilon. \tag{5.4}$$

The conceptual difficulties which analysis causes the beginner stem in large measure from such pileups of quantifiers as occur 45 here. In (5.4) we have three successive quantifiers, the order of which is essential. For example, the statement

$$\bigvee_{N}\bigwedge_{\varepsilon}\bigwedge_{m,n}.m > N \wedge n > N \dot{\to} |r_m - r_n| < \varepsilon., \tag{5.5}$$

got by interchanging two quantifiers, is not by any means satisfied by all decimal expansions, but only by those of finite length. For, since there is supposed to be an N such that for all $m > N$ and $n > N$

$$\bigwedge_{\varepsilon}|r_m - r_n| < \varepsilon,$$

it must be the case that $r_m = r_n$. Thus a sequence which fulfills (5.5) looks like the following:

$$r_1, \quad r_2, \ldots, \quad r_N, \quad r_{N+1}, \quad r_{N+1}, \ldots\ldots$$

From a certain subscript on it is constant. In such a situation we say for short that the sequence is "*finally*" *constant.*

All finally constant sequences fulfill the condition (5.4), but not vice versa: every infinitely long decimal fraction is a sequence that fulfills (5.4) but is not finally constant. A sequence which satisfies (5.4) is not necessarily finally constant, but only approaches final constancy with arbitrarily small tolerance; i.e., for any tolerance ε we finally get (for all $m > N$, $n > N$ for an appropriate N)

$$|r_m - r_n| < \varepsilon.$$

The property (5.4) of decimal expansions and its importance for analysis were first discovered by Cauchy (early nineteenth century). Sequences with that property are therefore called Cauchy sequences. Recently the expression "*concentrated*" has also gained currency for such sequences.

DEFINITION 5.1 A rational sequence r_* is *concentrated* if

$$\bigwedge_{\varepsilon}\bigvee_{N}\bigwedge_{m,n}.m > N \wedge n > N \dot{\to} |r_m - r_n| < \varepsilon.$$

In the discussion leading to this definition, we claimed that infinite decimal expansions are concentrated. That has yet to be proved, however. Every decimal expansion $0.a_1a_2\ldots\ldots$ yields an *increasing* sequence of rational numbers

$$r_1 \leqslant r_2 \leqslant r_3 \leqslant \ldots\ldots,$$

if we set $r_n = 0.a_1a_2\ldots a_n$.

Now we have

$$r_n \leqslant \frac{9}{10} + \frac{9}{10^2} + \ldots + \frac{9}{10^n} = \frac{9}{10}\left(1 + \frac{1}{10} + \ldots + \frac{1}{10^{n-1}}\right)$$
$$< \frac{9}{10}\,\frac{1}{1 - 1/10} = 1$$

$$\left(\text{for } 1 + q + q^2 + \ldots + q^{n-1} = \frac{1 - q^n}{1 - q} < \frac{1}{1 - q} \text{ when } 0 < q < 1\right).$$

Thus the sequence r_* is also *bounded*; i.e., 46

$$\bigvee_c \bigwedge_n |r_n| \leqslant c. \tag{5.6}$$

In this case "c" can be a variable for natural numbers.

Rather than prove concentration for the sequences associated with decimal fractions, we shall now demonstrate the more general

THEOREM 5.1 Every increasing bounded sequence is concentrated.

We preface the proof with a definition of the *subsequences* of a given sequence r_*. Let m_* be a *properly increasing* sequence of natural numbers

$$m_1 < m_2 < m_3 < \ldots\ldots$$

Then $r_{m_*} = r_{m_1}, r_{m_2}, r_{m_3}, \ldots\ldots$ is a *subsequence* of r_*.

Such properly increasing sequences of natural numbers as we have here are called for short "*natural sequences.*" They are the subsequences of the identity sequence 1, 2, 3, ... of the natural numbers. It follows from this definition of subsequence that the subsequences of any increasing or bounded sequence are themselves likewise increasing or bounded.

To prove the above theorem, we assume that r_* is an unconcentrated increasing sequence. We want to prove that in that case there is an unconcentrated subsequence of r_*. If r_* is unconcentrated, then for some ε there is for every N an m and n such that $N < m < n$ and $r_n - r_m > \varepsilon$. For each N we select the smallest natural numbers m, n such that $r_n - r_m > \varepsilon$. For $N = 1$ let these be m_1, n_1; for $N = n_1$: m_2, n_2; for $N = n_2$: m_3, n_3; etc. Altogether we get

$$r_{m_1} < r_{n_1} \leqslant r_{m_2} < r_{n_2} \leqslant r_{m_3} < r_{n_3} \leqslant \ldots\ldots,$$

and it is furthermore the case that

$$\bigwedge_i r_{n_i} - r_{m_i} > \varepsilon$$

(where i ranges over natural numbers).

In this way we have inductively defined two subsequences by the use of μ-terms. It follows from the above inequalities for the subsequence r_{m_*} that

$$r_{m_{i+1}} - r_{m_1} > i\varepsilon.$$

Hence the subsequence r_{m_*} is not bounded.

In just the same way as Theorem 5.1, we can of course also prove that every *decreasing* bounded sequence $r_1 \geqslant r_2 \geqslant r_3 \geqslant \ldots\ldots$ is concentrated.

If we call increasing and decreasing sequences together "*monotonic*," then we can summarize by saying that every bounded monotonic sequence is concentrated.

47 A bounded but nonmonotonic sequence is not necessarily concentrated. It does, however, hold that

THEOREM 5.2 Every bounded sequence has a concentrated subsequence.

This theorem follows directly from Theorem 5.1 together with the following

THEOREM 5.3 Every sequence has a monotonic subsequence.

Proof. In any sequence r_* we call a member r_n an *upper crest* if every succeeding member $r_m \leqslant r_n$ (where $m > n$). If there are infinitely many upper crests of which r_{n_1} is the first, r_{n_2} the second, etc., then these crests clearly constitute a decreasing subsequence.

If, on the other hand, there are not infinitely many upper crests, then there is a subscript N such that for all $n > N$ the member r_n is not an upper crest. We then have

$$\bigwedge_{n > N} \bigvee_{m > n} r_m > r_n,$$

which immediately entails the existence of a properly increasing subsequence of r_*.

The inductive definitions of the subsequences claimed here to exist may easily be written out by means of μ-terms (cf. §3).

Decimal calculations (the basis for which will be laid in §6) always proceed in such a way that the arithmetical operations are executed member by member (place by place). For example if r_* is to be added to s_*, then the sequence $r_* + s_*$ is formed from the members

$$r_1 + s_1, \quad r_2 + s_2, \quad r_3 + s_3, \ldots\ldots$$

Such member-by-member figuring leads in the case of subtraction and division to sequences that are no longer increasing.

How, e.g., are we to carry out the division operation

$$1 \div 0.11010010001\ldots\ldots$$

on the nonterminating decimal indicated (one more 0 is inserted each time)? The members of the sequence are

$$0.1, \quad 0.11, \quad 0.110, \quad 0.1101, \ldots\ldots$$

i.e.

$$\frac{1}{10}, \quad \frac{11}{100}, \quad \frac{11}{100}, \quad \frac{1101}{10\,000}, \ldots\ldots$$

Member-by-member division yields the sequence

$$10, \quad \frac{100}{11}, \quad \frac{100}{11}, \quad \frac{10\,000}{1101}, \ldots\ldots$$

i.e.

$$10, \quad 9\frac{1}{11}, \quad 9\frac{1}{11}, \quad 9\frac{91}{1101}, \ldots\ldots$$

This sequence s_* is decreasing; i.e., 48

$$s_1 \geqslant s_2 \geqslant s_3 \geqslant \ldots\ldots$$

Later on we shall undertake the task of defining the sense in which this sequence represents a number. We already see right here, however, that there is something arbitrary about giving priority to the increasing sequences.

As we shall show in the sequel, the class of concentrated sequences is closed under the arithmetical operations; i.e., member-by-member execution of those operations does not take us outside that class. That is what makes concentrated sequences preeminently fit to define the real numbers.

Among the nonterminating decimals are some which we regard as representing rational numbers. They are the repeating decimals, e.g.,

$$\frac{1}{3} = 0.\overline{3} = 0.333\cdots\cdots$$

The members of the sequence r_* here are

$$r_1 = 0.3, \quad r_2 = 0.33, \quad r_3 = 0.333, \ldots\ldots$$

i.e.

$$\frac{3}{10}, \quad \frac{33}{100}, \quad \frac{333}{1000}, \ldots\ldots$$

The repeating decimal $0.\overline{3}$ is said to represent the number $\frac{1}{3}$, because the members of the sequence r_* approach "arbitrarily close" to $\frac{1}{3}$ for "sufficiently large" subscripts.

This arbitrarily close approximation for sufficiently large subscripts is again a property that requires three quantifiers for its mathematical formulation: for every ε there is an appropriate N such that for all $n > N$, $|r_n - s| < \varepsilon$ (in this case for $s = \frac{1}{3}$).

In logical notation we get

$$\wedge_\varepsilon \vee_N \wedge_n . n > N \rightarrow |r_n - s| < \varepsilon. \tag{5.7}$$

DEFINITION 5.2 A sequence r_* is *convergent* if there is a number s for which (5.7) holds.

We also say that r_* converges to s: $r_* \rightsquigarrow s$; or that s is the *limit* of r_*:

$$\lim r_* = s.$$

The possibility of writing $r_* \rightsquigarrow s$, i.e., the convergence of r_* to s, as an equation, $\lim r_* = s$, rests of course in the fact that the condition (5.7) uniquely determines the number s. For we have

49 THEOREM 5.4 $r_* \rightsquigarrow s \wedge r_* \rightsquigarrow s' \dot{\rightarrow} s = s'$.

Proof. Were it the case that $r_* \rightsquigarrow s$ and $r_* \rightsquigarrow s'$ and $s \neq s'$, then $\varepsilon = |s - s'|$ would be > 0. Thus there would be an n such that $|r_n - s| < \frac{\varepsilon}{2}$ and $|r_n - s'| < \frac{\varepsilon}{2}$. That, however, would lead to

$$|s - s'| \leq |s - r_n| + |r_n - s'| < \varepsilon,$$

which would contradict $|s - s'| = \varepsilon$.

We have here another example of a so-called indirect proof, as we did in (4.11). The task there was to prove a statement of the form "for some n, $A(n)$," and that was done by proving

$$\neg \wedge_n \neg A(n).$$

The indirect proof was based on the convention of classical logic which treats the two statements as logically equivalent.

In the proof of Theorem 5.4 the situation is different. Here it is simply a case of a (tacit) application of the logically true subjunction (4.10). For any two rational numbers s, s', either $s = s'$ or $s \neq s'$. The proof of Theorem 5.4 shows that $s \neq s'$ does not hold. The two together, "$s = s'$ or $s \neq s'$" and "$\neg\, s \neq s'$," then yield the desired $s = s'$ by (4.10).

The celebrated indirect arguments of Eudoxos (Euclid V) and Archimedes in infinitesimal problems are in essence always arguments that prove the equality of two quantities by virtue of the fact that both are limits of the same sequence. Modern analysis replaces those indirect proofs by the indefinite Theorem 5.4. This theorem does indeed hold for "all" sequences.

We now have the alternative of construing a repeating decimal like $0.\overline{3}$ as a symbol for the limit of the sequence $r_* = 0.3, 0.33, 0.333, \ldots$. The equation $\frac{1}{3} = 0.\overline{3}$ then means that $\lim r_* = \frac{1}{3}$. We have yet to prove this equation. First of all,

$$r_1 = \frac{3}{10}, \quad r_2 = \frac{3}{10} + \frac{3}{100}, \ldots\ldots$$

and hence

$$r_n = \frac{3}{10}\left(1 + \frac{1}{10} + \ldots + \frac{1}{10^{n-1}}\right).$$

Therefore

$$r_n = \frac{3}{10}\,\frac{1 - \dfrac{1}{10^n}}{1 - \dfrac{1}{10}} = \frac{3}{10}\,\frac{1}{1 - \dfrac{1}{10}} - \frac{3}{10}\,\frac{1}{1 - \dfrac{1}{10}}\,\frac{1}{10^n},$$

i.e.,

$$\frac{1}{3} - r_n = \frac{1}{3}\,\frac{1}{10^n}.$$

Thus all we have left to prove is that $\frac{1}{10^*} \rightsquigarrow 0$. With that, 50
$r_* - \frac{1}{3} \rightsquigarrow 0$ then follows, whence $\lim r_* = \frac{1}{3}$.

For $\frac{1}{10^*} \rightsquigarrow 0$ it is sufficient to show that

$$\bigwedge_n \bigvee_N \frac{1}{10^N} < \frac{1}{n},$$

i.e.,

$$\bigwedge_n \bigvee_N 10^N > n. \tag{5.8}$$

But by induction we immediately get

$$2^n > n,$$

for $2^1 > 1$, and from $2^n > n$ it follows that $2^{n+1} > 2n \geqslant n + 1$. Hence also

$$10^n > n.$$

For (5.8), therefore, all we have to do is to set $N = n$.

As this example shows, the case $\lim r_* = 0$ is especially important, for by Definition 5.2, $\lim r_* = s \leftrightarrow \lim (r_* - s) = 0$.

DEFINITION 5.3 Sequences which converge to 0 are called *null sequences*.

In the case of decreasing sequences of nonnegative numbers, the condition for being a null sequence can be simplified to

$$\wedge_\varepsilon \vee_n r_n < \varepsilon.$$

We already made use of this in the case of $\frac{1}{10^*} \rightsquigarrow 0$.

The convergent sequences are frequently considered to include also such sequences as $r_* = 2^*$, for which

$$\wedge_m \vee_N \wedge_n . n > N \rightarrow r_n > m. \tag{5.9}$$

We then write $r_* \rightsquigarrow \infty$ or $\lim r_* = \infty$, and we call r_* *improperly convergent.*

If the sequence $-r_*$ satisfies condition (5.9), then we write $r_* \rightsquigarrow -\infty$ or $\lim r_* = -\infty$. Such a sequence is likewise called *improperly convergent.*

Beyond the introduction of this notation, the symbols ∞ and $-\infty$ can also be taken to denote "improper" numbers, providing that appropriate rules are laid down for their arithmetical treatment. (The symbol ∞ derives from the Etrusco-Roman symbol ↀ for "1000"; it was introduced into mathematics by Wallis in the seventeenth century.) First of all, we stipulate for positive numbers ε and ∞ $(= +\infty)$ that

51

$$\left.\begin{array}{l} \varepsilon + \infty = \infty + \varepsilon = \infty + \infty = \infty \\ \varepsilon \cdot \infty = \infty \cdot \varepsilon = \infty \cdot \infty = \infty. \end{array}\right\} \tag{5.10}$$

Thus the subtraction $\infty - \varepsilon = \infty$ and the division $\frac{\infty}{\varepsilon} = \infty$ can be carried out, whereas $\infty - \infty$ and $\frac{\infty}{\infty}$ cannot [by (5.10) these expressions do not determine any number]. $\infty - \infty$ and $\frac{\infty}{\infty}$, like $\frac{0}{0}$ and $\frac{\pm\varepsilon}{0}$, are called "*indeterminate*" expressions.

The arithmetical laws (associativity, commutativity, and distributivity) remain valid. Upon the introduction of

$$-\infty = (-1) \cdot \infty$$

those laws govern calculations involving $\pm\infty$ and positive and negative numbers.

If ∞ is substituted for ε in $0 \pm \varepsilon = \pm\varepsilon$, or 0 for ε in $\varepsilon \pm \infty = \pm\infty$, either case gives rise to

$$0 \pm \infty = \pm\infty.$$

For multiplication we have so far established that $0 \cdot \varepsilon = 0$ and $\varepsilon \cdot \infty = \infty$. Thus no convention compatible with the arithmetical laws can be established for $0 \cdot \infty$. $0 \cdot \infty$ and $\infty \cdot 0$ are also indeterminate expressions.

According to these definitions, $\frac{0}{\infty}$ and $\frac{\pm\varepsilon}{\infty}$ ought likewise to be indefinite. Nevertheless, it is advisable to stipulate for every rational number r that

$$\frac{r}{\infty} = 0.$$

To see the justification for this, take two sequences r_* and s_* such that

$$\lim r_* = r, \quad \lim s_* = \infty$$

and form the sequence $\frac{r_*}{s_*}$ (replacing members $s_n = 0$ in any random manner by members $\neq 0$; because $\lim s_* = \infty$, there are at most finitely many such members). It is then easily proved that $\lim \frac{r_*}{s_*} = 0$: for every ε there is an N such that for all $n > N$, $|r_n - r| < 1$ and $s_n > \frac{1}{\varepsilon}$, whence $\left|\frac{r_n}{s_n}\right| < (|r| + 1)\varepsilon$.

These considerations do not, to be sure, justify setting $\frac{r}{0}$ or $\frac{\infty}{0} = \infty$: from $\lim s_* = 0$ it does not follow that $\lim \frac{1}{s_*} = \infty$, but only that $\lim \left|\frac{1}{s_*}\right| = \infty$. Thus the most we can do is to set $\left|\frac{r}{0}\right| = \left|\frac{\infty}{0}\right| = \infty$. 52

This plethora of expressions and stipulations is rather bewildering at first glance. If we confine ourselves to positive numbers for the time being, adding only ∞ as an improper number, then it is enough to note that

$$\infty - \infty \quad \text{and} \quad \frac{\infty}{\infty}$$

are indefinite. For $0 = \frac{1}{\infty}$, these in turn establish the indeterminacy of $0 \cdot \infty$ and $\frac{0}{0}$.

The old controversy over whether ∞ is a number or whether the symbol "∞" is only introduced to permit a briefer formulation of certain sentences about numbers proper is rather pointless. Even sentences about positive rational numbers can be construed as a "manner of speaking," i.e., as sentences that are meant only to serve as a convenient means of communicating other sentences (viz., sentences about the natural numbers). If we admit 0, division is thereafter only restrictedly applicable

$\left(\text{indeterminacy of } \frac{0}{0}\right)$. If we add ∞, then subtraction and division are only restrictedly applicable $\left(\text{indeterminacy of } \infty - \infty, \frac{\infty}{\infty}\right)$. Upon admission of both 0 and ∞, even the applicability of multiplication is restricted (indeterminacy of $0 \cdot \infty$).

All sentences in ∞ can be construed as formulations of sentences not involving ∞. That holds just as well for 0 and even for $\frac{1}{2}$ or -1. Only the case of the natural numbers is different. It is only the construction of the natural numbers and the (inductive) definition of the arithmetical operations that—as Kant would say—make the sentences of arithmetic possible in the first place.

§6. Definition of the Real Numbers

We saw in §5 that nonterminating decimal fractions are a special kind of concentrated sequence of rational numbers. Some concentrated sequences (e.g., the repeating decimals) converge to rational numbers. From this it follows logically that some convergent sequences are concentrated. In fact, however,

53 THEOREM 6.1 All convergent sequences are concentrated.

Proof. If $r_* \rightsquigarrow r$, then for every ε there is an N such that

$$|r_n - r| < \frac{\varepsilon}{2} \quad \text{for all } n > N.$$

For $m > N$ and $n > N$ it is therefore the case that

$$|r_m - r_n| \leqslant |r_m - r| + |r - r_m| < \varepsilon.$$

Theorem 6.1 cannot be reversed: not all concentrated sequences converge to a rational number. Thus, in terms so far available, we should have to say that some concentrated sequences are not convergent at all. Let us rather say, however, that they are not *rationally convergent.* As an example, take any nonrepeating decimal or, say, a sequence of closer and closer approximate solutions of the equation $x^2 = 2$. These examples make use of the addition and multiplication of rational numbers. However, recourse to ordering relations is sufficient.

Let $s_1, s_2, s_3, \ldots\ldots$ be an enumeration of all rational numbers in the *closed interval* $[0|1]$, i.e., $\epsilon_r\ 0 \leqslant r \leqslant 1$.

In order of increasing denominators (or increasing numerators in the case of identical denominators), we get the enumeration

$$0,\ 1,\ \frac{1}{2},\ \frac{1}{3},\ \frac{2}{3},\ \frac{1}{4},\ \frac{3}{4},\ \frac{1}{5},\ \frac{2}{5},\ \frac{3}{5},\ \frac{4}{5},\ \frac{1}{6},\ \ldots\ldots$$

For any two distinct members s_m, s_n of this sequence there is another member that lies in the *open interval* $(s_m \mid s_n)$, i.e., $\epsilon_r\ s_m < r < s_n$, where $s_m < s_n$; e.g.,

$$\frac{1}{2}(s_m + s_n).$$

The sequence s_* is therefore said to be "*dense.*"

We define inductively a subsequence s_{k_*} of s_*. Let $k_1 = 1$, $k_2 = 2$, and hence $s_{k_1} = 0$, $s_{k_2} = 1$.

Furthermore, let k_{n+1} be the smallest subscript k such that s_k lies between $s_{k_{n-1}}$ and s_{k_n}:

$$k_{n+1} = \mu_k\ s_k \in (s_{k_{n-1}} \mid s_{k_n}).$$

Accordingly, we have

$$s_{k_3} = \frac{1}{2},\quad s_{k_4} = \frac{2}{3},\quad s_{k_5} = \frac{3}{5},\quad s_{k_6} = \frac{5}{8},\quad s_{k_7} = \frac{8}{13},\ \ldots\ldots$$

The denominators of these fractions form the so-called Fibonacci sequence,

$$q_* = 1, 1, 2, 3, 5, 8, 13, \ldots\ldots,$$

such that $q_{n+1} = q_{n-1} + q_n$. We have $s_{k_n} = \dfrac{q_{n-1}}{q_n}$. That the above definition yields just this sequence is easily proved by induction. 54

Upon constructing the subsequence s_{k_*} we get

$$s_{k_1} < s_{k_3} < s_{k_5} < \ldots\ldots < s_{k_6} < s_{k_4} < s_{k_2}.$$

The sequence of members $s_{k_1}, s_{k_3}, s_{k_5}, \ldots\ldots$ is thus increasing and bounded. The members $s_{k_2}, s_{k_4}, \ldots\ldots$ correspondingly form a decreasing bounded sequence.

These sequences are therefore concentrated. However, none of the members s_n (i.e., no rational number, since s_* includes *all* rational numbers in the interval $[0 \mid 1]$) is the limit of the sequences. There is exactly one i such that $k_i \leqslant n \leqslant k_{i+1}$. For this i, $s_n \notin (s_{k_{i-1}} \mid s_{k_i})$ holds by the definition of k_{i+1}, so that s_n is neither the limit of $s_{k_1}, s_{k_3}, \ldots\ldots$ nor of $s_{k_2}, s_{k_4}, \ldots\ldots$ [Both sequences have the same real number ρ as limit such that $\rho = \dfrac{1}{1+\rho}$, i.e., $\rho = \dfrac{1}{2}(\sqrt{5} - 1)$—as is shown by the representation

$$s_{k_{n+1}} = \frac{q_n}{q_{n+1}} = \frac{q_n}{q_{n-1} + q_n} = \frac{1}{1 + \dfrac{q_{n-1}}{q_n}} = \frac{1}{1 + s_{k_n}};$$

though we are not yet in a position to prove this.]

The irreversibility of Theorem 6.1 can also be expressed as follows: the limit operation (the move from r_* to lim r_*) is not applicable to every concentrated sequence. The situation is like that prior to the introduction of the positive rational numbers. Division failed of universal applicability to the natural numbers. Now the limit operation fails of universal applicability to the rational numbers (with the obvious restriction to concentrated sequences, of course).

If we are to proceed to extend the rational numbers so that "all" concentrated sequences will be convergent, we must bear in mind that the concentrated sequences (of rational numbers) do not form a definite set, but only a class. The best we can do is to initiate an indefinite process of extension which, to begin with, will yield limits for certain definite sequences, but which can always be extended to such further definite sequences as may crop up.

The desired limits are again obtained by abstraction. Let r_* and s_* be two concentrated sequences (of rational numbers). We call r_* and s_* *equivalent* (and write $r_* \sim s_*$) if the *difference sequence* $r_* - s_*$ with members $r_1 - s_1, r_2 - s_2, \ldots\ldots$ is a null sequence.

DEFINITION 6.1 $r_* \sim s_* \leftrightharpoons r_* - s_*$ is a null sequence.

The relation $\sim$ is indeed an equivalence relation:

$$r_* \sim r_* . \tag{6.1}$$

55 For it is, after all, the case that

$$r_* - r_* = 0, 0, 0, \ldots\ldots$$

$$r_* \sim t_* \wedge s_* \sim t_* \dot{\rightarrow} r_* \sim s_* . \tag{6.2}$$

Proof. By hypothesis there is for every ε an M and N such that

$$\bigwedge_{m>M} |r_m - t_m| < \varepsilon, \quad \bigwedge_{n>N} |s_n - t_n| < \varepsilon.$$

Now if, say, $M \leqslant N$, then it will follow for all $n > N$ that

$$|r_n - s_n| \leqslant |r_n - t_n| + |t_n - s_n| < 2\varepsilon,$$

i.e., $r_* \sim s_*$; and similarly for $N \leqslant M$.

By abstraction with respect to $\sim$ we now form "real numbers" from concentrated sequences of rational numbers. That is, we restrict our statements about sequences to those which are invariant (with respect to $\sim$)—and we then write the statements as statements about new abstract objects, *the real numbers*.

If $A(t_*)$ is an invariant statement about t_*, i.e., if

$$r_* \sim s_* \wedge A(r_*) \dot{\to} A(s_*) \tag{6.3}$$

(an indefinite condition), then we write $A(\lim t_*)$ in place of $A(t_*)$.
In particular, however, we write

$$\lim r_* = \lim s_* \quad \text{for} \quad r_* \sim s_*.$$

The expressions $\lim t_*$ or $\lim_n t_n$, or even more explicitly $\lim_{n \to \infty} t_n$, are called *terms for real numbers.*

For the real numbers we use the variables $\xi, \eta, \zeta, \ldots$

Next we have to define the *arithmetical operations* in an invariant manner for the real numbers, i.e., for concentrated sequences. We stipulate that

$$\begin{aligned} \lim r_* + \lim s_* &= \lim(r_* + s_*) \\ \lim r_* \cdot \lim s_* &= \lim(r_* \cdot s_*), \end{aligned} \tag{6.4}$$

where the sequences on the right are to be composed member by member.

For these definitions to be invariant, it clearly suffices that

$$\begin{aligned} r_* \sim r_*' &\to r_* + s_* \sim r_*' + s_* \\ r_* \sim r_*' &\to r_* \cdot s_* \sim r_*' \cdot s_*. \end{aligned} \tag{6.5}$$

The first of these is trivial. For the second, we must prove that $r_* \cdot s_* - r_*' \cdot s_*$ is a null sequence if $r_* - r_*'$ is. This follows from

$$|r_n \cdot s_n - r_n' \cdot s_n| = |r_n - r_n'| \cdot |s_n|,$$

56

for the factor $|s_n|$ here is bounded; i.e.,

$$\vee_\delta \wedge_n |s_n| < \delta.$$

Its boundedness follows in turn directly from the concentration of s_*. This is our first use of the latter assumption.

For the sum and the product of real numbers to be real numbers themselves means that $r_* + s_*$ and $r_* \cdot s_*$ must be concentrated sequences if r_* and s_* are. This is in fact the case. If

$$\wedge_{\substack{m_1, m_2 \\ >M}} |r_{m_1} - r_{m_2}| < \varepsilon, \quad \wedge_{\substack{n_1, n_2 \\ >N}} |s_{n_1} - s_{n_2}| < \varepsilon,$$

then (providing $M \leqslant N$)

$$\wedge_{\substack{n_1, n_2 \\ >N}} |(r_{n_1} + s_{n_1}) - (r_{n_2} + s_{n_2})| < 2\varepsilon$$

and

$$\bigwedge_{\substack{n_1,n_2\\>N}} |r_{n_1} \cdot s_{n_1} - r_{n_2} \cdot s_{n_2}| \leqslant |(r_{n_1} - r_{n_2}) \cdot s_{n_1} + r_{n_2}(s_{n_1} - s_{n_2})|$$
$$\leqslant (|s_{n_1}| + |r_{n_2}|)\varepsilon.$$

Thus, since r_* and s_* are bounded, $r_* \cdot s_*$ is concentrated too.

If we are given an arbitrary definite set of rational sequences for the production of real numbers, that set need not contain $r_* + s_*$ and $r_* \cdot s_*$ along with r_* and s_*. But it is of course always possible to extend the set of sequences in such a (definite) way that it becomes *closed* under addition and multiplication. This state of affairs can be expressed by saying that the class of "all" sequences is *closed* under addition and multiplication.

The laws of commutativity, associativity, and distributivity immediately extend from the rational to the real numbers by virtue of (6.4).

Subtraction again turns out to be applicable without restriction. The equation

$$\lim r_* + \xi = \lim s_*$$

is solved by $\xi = \lim(s_* - r_*)$. We may thus assume that the difference sequence is contained in the set of sequences in question.

The applicability of division to real numbers, $\frac{\lim r_*}{\lim s_*}$, holds under the condition that

$$\lim s_* \neq \bar{0}.$$

$\bar{0}$ here refers to the limit of the (concentrated) sequence 0, 0, 0, ... $\bar{0}$ is the identity element for the addition of real numbers:

57

$$\zeta + \bar{0} = \bar{0} + \zeta = \zeta.$$

$\lim s_* \neq \bar{0}$ means in view of $s_n - 0 = s_n$ that s_* is not a null sequence; i.e.,

$$\bigvee_\varepsilon \bigwedge_N \bigvee_{\substack{n\\>N}} |s_n| > \varepsilon. \tag{6.6}$$

Since s_* is also concentrated—only here do we fully exploit this assumption—we even get

$$\bigvee_\varepsilon \bigvee_M \bigwedge_{\substack{m\\>M}} |s_m| > \varepsilon. \tag{6.7}$$

For if $|s_m - s_n| < \frac{\varepsilon}{2}$ and $|s_n| > \varepsilon$ hold for all $m, n > M$, then it follows that $|s_m| > \frac{\varepsilon}{2}$.

Thus it follows in any case from $\lim s_* \neq \bar{0}$ that

$$\bigvee_m \bigwedge_{m > M} s_m \neq 0.$$

All members of the sequence s_* are finally $\neq 0$.

If the finitely many members of s_* that $= 0$ are replaced by, say, 1, then a concentrated sequence s_*' results such that $s_* \sim s_*'$. All members of s_*' are $\neq 0$.

The solution of the equation

$$\zeta \cdot \lim s_* = \lim r_*$$

can now be given as

$$\zeta = \lim \frac{r_*}{s_*'}.$$

The sequence $\frac{r_*}{s_*'}$ is concentrated, since

$$\left|\frac{r_m}{s_m'} - \frac{r_n}{s_n'}\right| = \frac{|r_m s_n' - r_n s_m'|}{|s_m' s_n'|} = \frac{|(r_m - r_n)s_n' + r_n(s_n' - s_m')|}{|s_m' s_n'|}$$

$$\leq \frac{|r_m - r_n||s_n'| + |r_n||s_n' - s_m'|}{|s_m'||s_n'|}.$$

The right-hand factors s_n', r_n, $\frac{1}{s_m'}$, $\frac{1}{s_n'}$ are all bounded, particularly in view of (6.7).

With that we have established that the real numbers form a *field* (and remember: a class!).

To define an *ordering relation*, we first stipulate that 58

$$\lim r_* > \overline{0} \leftrightharpoons \bigvee_\varepsilon \bigvee_N \bigwedge_{n > N} r_n > \varepsilon. \tag{6.8}$$

Thus all members of r_* are to be finally $> \varepsilon$ (for an appropriate ε). The real number $\lim r_*$ in such a case is called *positive*.

The statement of positivity is invariant:

$$r_* \sim r_*' \wedge \lim r_* > \overline{0} \rightarrow \lim r_*' > \overline{0}.$$

Proof. If for all sufficiently large n, $r_n > \varepsilon$ and $|r_n - r_n'| < \varepsilon/2$, then it is also true for all sufficiently large n that $r_n' > \varepsilon/2$.

We now posit

$$\lim r_* > \lim s_* \leftrightharpoons \lim (r_* - s_*) > \overline{0}. \tag{6.9}$$

For $s_* = 0, 0, 0, \ldots\ldots$ (6.9) reverts to (6.8). We use $<$, $\leq$, $\geq$ as with the rationals.

For the ordering relation $>$ we must prove the order laws (totality, transitivity, and monotony). To this end it is enough to prove the following statements concerning the positivity of the real numbers:

$$\begin{array}{ll} (1) & \xi = \bar{0} \dot{\rightarrow} \xi > \bar{0} \vee -\xi > \bar{0} \\ (2) & \xi > \bar{0} \wedge \eta > \bar{0} \dot{\rightarrow} \xi + \eta > \bar{0} \\ (3) & \xi > \bar{0} \wedge \eta > \bar{0} \dot{\rightarrow} \xi \cdot \eta > \bar{0}. \end{array} \tag{6.10}$$

For (1) we have only to derive a slightly strengthened form of (6.7) from the assumption (6.6). (2) and (3) follow immediately from the monotony of the ordering relation for the rational numbers.

In view of (6.10), absolute value may be defined for real numbers in the same way as for rationals:

$$|\xi| = \begin{cases} \xi & \text{if} \quad \xi > \bar{0} \\ \bar{0} & \text{if} \quad \xi = \bar{0} \\ -\xi & \text{if} \quad \xi < \bar{0}. \end{cases}$$

The ordering furthermore satisfies the "*Archimedean*" property (which is trivial for rational numbers):

$$\bigwedge_{\substack{\xi,\eta \\ >\bar{0}}} \bigvee_n n\xi > \eta. \tag{6.11}$$

Since $\xi > 0$ and $\eta > 0$ imply that $\xi/\eta > 0$, too, (6.11) reduces to

$$\bigwedge_{\substack{\zeta \\ >\bar{0}}} \bigvee_n n\,\zeta > \bar{1}. \tag{6.12}$$

59 $\bar{1} = \lim 1, 1, 1, \ldots\ldots$ here is the *identity element* for multiplication.

Since the real number $n\zeta$ is furthermore defined for $\zeta = \lim s_*$ as the limit of the sequence ns_* (with members $ns_1, ns_2, \ldots\ldots$), (6.12) follows directly from the corresponding property of the rational numbers.

The class of all real numbers is thus an *Archimedean field.*

§7. The Completeness of the Field of Real Numbers

The real numbers include in particular the limits $\bar{r}$ of the constant sequences $r, r, r, \ldots\ldots$

Of these special real numbers, the following are true:

$$\begin{aligned} \bar{r} = \bar{s} &\leftrightarrow r = s \\ \bar{r} > \bar{s} &\leftrightarrow r > s \\ \bar{r} + \bar{s} &= \overline{r + s} \\ \bar{r} \cdot \bar{s} &= \overline{r \cdot s}. \end{aligned}$$

Therefore we shall simply write r from now on when we mean the real number $\bar{r}$.

The (definite) field of the rational numbers thereby comes to be seen as a subfield of the (indefinite) field of reals.

Nevertheless, it is still advisable to retain the notation $\bar{r}$ in place of r for the following theorem about concentrated sequences of rationals:

$$\lim \bar{r}_* = \lim r_*. \tag{7.1}$$

We have here on the left a sequence $\bar{r}_*$, i.e., a sequence of real numbers. (7.1) asserts that this sequence is convergent and that the real number $\lim r_*$ is the limit of the real sequence $\bar{r}_1$, $\bar{r}_2, \ldots$ . .

The general definition of convergence for real sequences is analogous to Definition 5.2:

A real sequence ξ_* is *convergent* if there is a real number η such that

$$\bigwedge_{\varepsilon} \bigvee_N \bigwedge_{n>N} |\xi_n - \eta| < \bar{\varepsilon}. \tag{7.2}$$

To get (7.1) we must therefore prove that

$$\bigwedge_{\varepsilon} \bigvee_N \bigwedge_{n>N} |\bar{r}_n - \lim r_*| < \bar{\varepsilon}.$$

The inequality $|\bar{r}_n - \lim r_*| < \bar{\varepsilon}$ means that

$$\bar{\varepsilon} \pm (\bar{r}_n - \lim r_*) > \bar{0}.$$

The two real numbers on the left are the limits of the sequences $\varepsilon \pm (r_n - r_*)$.

Thus, by (6.8), we have to prove

60

$$\bigvee_{\delta} \bigvee_M \bigwedge_{m>M} \varepsilon \pm (r_n - r_m) > \delta.$$

For $\delta = \frac{\varepsilon}{2}$ this holds for all sufficiently large n, as the concentration of r_* directly implies.

(7.1) guarantees that a concentrated sequence of rational numbers (or, more precisely, of the special reals $\bar{r}$) will have a real number as limit.

With due attention to the indefiniteness of the class of all sequences, we can also say that "all" concentrated sequences of rational numbers converge. However, that can mean nothing more than that appropriate real limits can be constructed for every definite set of such sequences.

The natural question of whether "all" concentrated sequences of *real* numbers converge (the converse is trivial) must likewise first of all be interpreted. Since every real number ξ is to be represented as the limit of a rational sequence r_*, the representation of a real sequence $\xi_1, \xi_2, \xi_3, \ldots\ldots$ requires a term $S(m)$ that represents a rational sequence for every m:

$$S(m) = T(m, n).$$

That is, we need a 2-place rational term.

The term $T(m, n)$ represents a sequence r_{m*} for every m, and hence represents the real number $\xi_m = \lim r_{m*}$. We can also say that the real sequence $\xi_\dagger$ is represented by the double rational sequence $r_{\dagger *}$.

Now let the real sequence $\xi_\dagger$ be concentrated (i.e., let

$$|\lim r_{m*} - \lim r_{n*}| < \varepsilon$$

hold for every ε and for all sufficiently large m, n) and let the sequences r_{n*} be concentrated for every n. By means of a null sequence δ_*, say $\delta_n = \frac{1}{n}$, we define a rational sequence s_* as follows:

Let k_n be the smallest k such that $|r_{nk} - \xi_n| < \delta_n$, i.e.,

$$|r_{nk_n} - \xi_n| < \delta_n .$$

We set $s_n = r_{nk_n}$ (Cauchy diagonal procedure).

The sequence s_* is concentrated, since

$$|s_m - s_n| \leqslant |s_m - \xi_m| + |\xi_m - \xi_n| + |\xi_n - s_n|,$$

where, for every ε, the three right-hand terms become less than $\frac{\varepsilon}{3}$ for all sufficiently large m, n.

61 For the real number $\xi = \lim s_*$ it follows that $\xi = \lim \xi_\dagger$, for in

$$|\xi - \xi_n| \leqslant |\xi - s_n| + |s_n - \xi_n|$$

the three right-hand terms again become, for every ε, less than $\frac{\varepsilon}{2}$ for all sufficiently large n.

From each definite term representing a double sequence $r_{\dagger *}$ we have constructed a new definite term that represents a sequence s_*. In this way we can construct a real number $\xi = \lim \xi_\dagger$ for every real concentrated sequence $\xi_\dagger$; i.e., we can construct a term representing such a real number.

This state of affairs may be formulated as the "*Cauchy convergence criterion*": that every concentrated real sequence converges to a real number (and vice versa). Because of the indefiniteness of the class of all real numbers, however, this does not mean that a *definite* set of real numbers could be constructed that would be *complete*, i.e., such that every concentrated sequence of numbers in that set would have a limit within that set.

In order to prove the incompleteness of any definite set of real numbers, we must first agree on what we mean by a definite set of real numbers. Since the real numbers are represented by terms (for rational sequences), what we must construct are sets of terms. Let us now speak of a definite set only if a construction rule is given such that every term to be constructed is a finite combination of finitely many specified individual symbols.

The totality of (finite) combinations is denumerable; i.e., by an inductive definition a unique correlation can be established between the natural numbers and the symbol combinations.

From this it follows that every definite set of real numbers can be mapped one-to-one into the set of natural numbers: every real number in the definite set is correlated with exactly one natural number.

Now, every set of natural numbers is at most denumerable. (Take the smallest number in the set as the first one, the next larger as the second one, etc. Then if n occurs at all, it will be reached in n steps at the latest.) Every definite set of real numbers is thereby seen to be denumerable. Such a set can be represented as the set of the members of a real sequence ξ_1, ξ_2, ξ_3, ... We shall refer to this set $\{\xi_1, \xi_2, \xi_3, \ldots\ldots\}$ as the *set associated* with the sequence $\xi_1, \xi_2, \xi_3, \ldots\ldots$

To be sure, there are sequences of real numbers that are "*complete*" in the sense that *every* concentrated subsequence converges to a member of the sequence. An example is any convergent sequence which contains its limit as a member.

But even the property of *density* is incompatible with completeness. A set K of real numbers is called dense if for any two members $\xi, \eta \in K$ such that $\xi < \eta$ there is always a $\zeta \in K$ such that $\xi < \zeta < \eta$. *62*

If a set K is dense, then clearly $\underset{K}{\in_\zeta}(\xi < \zeta < \eta)$ will always be infinite for $\xi, \eta \in K$ such that $\xi < \eta$.

For sequences whose associated set is dense (and is not trivially a unit set), we now have

THEOREM 7.1 For every nonconstant sequence whose associated set is dense there is a concentrated subsequence the limit of which is not in the sequence.

We already established a special case of this theorem to prove the existence of rational concentrated sequences without rational limits. The proof strategy followed in the special case extends immediately to the general theorem. (Cf. pp. 54–55.)

If a sequence of real numbers includes all rational numbers—or at least all the rational numbers of an interval in which the members of the sequence lie—then we know for certain that the sequence is dense: between any two real numbers there is always a rational number.

Proof. If $\xi = \lim r_*$ and $\eta = \lim s_*$ and, say, $\xi < \eta$, then

$$\lim (s_* - r_*) > 0;$$

i.e.,

$$\bigvee_{\varepsilon} \bigwedge_{N} \bigvee_{\substack{n \\ >N}} s_n - r_n > \varepsilon.$$

From this it follows immediately that

$$\eta - \xi > \varepsilon \quad \text{for some } \varepsilon.$$

Now if, say,

$$|r_n - r_N| < \frac{\varepsilon}{3} \quad \text{for all} \quad n > N,$$

then $r_N + \frac{2}{3}\varepsilon$ is a rational number such that

$$\xi < r_N + \frac{2}{3}\varepsilon < \eta.$$

For we have

$$|\xi - r_N| \leqslant \frac{\varepsilon}{3},$$

i.e.,

$$\xi \leqslant r_N + \frac{\varepsilon}{3} \quad \text{and} \quad r_N \leqslant \xi + \frac{\varepsilon}{3},$$

whence

$$r_N + \frac{2}{3}\varepsilon \leqslant \xi + \varepsilon < \eta.$$

63 Thus the incompleteness of the rational numbers with respect to the limit operation (as applied to concentrated sequences) cannot be disposed of in the same way as, say, the incompleteness of the positive rational numbers with respect to subtraction. It is possible so to construct the field of rational numbers that all subtractions can be carried out, whereas there is no construction of real numbers starting from the set of rational numbers so that every (concentrated) sequence of the numbers constructed has a limit within the constructed extension. Rather, any definite extension by means of real numbers can in turn be extended all over again. It changes nothing to express this state of affairs by saying of the (indefinite) class of all real numbers that it is *complete.*

Most of the current textbooks on analysis posit an axiom system for real numbers at the outset, according to which the real numbers form an Archimedean field which is supposed to be "complete" in the sense that *every* concentrated sequence of real numbers converges to a real number.

As we have seen, such an axiom system is not satisfiable under the construction of real numbers described (viz., an abstraction from concentrated rational sequences that "equates" two sequences when their difference sequence is a null sequence).

Whether the axiom system can be satisfied at all upon appropriate extension to include axioms for the set or sequence concept utilized, indeed, whether it is even consistent—these questions can be dispensed with so far as figuring with differentials and integrals is concerned.

From the constructive standpoint, however, the significance of the notion "Archimedean field" for real-number theory can be elucidated as follows. Every Archimedean field F is isomorphic with a field of real numbers (i.e., with a field the elements of which are contained in the class of all real numbers). "*Isomorphic*" is here understood to mean that each element of the field F can be correlated with exactly one real number in such a way that the sum or product of two elements is always correlated with the sum or product of the two correlative numbers.

To prove this, we can proceed as follows. As F is a field, it includes a null element o and a unit element e. These elements are correlated with the numbers 0 and 1. As F is to be an ordered field, all elements $e, e + e, e + e + e, \ldots\ldots$ must be distinct (F has the characteristic 0, as one would say in algebra). These elements are correlated with the natural numbers 1, 2, 3, ... At this point, we immediately see that F also contains elements that must be correlated with the rest of the rational numbers. F therefore contains a subfield F_0 that is isomorphic with the field of rational numbers. F_0 consists of the elements re. If now $F = F_0$, then F is itself isomorphic with the field of rational numbers, and is hence a field of real numbers. If, on the other hand, $F_0 \subset F$, then there must be certain elements x of F which are not in F_0. We have to show how to correlate each x with *64* exactly one real number. At this point, the Archimedean property assumes a crucial role. Let $x > 0$.

For every n there is an m, hence also a smallest m, such that

$$me > nx.$$

Thus there is a smallest rational number r_n $\left(\text{viz.}, \frac{m}{n}\right)$ representable with the denominator n for which $r_n e > x$.

It is furthermore true of r_n that

$$r_n e > x \geqslant \left(r_n - \frac{1}{n}\right)e,$$

whence

$$0 < r_n e - x \leqslant \frac{1}{n} e.$$

Next we consider the definite sequence r_*, which is easily proved to be concentrated. From $|x - r_n e| \leqslant \frac{1}{n} e$ it follows that

$$|(r_m - r_n)e| \leqslant \frac{2}{m} e \quad \text{for} \quad m < n,$$

and hence that

$$|r_m - r_n| \leqslant \frac{2}{m}.$$

The sequence $r_* e$ is likewise concentrated, and within F it is the case that

$$x = \lim (r_* e).$$

We correlate the element x with the real number $\xi = \lim r_*$ and set $x = \xi e$.

If $y \neq x$ is another element of F, and if y is correlated in the way just described with the real number η $\left(\eta = \lim s_*, \text{ where } 0 < s_n e - y \leqslant \frac{1}{n} e\right)$, then η will also be $\neq \xi$.

Proof. If y is, say, $> x$, then the Archimedean property entails the existence of an n such that $y - x > \frac{1}{n} e$. From $x = \lim (r_* e)$ and $y = \lim (s_* e)$ it follows further that $y - x = \lim (s_* - r_*)e$. Thus $s_* - r_*$ cannot be a null sequence.

65 The fact that this correlation of the elements of F with the real numbers is an isomorphism (with respect to addition and multiplication) follows from

$$\lim r_* + \lim s_* = \lim (r_* + s_*)$$
$$\lim r_* \cdot \lim s_* = \lim (r_* \cdot s_*).$$

We can therefore say that every Archimedean field is contained in the class of all real numbers up to isomorphisms (i.e., upon abstraction with respect to the equivalence relation given by an appropriate isomorphic mapping). Conversely, the indefinite field of all real numbers contains (up to isomorphisms) all Archimedean fields. In this sense it is "the" largest Archimedean field.

The completeness we have been discussing of the indefinite field of all real numbers (every concentrated sequence of real numbers has a real number as its limit) is frequently expressed in a different though equivalent manner; e.g.,

Every bounded nonempty set of real numbers has a real number as its least upper bound.

The notions referred to here must first of all be elucidated. A set K is said to be *bounded* if

$$\bigvee_c \bigwedge_\xi . \xi \in K \rightarrow |\xi| < c.$$

(Whether c is taken here to be a real, a rational, or a natural number is immaterial.)

c is an *upper* {*lower*} *bound* of K if

$$\bigwedge_\xi . \xi \in K \rightarrow \xi \leqslant c.$$
$$\{\bigwedge_\xi . \xi \in K \rightarrow \xi \geqslant c.\}.$$

A lower bound γ of a set K is called the *greatest lower bound* ("*glb*") if all lower bounds are $\leqslant \gamma$. There is clearly at most one greatest lower bound. We express this as $\gamma = \text{glb } K$. The *least upper bound*, lub K, is defined analogously.

We may wish to show that if a set K has an upper bound at all, then it must also have a lub. It is natural to consider for each n the least upper bound r_n of K which is representable in the form $\frac{m}{n}$, and then to try forming $\lim r_*$. However, if we write out the definition of this sequence r_*,

66

$$r_n = \frac{T(n)}{n}$$
$$T(n) = \mu_m \bigwedge_\xi . \xi \in K \rightarrow \xi \leqslant \frac{m}{n}., \tag{7.3}$$

then we see that the term representating the numerator of r_n contains an indefinite quantifier. Thus the sequence r_* would not necessarily be definite.

We must therefore impose a definiteness requirement on the representation of K. Let this be done as follows. We replace the set K by the set of rational numbers r such that $r < \xi \in K$ for some ξ. Call this set the *left class* of K. The set K and its left class have the same numbers as upper bounds and hence—if there is one—the same lub. We now require that the left class be definite, i.e., that it be representable as $\in_r A(r)$ with a definite statement form $A(r)$ (even though $\bigvee_\xi r < \xi \in K$ is indefinite).

This statement form $A(r)$ is adequate to characterize the upper bounds of K. From $\bigwedge_\xi . \xi \in K \rightarrow \xi \leq \xi_0$. it immediately follows that $\wedge_r . A(r) \rightarrow r \leq \xi_0$., since $A(r) \rightarrow \bigvee_\xi r < \xi \in K$; and from $\wedge_r .A(r) \rightarrow r \leq \xi_0$. it follows that $\bigwedge_\xi . \xi \in K \rightarrow \xi \leq \xi_0$., since $\xi > \xi_0 \rightarrow \vee_r \xi > r > \xi_0$.

We can formulate the following theorem:

Every nonempty set of real numbers with an upper bound and a definite left class has a real number as lub.

Proof. The numbers r_n defined above now form—as is easily seen—a definite concentrated sequence r_*. Therefore $\eta = \lim r_*$ is a real number. η is the lub of K. From $\xi \leq r_n$ for all $\xi \in K$ it follows in any case that $\xi \leq \eta$ for all $\xi \in K$, so that η is an upper bound. Were ζ some upper bound such that $\zeta < \eta$, then $\eta - \zeta$ would be $> \varepsilon$ for some ε. Furthermore, $r_n > \eta - \frac{\varepsilon}{2}$ for all sufficiently large n. For these n it follows that $r_n - \frac{\varepsilon}{2} > \zeta$; i.e., $r_n - \frac{\varepsilon}{2}$ would still be an upper bound. For those n such that $\frac{1}{n} < \frac{\varepsilon}{2}$, that contradicts the definition of r_n.

The limit of a bounded increasing sequence is also the lub of the associated set.

Proof. First of all, the limit ξ is an upper bound. It is furthermore the lub, since for every ε all members of the sequence will finally lie between ξ and $\xi - \varepsilon$.

Corresponding theorems hold for the glb.

Our earlier proof that dense definite sets of real numbers are incomplete, or contain concentrated sequences with no limit (inside the set), thus shows that such a set is also incomplete in the following sense: it includes bounded subsets with no lub in the set. An example is the set associated with the increasing sequence we constructed.

67

Nevertheless, we can still say that every nonempty bounded set (with definite left class) in the indefinite class of the real numbers has a lub. For all definite extensions of the rational numbers by real numbers, the appropriate lub of any nonempty bounded subset (with definite left class) can be annexed.

Just as every real number can be represented as the limit of a definite sequence of rationals, so it can also be represented as the glb or lub of a bounded set of rationals.

The historically most important representation, the decimal fraction, is a particular case of representation by concentrated sequences. The advantage of the sequential representation is primarily that the statement $\xi = \lim r_*$ can always be supplemented by an estimate

$$|\xi - r_n| < \varepsilon_n,$$

where ε_* is a null sequence. $\left(\text{For decimals, } \varepsilon_n \text{ is } \frac{1}{10^n}\text{, and hence is defined independently of } \xi \text{ and } r_*.\right)$

For the representation of real numbers by sets $\xi = \text{glb } K$, no such supplementation is possible. We therefore gave preference to representation by means of sequences $\xi = \lim r_*$, and used the notions of glb and lub only to simplify certain proof procedures. In particular, the existence of a lub yields a general principle for proving certain statements about real numbers. The principle closely resembles the arithmetical principle of induction: it may be called the *analytic induction principle.* We consider *"left-hereditary"* statement forms (we could also speak of properties or classes), i.e., statements $A(\xi)$ for which $\xi_1 < \xi_2 \dot\to A(\xi_2) \to A(\xi_1)$.

Analytic Induction Principle. For left-hereditary definite statement forms $A(\xi)$ it always follows from

(1) $A(\alpha)$

(2) $\bigwedge_{\xi_0} . \bigwedge_{\xi} . \xi < \xi_0 \to A(\xi) . \to \bigvee_{\delta} A(\xi_0 + \delta).$

that

(3) $\bigwedge_{\xi} A(\xi).$

Thus the analytic induction principle permits the following sort of inference. From the fact that a definite left-hereditary formula (1) is true of a number and (2) holds in an interval to the right of ξ_0 whenever it holds everywhere to the left of ξ_0, we can conclude that the formula is universally valid. To prove this induction principle, it is enough in view of left heredity to show that the set $\epsilon_r A(r)$ has no upper bound. In view of (1), $A(r)$ holds of all $r < \alpha$, so the set $\epsilon_r A(r)$ is not empty. We now proceed by indirect proof. If $\epsilon_r A(r)$ had an upper bound, then it would have a lub, *68*

$$\xi_0 = \text{lub } \epsilon_r\, A(r).$$

It follows that

$$\wedge_r . r < \xi_0 \rightarrow A(r).,$$

and hence also that

$$\wedge_\xi . \xi < \xi_0 \rightarrow A(\xi).$$

Thus, by (2), $A(\xi_0 + \delta)$ holds of some δ, and in particular of several $r > \xi_0$, contrary to the definition of ξ_0.

The induction principle is usually applied to statement forms of the type

$$\alpha \leqslant \xi \rightarrow B(\xi).$$

Here $A(\xi)$ is trivially true of all $\xi < \alpha$.

We shall have frequent occasion to employ the induction principle. As the proof shows, it is only available when the statement form $A(r)$ is definite. Otherwise no lub can be formed.

III. ONE-PLACE FUNCTIONS

§8. Continuous Functions

In §3 we defined functions with natural or rational numbers as arguments. For example, a function of rationals is represented by a term $T(r)$ with a variable r ranging over rational numbers. The domain of arguments, i.e., the set of numbers that can be substituted for r, is in such a case constructible.

The peculiarity of the real functions—by which we mean in this book functions from reals to reals—is that the argument domain is not in general constructible, but is rather an indefinite class.

It is nevertheless a quite simple matter to define functions taking *all* real numbers as arguments, e.g., by means of the term 3ξ. Every real number ξ is correlated as an argument with the real number 3ξ as value. If we have a representation $\xi = \lim r_*$ for the argument ξ, then for the value 3ξ we have the representation $3\xi = \lim 3r_*$. All we have to do is to show—as we did in §6—that the various possible representations of the argument always give rise to the same value for the function.

If we use unique representations, e.g., decimals

$$\xi = \sum_{0}^{\infty}{}_n \frac{a_n}{10^n}$$

with a sequence a_* of integers (where $0 \leqslant a_n \leqslant 9$ for $n > 0$ and not finally all $a_n = 9$, i.e., $\bigwedge_N \bigvee_{m>n} a_n \neq 9$), then we may disregard the question of invariance. We can define, e.g.,

$$\phi\xi = \begin{cases} 0 \text{ when only for finitely many } n, a_n = 7 \\ 1 \text{ when for infinitely many } n, a_n = 7. \end{cases}$$

The right side here could be replaced by an ι-term (cf. §3).

We now generalize the notion of function by ceasing to require that *all real* numbers be substituends for the variable ξ in the representing term $T(\xi)$. We shall call $\phi = \daleth_\xi T(\xi)$ a function even when $T(\xi)$ is a term (e.g., a description rather than a pseudodescription) only for ξs from a certain set K. The set K is then called the *argument domain* (or simply the *domain*) of ϕ. If ξ 70 belongs to the argument domain of a function ϕ, then we say for short that ϕ is *applicable* to ξ. The *application* of ϕ to ξ yields the function value ϕ℩ξ. ℩ may thus also be called the *application operator* (or the application function: its arguments are pairs composed of a function and an argument thereof).

We avoid the common formulation "ϕ is defined for ξ" because this conflicts with the usage of §4 according to which the term $\phi\xi$ or the function symbol ϕ is the sort of thing that can be defined. Thus "definition," like "proof," is only employed metamathematically.

For many applications, e.g., for "describing" empirical dependencies, the *continuous* functions are the most important. Suppose we conjecture that in some physical system a certain measurable quantity B (e.g., the pressure at some point) depends solely upon another quantity A (e.g., the temperature). This conjecture can be confirmed only by arbitrarily varying the quantity A and then measuring A and B each time. The measurement values will always be rational numbers, 6-place decimals for instance. Only when improved precision in the measurement of A results in more and more precise measurement values for B can we say that B depends solely on A. Were B to vary measurably although A remained constant within the range of tolerance, then we should suspect that the quantity B did *not* depend on A alone. We should then have to take the variation of other quantities A', A'', . . . (e.g., the external air pressure) into account until B finally turned out to depend only on A, A', A'', . . . The closer A, A', A'' are approximated to particular values, the closer a definite value of B must be measured. Now assume that B has proved empirically to be dependent on A alone. B is then said to be a "function" of A. In this physical usage of the word "function," a function is a quantity that depends on another quantity.

In mathematics, however, the word "function" is used differently. Functions here are abstract objects representable by terms. The connection with dependency relationships between quantities is the following: mathematical functions serve to "describe" the dependency relationships.

If B is dependent on A, then to describe the dependency on A a function (i.e., a term in an appropriate language) is sought which yields as values the measurements of B corresponding to measurement values of A taken as arguments.

The "describing" function (it is really a matter here of something entirely different from a description, such as of a landscape *71* or a trip) is subject to a condition arising from the requirement that increased precision in the settings of A shall make any desired precision possible in the measurements of B. The resulting condition on the describing function ϕ is that of *uniform continuity*. If a maximum tolerance ε is stipulated for the measurement of B, then there must be a maximum tolerance δ for A such that

$$\bigwedge_{\xi_1, \xi_2} \cdot |\xi_1 - \xi_2| < \delta \rightarrow |\phi \xi_1 - \phi \xi_2| < \varepsilon.$$

The arguments ξ_1, ξ_2 here must of course be restricted to the possible settings of A.

It will turn out that a uniformly continuous function is uniquely determined by the values of the function for rational arguments. (Even for rational arguments the exact values of the function cannot be ascertained by measurement, however. The measurement results are rather "idealized" by the describing function.)

In order to investigate this relationship—independently of any applications for the time being—we first of all define what it is for a function to be uniformly continuous in its domain.

Let the function ϕ have as domain a set K of real numbers.

DEFINITION 8.1 ϕ is *uniformly continuous* in $K \Leftrightarrow$

$$\bigwedge_{\varepsilon} \bigvee_{\delta} \bigwedge_{\substack{\xi_1, \xi_2 \\ K}} \cdot |\xi_1 - \xi_2| < \delta \rightarrow |\phi \xi_1 - \phi \xi_2| < \varepsilon. \qquad (8.1)$$

THEOREM 8.1 Let a definite function ϕ_0 be uniformly continuous for the domain of *rational* numbers in a closed interval $[\alpha | \beta]$. Then there is exactly one uniformly continuous function ϕ which has *all real* numbers in the interval $[\alpha | \beta]$ as arguments and which coincides with ϕ_0 for rational arguments.

Proof. (8.1) immediately entails that, for a uniformly continuous function ϕ, $\xi = \lim r_*$ always implies $\phi \xi = \lim \phi r_*$. Accordingly, if ϕ_0 is given, let us define for $\xi = \lim r_*$

$$\phi \xi = \lim \phi_0 r_*.$$

If r_* is a definite sequence, then so is $\phi_0 r_*$ because of the definiteness of ϕ_0. We must also show that $\phi_0 r_*$ is a concentrated sequence when r_* is:

$$\bigwedge_{\varepsilon} \bigvee_N \bigwedge_{\substack{m,n \\ > N}} |\phi_0 r_m - \phi_0 r_n| < \varepsilon.$$

This follows from (8.1) and

$$\bigwedge_{\delta} \bigvee_N \bigwedge_{\substack{m,n \\ > N}} |r_m - r_n| < \delta.$$

Finally, we have $\lim r_* = \lim s_* \rightarrow \lim \phi_0 r_* = \lim \phi_0 s_*$, for

72

$$\bigwedge_\varepsilon \bigvee_N \bigwedge_{n>N} |\phi_0 r_n - \phi_0 s_n| < \varepsilon$$

again follows from (8.1) and

$$\bigwedge_\delta \bigvee_N \bigwedge_{n>N} |r_n - s_n| < \delta.$$

With that we have defined a function ϕ with the entire interval $[\alpha | \beta]$ as its domain. For every rational number r we get $\phi r = \phi_0 r$.

This function ϕ is uniformly continuous in $[\alpha | \beta]$, because all δ, ε such that, for all rational $r_1, r_2 \in [\alpha | \beta]$,

$$|r_1 - r_2| < \delta \rightarrow |\phi r_1 - \phi r_2| < \varepsilon$$

also satisfy

$$|\xi_1 - \xi_2| < \delta \rightarrow |\phi\xi_1 - \phi\xi_2| \leqslant \varepsilon$$

for all real $\xi_1, \xi_2 \in [\alpha | \beta]$. This follows from the fact that the real numbers ξ_1, ξ_2 such that $|\xi_1 - \xi_2| < \delta$ can be represented as

$$\xi_1 = \lim r_{1*}, \qquad \xi_2 = \lim r_{2*},$$

where $|r_{1m} - r_{2n}| < \delta$ for all m, n. Hence we have

$$|\phi r_{1m} - \phi r_{2n}| < \varepsilon,$$

and passage to the limits $m \rightsquigarrow \infty$, $n \rightsquigarrow \infty$ yields

$$|\phi\xi_1 - \phi\xi_2| \leqslant \varepsilon.$$

A uniformly continuous function is thus already uniquely determined by its *rational subfunction* (by which we mean the function as restricted to rational arguments). If the rational subfunction is uniformly continuous, we call the function "*rationally uniformly continuous.*" In what follows we shall confine ourselves to real functions with an interval as argument domain. *We shall always presuppose that the rational subfunctions of those functions are definite functions.* This "*rational definiteness*" will not be expressly mentioned again, since we shall have to do exclusively with real functions which fulfill this condition.

If we consider a uniformly continuous function only in some *neighborhood* of a point ξ_0, i.e., for those ξ such that $|\xi - \xi_0| < \delta_0$ for some δ_0, then there too we shall have

$$\bigwedge_\varepsilon \bigvee_\delta \bigwedge_\xi . |\xi - \xi_0| < \delta \rightarrow |\phi\xi - \phi\xi_0| < \varepsilon. \tag{8.2}$$

73 The number $\eta_0 = \phi\xi_0$ is in fact already uniquely determined by the (somewhat weaker) condition

$$\bigwedge_\varepsilon \bigvee_\delta \bigwedge_\xi . 0 < |\xi - \xi_0| < \delta \rightarrow |\phi\xi - \eta_0| < \varepsilon. \tag{8.3}$$

If (8.3) holds, then η_0 is called the *limit* of ϕ at ξ_0. That does not mean that ξ_0 must necessarily belong to the domain of ϕ. It is enough for a *"deleted neighborhood"* of ξ_0, i.e., a neighborhood $\epsilon_\xi\, |\xi - \xi_0| < \delta_0$ surrounding but not including the point ξ_0, to belong to the argument domain. (The arguments are often referred to as "points.")

We write this as

$$\lim_{\xi \rightarrow \xi_0} \phi\xi = \eta_0 \tag{8.4}$$

or, more briefly,

$$\lim_{\xi_0} \phi = \eta_0.$$

DEFINITION 8.2 A function ϕ applicable in a neighborhood of ξ_0 is *continuous at* ξ_0 if (1) $\lim\limits_{\xi_0} \phi$ exists; and (2) this limit is equal to $\phi\xi_0$; i.e., if (8.2) holds.

If the domain of ϕ is a closed interval $[\alpha \,|\, \beta]$, then ϕ is only defined in a *one-sided* neighborhood of the end points, e.g., for $\xi \geqslant \alpha$.

The condition $\bigwedge_\varepsilon \bigvee_\delta \bigwedge_{\substack{\xi \\ \xi > \xi_0}} . \xi - \xi_0 < \delta \rightarrow |\phi\xi - \eta_0| < \varepsilon.$ also determines the number η_0 uniquely (if there is any such number at all). We call η_0 the *right-hand limit* of ϕ at ξ_0, and we write

$$\lim_{\xi \overset{\rightarrow}{>} \xi_0} \phi\xi = \eta_0 \quad \text{or} \quad \lim_{> \xi_0} \phi = \eta_0.$$

The *left-hand limit* at ξ_0 is characterized exactly analogously for functions defined in a left-hand neighborhood (i.e., for those ξ such that $\xi < \xi_0$ and $\xi_0 - \xi < \delta_0$ for some δ_0). We then write

$$\lim_{\xi \overset{\rightarrow}{<} \xi_0} \phi\xi = \eta_0 \quad \text{or} \quad \lim_{< \xi_0} \phi = \eta_0.$$

If a function ϕ is continuous at ξ_0, the two one-sided limits exist and are equal to $\phi\xi_0$. (The existence of the two-sided limit itself follows from that of the two one-sided limits if and only if the two one-sided limits are the same.)

If $\lim\limits_{<\xi_0} \phi = \phi\xi_0$ $\{\lim\limits_{>\xi_0} \phi = \phi\xi_0\}$, then ϕ is said to be *left-*{*right-*} *continuous*.

DEFINITION 8.3 A function ϕ is *continuous in an interval* if it is continuous at every point of the interval. Should the interval include end points, then the function must be one-sidedly continuous there.

74 We have seen that in the case of functions uniformly continuous in a closed interval all that matters is the rational subfunction. Any rational sequence r_* such that $r_* \rightharpoonup \xi_0$ determines the value of such a function at an arbitrary point ξ_0 as follows:

$$\phi\xi_0 = \lim \phi r_*.$$

Continuity at ξ_0, i.e., $\lim\limits_{\xi_0} \phi = \phi\xi_0$, entails more generally for every real sequence ξ_* that

$$\xi_* \rightharpoonup \xi_0 \rightarrow \phi\xi_* \rightharpoonup \phi\xi_0. \tag{8.5}$$

Proof. For $|\phi\xi_n - \phi\xi_0| < \varepsilon$ to hold, it is sufficient in view of the continuity of ϕ that $|\xi_n - \xi_0| < \delta$ for some appropriate δ. This follows for all $n > N$ (for an appropriate N) from the convergence of ξ_* to ξ_0.

In order to obtain (8.5) for rational sequences alone, it is enough to take account of the rational arguments in a neighborhood of ξ_0 for the continuity of ϕ at ξ_0. The limit is then called a "*rational limit,*" written as

$$\lim_{r \rightharpoonup \xi_0} \phi r = \eta_0$$

for

$$\wedge_\varepsilon \vee_\delta \wedge_r . 0 < |r - \xi_0| < \delta \rightarrow |\phi r - \eta_0| < \varepsilon. \tag{8.6}$$

The one-sided rational limit is defined similarly. If $\lim\limits_{r \rightharpoonup \xi_0} \phi r = \phi\xi_0$, then ϕ is said to be *rationally continuous at* ξ_0.

In expounding the notion of limit for functions, textbooks typically prove the equivalence of the two conditions

(1) $$\lim_{\xi \rightharpoonup \xi_0} \phi\xi = \eta_0$$

and

(2) $$\xi_* \rightharpoonup \xi_0 \rightarrow \phi\xi_* \rightharpoonup \eta_0 \quad \text{for all sequences } \xi_*.$$

(2) may be inferred from (1) as in the proof of (8.5)

We shall do without the inference of (1) from (2), since whenever we shall be able to prove (2), we shall already have been able to prove (1).

Even so, let us take a look at the proof of (1) from (2), since it makes use of the so-called *principle of choice.*

To begin with, we confine ourselves to rational arguments; i.e., we replace the variables ξ and ξ_* in (1) and (2) by variables r and r_*. We then proceed by indirect proof. Assume that (1) is false. The negation of (8.6) implies that for some ε,

$$\wedge_\delta \vee_r .0 < |r - \xi_0| < \delta \wedge |\phi r - \eta_0| \geq \varepsilon.$$

For a null sequence δ_* it follows in particular that

$$\wedge_n \vee_r .0 < |r - \xi_0| < \delta_n \wedge |\phi r - \eta_0| \geq \varepsilon.$$

Now, by the *principle of choice* one concludes generally from a statement of the form $\wedge_n \vee_r A(n, r)$ that $\vee_{r_*} \wedge_n A(n, r_n)$. As the range of values of "r" is denumerable, this presents no difficulty. If for every n there is an r such that $A(n, r)$, and if R_* is a sequence containing *all* rational numbers as members, then there is an m such that $A(n, R_m)$. Now let $\mu(n)$ be the least m such that $A(n, R_m)$: 75

$$\mu(n) = \mu_m A(n, R_m).$$

If r_n is then set $= R_{\mu(n)}$, $r_* = R_{\mu(*)}$ is indeed a sequence which satisfies $\wedge_n A(n, r_n)$; i.e.,

$$\vee_{r_*} \wedge_n A(n, r_n).$$

In the above situation, this principle gives us

$$\vee_{r_*} \wedge_n .0 < |r_n - \xi_0| < \delta_n \wedge |\phi r_n - \eta_0| \geq \varepsilon.$$

For the sequence r_* it is true that $r_* \rightharpoonup \xi_0$, but not that $\phi r_* \rightharpoonup \eta_0$. (2) is therefore false. Thus from (2) it follows (constructively) that *not not* (1), and thence (classically) the definite statement (1).

If we do *not* confine ourselves to rational arguments, then we *cannot* deal analogously with a sequence containing *all* real numbers as members. The way out is to treat the formula

$$\wedge_x \vee_y A(x, y) \rightarrow \vee_f \wedge_x A(x, fx)$$

(where f is a function whose range is included in that of "y") as an "axiom" (sometimes as a "logical axiom," sometimes as a "set-theoretical axiom"). So far, however, no one has shown what implications such axiomatic considerations (in which, of course, anyone can indulge at his pleasure) might have for constructive analysis. Such considerations can thus—so far—prove nothing for our purposes.

It pays to introduce the concept of continuity along with that of uniform continuity, because continuity is frequently easier to prove. To that end the function need only, of course, be considered "*locally*" (i.e., in the neighborhood of a point).

On the other hand, uniform continuity can be inferred with the help of the following theorem:

THEOREM 8.2 If a function ϕ is continuous in a *closed* interval $[\alpha | \beta]$, then ϕ is uniformly continuous in $[\alpha | \beta]$.

Note. A sufficient condition is the rational continuity of ϕ for all ξ_0 in $[\alpha|\beta]$, though not continuity merely for all rational arguments.

Proof. We have to show for arbitrary ε that

$$\bigvee_\delta \bigwedge_{\substack{r_1, r_2 \\ [\alpha|\beta]}} . |r_1 - r_2| < \delta \rightarrow |\phi r_1 - \phi r_2| < \varepsilon.$$

76 To that end we consider the following statement form $A(\xi)$ for $\xi \leqslant \beta$:

$$\bigvee_\delta \bigwedge_{\substack{r_1, r_2 \\ [\alpha|\xi]}} . |r_1 - r_2| < \delta \rightarrow |\phi r_1 - \phi r_2| < \varepsilon.$$

For $\xi > \beta$ let $A(\xi) \leftrightharpoons A(\beta)$. To establish $A(\beta)$, we show by the analytic induction principle that $\bigwedge_\xi A(\xi)$:

(1) $$\bigwedge_\xi . \xi < \xi_0 < \beta \rightarrow A(\xi). \rightarrow \bigvee_{\delta_0} A(\xi_0 + \delta_0)$$

(2) $$\bigwedge_\xi . \xi < \beta \rightarrow A(\xi). \rightarrow A(\beta).$$

(2) here arises from (1) by the substitution of $\xi_0 \geqslant \beta$ for $\xi_0 < \beta$. Since $A(\xi)$ is left-hereditary and $A(\alpha)$ clearly holds, it follows from (1) and (2) that $\bigwedge_\xi A(\xi)$.

A sufficient condition for (1) is rational continuity at ξ_0:

$$\bigvee_{\delta_0} \bigwedge_r . |r - \xi_0| < \delta_0 \rightarrow |\phi r - \phi \xi_0| < \frac{\varepsilon}{2}.$$

For let

$$r_0 \in (\xi_0 - \delta_0 | \xi_0).$$

By hypothesis we have $A(r_0)$, whence for some $\delta < r_0 - \xi_0 + \delta_0$

$$\bigwedge_{\substack{r_1, r_2 \\ [\alpha|r_0]}} . |r_1 - r_2| < \delta \rightarrow |\phi r_1 - \phi r_2| < \varepsilon.$$

For $r_1 < r_2 < \xi_0 + \delta_0$ we must distinguish the cases $r_2 \leqslant r_0$ and $r_0 < r_2$.

It follows in both cases that

$$r_2 - r_1 < \delta \rightarrow |\phi r_1 - \phi r_2| < \varepsilon.$$

The proof of (2) proceeds similarly, using the left continuity of ϕ at β.

Uniform continuity directly entails boundedness:

THEOREM 8.3 Every function continuous in $[\alpha|\beta]$ is bounded therein.

Proof. If ϕ is continuous in $[\alpha|\beta]$, it follows from its uniform continuity that there is, e.g., a δ such that

$$\bigwedge_{\xi_1,\xi_2}.|\xi_1 - \xi_2| < \delta \rightarrow |\phi\xi_1 - \phi\xi_2| < 1.$$

If $\beta - \alpha < n\delta$, then it is true of every $\xi \in [\alpha|\beta]$ that

$$|\phi\xi - \phi\alpha| < n,$$

and hence that

$$|\phi\xi| < |\phi\alpha| + n.$$

We now proceed to prove the general *intermediate-value theorem* for functions continuous in a closed interval. 77

THEOREM 8.4 For every function that is continuous in a closed interval, any number between two values of the function is itself a value of the function.

Proof. Let it be the case that, say, $\phi\alpha = a < c < b = \phi\beta$. We want a real number γ such that $\alpha < \gamma < \beta$ and $\phi\gamma = c$. We consider the set $\underset{[\alpha|\beta]}{\epsilon_r}\ \phi r \leqslant c$. This set has a definite left class and hence also a lub γ. Now let $\gamma = \lim r_*$, where $\bigwedge_n \phi r_n \leqslant c$. In the presence of continuity, this implies $\phi\gamma \leqslant c$. Were $\phi\gamma < c$, then in view of $\gamma < \beta$ (for $\gamma \in [\alpha|\beta]$ and $\gamma \neq \beta$ because $\phi\gamma \neq b$) and the continuity of ϕ, it would follow that there was an r such that $\gamma < r < \beta$ and $\phi r < c$—which contradicts our stipulation that γ is an upper bound of $\underset{[\alpha|\beta]}{\epsilon_r}\ \phi r \leqslant c$.

Just as monotonic sequences were of special significance among the concentrated sequences, so it is worthwhile to call attention to the continuous *monotonic* functions among continuous functions in general. They include—piecewise at any rate—all elementary functions (cf. §9) and even all analytic functions (cf. §12).

A function ϕ is *increasing* in its domain K if

$$\bigwedge_{\xi_1,\xi_2}.\xi_1 < \xi_2 \rightarrow \phi\xi_1 \leqslant \phi\xi_2.$$

Similarly, if this statement holds with $<$ $\{\geqslant;\ >\}$ in place of $\leqslant$, then ϕ is *properly increasing* $\{$*decreasing*; *properly decreasing*$\}$.

THEOREM 8.5 A function ϕ which is continuous in an interval K is (properly) increasing if and only if its rational subfunction ϕ_0 is (properly) increasing.

Proof. ("Only if" is trivial.) If $\xi_1 < \xi_2$, then there are rational numbers r_1, r_2 such that $\xi_1 < r_1 < r_2 < \xi_2$. Furthermore, $\xi_1 = \lim r_{1*}$ for a sequence r_{1*} such that $\bigwedge_n r_{1n} < r_1$, and $\xi_2 =$

lim r_{2*} for a sequence r_{2*} such that $\wedge_n r_2 < r_{2n}$. It follows for properly increasing ϕ_0 that

$$\phi\xi_1 \leqslant \phi r_1 < \phi r_2 \leqslant \phi\xi_2, \qquad \text{i.e.,} \qquad \phi\xi_1 < \phi\xi_2 .$$

Like the convergence of all bounded monotonic sequences, we immediately get the existence of the left-hand and right-hand limits for monotonic functions (in their argument interval). For instance, say that ϕ is increasing. We want to show that $\lim_{\xi \overset{\frown}{<} \xi_0} \phi\xi$ exists. The set of values ϕr corresponding to rational arguments $r < \xi_0$ is definite and bounded. Let η_0 be the lub of that set. Then η_0 is also the lub of the function values $\phi\xi$ for $\xi < \xi_0$; for it follows from $\xi < \xi_0$ that there is an r such that $\xi < r < \xi_0$, whence $\phi\xi \leqslant \phi r \leqslant \eta_0$. In view of monotony, however, the lub is also the left-hand limit of the function at ξ_0.

78

Monotony—like uniform continuity—is a *global* property. But here, too, monotony in an interval follows from *local* monotony, i.e., from the monotony of the function in an appropriate neighborhood of each point.

THEOREM 8.6 If ϕ is properly increasing locally in some interval, then it is so globally too.

Proof by analytic induction: By Theorem 8.5 it suffices to show that ϕ_0 is properly increasing. We consider an interval with the left end point α, and the formula

$$A(\xi) \leftrightharpoons \underset{[\alpha|\xi]}{\wedge_{r_1, r_2}} . r_1 < r_2 \rightarrow \phi r_1 < \phi r_2$$

which is obviously left-hereditary.

If $\bigwedge_{\xi} . \xi < \xi_0 \rightarrow A(\xi).$, then too

$$r_1 < r_2 < \xi_0 \rightarrow \phi r_1 < \phi r_2 .$$

Now, if ϕ is properly increasing locally at ξ_0, then there is some δ such that

$$\xi_0 - \delta < r_1 < r_2 < \xi_0 + \delta \rightarrow \phi r_1 < \phi r_2 .$$

Thus it follows from $r_1 < r_2 < \xi_0 + \delta$ that $\phi r_1 < \phi r_2$, at all events for $r_2 < \xi$ and also for $\xi - \delta < r_1$. But if $r_1 \leqslant \xi_0 - \delta$ and $\xi_0 \leqslant r_2$, then with the help of an r out of $(\xi_0 - \delta | \xi_0)$ it directly follows that $\phi r_1 < \phi r < \phi r_2$.

A function ϕ which is *properly monotonic* in its domain K (i.e., properly increasing or properly decreasing) is *invertible* in K; i.e.,

$$\underset{K}{\bigwedge_{\xi_1, \xi_2}} . \phi\xi_1 = \phi\xi_2 \rightarrow \xi_1 = \xi_2 .$$

For every number η in the range R of ϕ there is then exactly one number ξ such that $\phi\xi = \eta$.

The term $\iota_\xi\ \phi\xi = \eta$ ("the ξ such that $\phi\xi = \eta$") represents a new function ψ for which

$$\psi\eta = \xi \longleftrightarrow \phi\xi = \eta .$$

The domain of ψ is R and its range is K. ψ is called the *inverse* of ϕ.

An invertible function is not necessarily monotonic, but we do have

THEOREM 8.7 If a function is continuous in an interval $[\alpha \mid \beta]$ and is invertible, then it is properly monotonic in $[\alpha \mid \beta]$.

Proof. Let us say that $\phi\alpha < \phi\beta$. We have to show that ϕ is properly increasing. To begin with, it follows from $\alpha \leqslant \xi_1 < \beta$ that $\phi\xi_1 < \phi\beta$, for $\phi\alpha < \phi\beta \leqslant \phi\xi_1$ would contradict invertibility in view of the intermediate-value theorem ($\phi\xi = \phi\beta$ for some $\xi \leqslant \xi_1$, i.e., $\xi \neq \beta$). We then see that $\alpha \leqslant \xi_1 < \xi_2 \leqslant \beta$ always entails $\phi\xi_1 < \phi\xi_2$, for $\phi\xi_2 \leqslant \phi\xi_1 < \phi\beta$ together with the intermediate-value theorem would again contradict invertibility. *79*

It follows from this theorem that the inverse ψ of a definite continuous function will itself be definite. For

$$\xi = \psi r \leftrightarrow \phi\xi = r,$$

and, assuming ϕ to be, say, properly increasing, we get

$$\xi = \text{lub}\ \epsilon_s\ \phi s \leqslant r .$$

The inverse ψ of a properly monotonic function ϕ is of course also properly monotonic, and in the same sense: ψ is properly increasing if ϕ is. The domain of the inverse of a continuous properly monotonic function is itself an interval. This follows directly from the intermediate-value theorem.

The continuity of the inverse function results from the following

THEOREM 8.8 The inverse ψ of a function ϕ which is properly increasing in $[\alpha \mid \beta]$ is continuous at the interior points of the range of ϕ (i.e., at those points which belong to a neighborhood included in the range). (It is not presupposed that ϕ is continuous.)

Proof. Let $\phi\alpha < \phi\xi_0 = \eta_0$. For all ε there is an $\varepsilon_0 < \varepsilon$ such that

$$\alpha < \xi_0 - \varepsilon_0 < \xi_0 < \xi_0 + \varepsilon_0 < \beta .$$

It follows that

$$\phi(\xi_0 - \varepsilon_0) < \eta_0 < \phi(\xi_0 + \varepsilon_0) .$$

Hence there is a δ such that

$$\phi(\xi_0 - \varepsilon_0) < \eta_0 - \delta < \eta_0 < \eta_0 + \delta < \phi(\xi_0 + \varepsilon_0).$$

Since ψ too is properly increasing, it follows directly for every η such that

$$\eta_0 - \delta < \eta < \eta_0 + \delta$$

that

$$\xi_0 - \varepsilon_0 < \psi\eta < \xi_0 + \varepsilon_0$$

and thus also that

$$\psi\eta_0 - \varepsilon < \psi\eta < \psi\eta_0 + \varepsilon.$$

Inversion is an important means of getting continuous functions. Before we consider special continuous functions in §9 (the so-called elementary functions), let us take up here one more *80* general method of forming continuous functions, viz., *composition.* If ψ is a function with domain D and range K, and ϕ is a function with domain K and range R, then the composition of ϕ and ψ is defined as the function χ with domain D and range R such that

$$\chi\xi = \phi(\psi\xi).$$

We write $\chi = \phi\urcorner\psi$, for we then get

$$\phi\urcorner\psi\dot{\urcorner}\xi = \phi\dot{\urcorner}\psi\urcorner\xi.$$

For the composition of continuous functions, we have the following

THEOREM 8.9 If ψ is continuous in D and ϕ is continuous in the range of ψ, then $\phi\urcorner\psi$ is continuous in D.

For the proof one has only to reflect that the continuity of ϕ at $\psi\xi_0$ entails that $\lim_{\xi_0} \psi = \eta_0 \rightarrow \lim_{\xi_0} \phi\urcorner\psi = \phi\eta_0$.

Elementary arithmetical operations on continuous functions lead as a matter of course to continuous functions again.

For two functions ϕ and ψ that are continuous at ξ_0, it will first of all be the case that

$$\lim_{\xi \rightsquigarrow \xi_0} (\phi\xi \pm \psi\xi) = \phi\xi_0 \pm \psi\xi_0$$

and

$$\lim_{\xi \rightsquigarrow \xi_0} (\phi\xi \cdot \psi\xi) = \phi\xi_0 \cdot \psi\xi_0.$$

For if it always follows from $|\xi - \xi_0| < \delta$ that $|\phi\xi - \phi\xi_0| < \varepsilon$ and $|\psi\xi - \psi\xi_0| < \varepsilon$, then it will also be the case that

$$|(\phi\xi \pm \psi\xi) - (\phi\xi_0 \pm \psi\xi_0)| < 2\varepsilon$$

and

$$\begin{aligned} &|\phi\xi \cdot \psi\xi - \phi\xi_0 \cdot \psi\xi_0| \\ &= |\phi\xi \cdot \psi\xi - \phi\xi_0 \cdot \psi\xi + \phi\xi_0 \cdot \psi\xi - \phi\xi_0 \cdot \psi\xi_0| \\ &\leqslant |\psi\xi||\phi\xi - \phi\xi_0| + |\phi\xi_0||\psi\xi - \psi\xi_0| \\ &< (|\psi\xi_0| + |\phi\xi_0| + \varepsilon) \cdot \varepsilon . \end{aligned}$$

To establish the continuity of the quotient $\frac{\phi}{\psi} = \phi \cdot \frac{1}{\psi}$, we prove that the function $\frac{1}{\psi}$ is continuous at ξ_0 whenever $\psi\xi_0 \neq 0$ and ψ is continuous at ξ_0.

Continuity together with $\psi\xi_0 \neq 0$ entails the existence of ε_0, δ_0 such that

$$\bigwedge_\xi . |\xi - \xi_0| < \delta_0 \rightarrow |\psi\xi| > \varepsilon_0 .$$

Therefore, if

81

$$\bigwedge_\xi . |\xi - \xi_0| < \delta \rightarrow |\psi\xi - \psi\xi_0| < \varepsilon .$$

for $\delta \leqslant \delta_0$, then for every ξ such that $|\xi - \xi_0| < \delta$,

$$\left|\frac{1}{\psi\xi} - \frac{1}{\psi\xi_0}\right| = \left|\frac{\psi\xi_0 - \psi\xi}{\psi\xi \cdot \psi\xi_0}\right| < \frac{\varepsilon}{\varepsilon_0^{\,2}} .$$

Since $\lim_{\varepsilon \to 0} \frac{\varepsilon}{\varepsilon_0^{\,2}} = 0$, $\frac{1}{\psi}$ is thus continuous at ξ_0.

§9. The Elementary Functions

The distinction accorded in infinitesimal mathematics to certain functions known as "elementary" in contrast to all others is a result of historical accident. Simple rational and algebraic functions, as well as some circle functions, made their appearance in ancient times. The onset of modern mathematics brought the addition of exponential functions and logarithms. Thus it is that the following are customarily listed nowadays as *basic elementary functions* (along with the trivial *constant functions*, which take the same value for all arguments):

(1) power functions
(2) exponential and logarithmic functions

(3) circle functions (trigonometric and arc functions).

All functions obtainable from these basic ones by addition, multiplication, and composition are called *elementary functions*.

Elementary functions may be represented by *elementary terms*, which include

(1) the constants ±1
(2) the variables ξ, η, . . .
(3) if S and T, then also $S + T$, $S \cdot T$
(4) if $S(\xi)$ and T, then also $S(T)$.

Also, the following symbols may occur in elementary terms:

exp, ln, sin, cos, tan, arc sin, arc cos, arc tan.

Since $S \cdot T = \exp(\ln S + \ln T)$, we could do without $S \cdot T$, as well as cos, tan, arc cos, and arc tan. By the same token, powers are omitted in view of $S^T = \exp(T \ln S)$.

82 The elementary functions are discussed individually below. We intend not only to lay down definitions, but also to show the motivation for our choice of those definitions. For the circle functions we shall do this—in keeping with historical precedent—by recourse to elementary geometry, after which we shall provide a purely arithmetical foundation.

We begin with integral-rational functions, in particular with those of the first degree. They are represented by terms of the form $\alpha\xi + \beta$. We call them *general linear* functions; only for $\beta = 0$ do we speak of linear functions. The latter are characterized by *linearity*:

$$L(\alpha\xi + \beta\eta) = \alpha L(\xi) + \beta L(\eta).$$

The domain is throughout the class of all real numbers.

The linear functions include in particular the *identity function*, for which $\alpha = 1$. We use a special symbol for this function, namely I. Thus it is always the case that

$$I\xi = \xi,$$

i.e., argument and value are always identical.

The availability of I permits us to write the function ϕ, represented by the term $T(\xi)$, as $\phi = T(I)$ instead of the lengthier $\phi = \imath_\xi T(\xi)$; e.g.,

$$\phi = \frac{1}{I^2 + 1}.$$

Thus we could also construe the letter I as a place marker—like $*$, $\dagger$, $\nmid$ in the case of sequences.

The graphic representation of general linear functions is well known to be a straight line.

$\alpha = 0$ yields constant functions. For the linear functions L we get in particular

$$L(\gamma \xi) = \gamma L(\xi) \qquad (9.1)$$

This functional equation is all that is necessary to characterize linear functions.

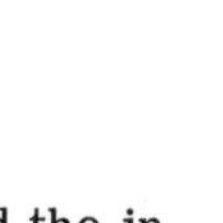

Fig. 1

For $\alpha \neq 0$ general linear functions are invertible, and the inverse is again general-linear.

The combination of constant and linear functions by means of addition and subtraction exclusively always yields general linear functions. If we throw in multiplication, the result is the *integral-rational* functions.   83

As terms representing integral-rational functions we may take the *polynomials*:

$$\alpha_0 + \alpha_1 \xi + \ldots + \alpha_n \xi^n.$$

The exponents occurring here are all natural numbers; such powers are already defined inductively in §4.

THEOREM 9.1 The integral-rational functions are everywhere continuous (i.e., in the class of all real numbers).

The proof requires only the consideration that the constants and the identity function are continuous, and that sums and products of continuous functions are themselves continuous.

If we also admit division for the formation of functions, there arise in addition *fractional-rational functions*. Any *rational function* can be represented by a polynomial quotient:

$$\frac{\alpha_0 + \alpha_1 \xi + \ldots + \alpha_m \xi^m}{\beta_0 + \beta_1 \xi + \ldots + \beta_n \xi^n}.$$

At least one of the constants $\beta_0, \ldots, \beta_n$ here must be $\neq 0$. As is taught in algebra, we can derive from any such given representation a uniquely determined representation with the smallest possible degrees m, n, the *reduced representation* or *reduction*.

A rational function takes as arguments all real numbers except for those which would yield 0 in the denominator of the reduction.

Two simple examples are $\eta = \frac{1}{\xi}$ and $\eta = \frac{1}{\xi^2}$, which may be represented graphically as in Figures 2 and 3.

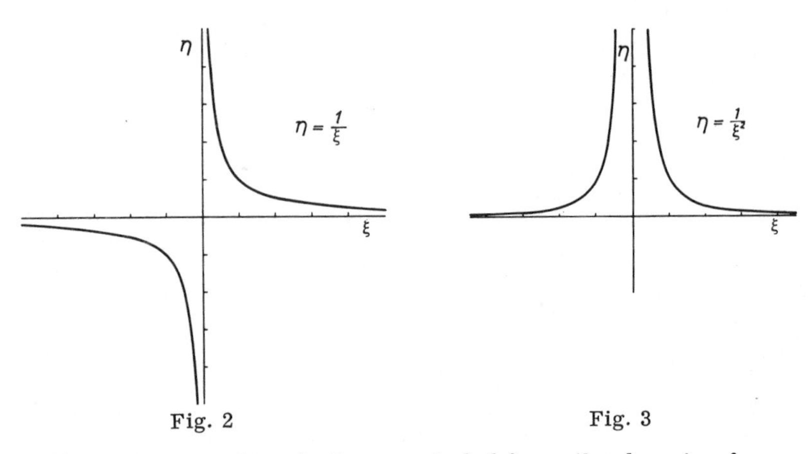

Fig. 2 Fig. 3

84 For every number ξ_0 thus excluded from the domain of a rational function ϕ we have

$$\lim_{\xi \to \xi_0} |\phi\xi| = \infty.$$

Such numbers are accordingly called *infinity points*. The infinity points are the only real numbers that do not belong to the domain of a rational function. The continuity of the rational functions in their domain follows immediately from the fact that the function $\frac{1}{\phi}$ is continuous at ξ_0 whenever $\phi\xi_0 \neq 0$ and ϕ is continuous at ξ_0.

Finally, we know from algebra that a polynomial always has at most finitely many substitution instances for which the denominator equals 0; hence every rational function has at most finitely many infinity points.

The rational functions include in particular the *power functions* I^n with integers as exponents. We set $\xi^0 = 1$, and we supplement this customary definition for $\xi \neq 0$ by stipulating the value 1 for the function when $\xi = 0$, too.

Furthermore, we set $\xi^{-n} = \frac{1}{\xi^n}$.

In what follows, we consider only arguments $\xi > 0$. In that argument domain of positive real numbers, the function I^n is properly increasing for $n > 0$, and properly decreasing for $n < 0$. All values are likewise positive. In particular, the inverse function exists in the domain under consideration for $n > 0$. By virtue of the intermediate-value theorem, the range consists of all positive real numbers, since $\lim_{\xi \to 0} \xi^n = 0$ and $\lim_{\xi \to \infty} \xi^n = \infty$ (after all, $\xi^n \geq \xi$ for $\xi \geq 1$).

If $\eta = \xi^n$, then we write $\xi = \eta^{\frac{1}{n}}$ (or $\xi = \sqrt[n]{\eta}$, using $\sqrt{\ }$ rather than $\sqrt[2]{\ }$).

If two fractions $\frac{m_1}{n_1}$ and $\frac{m_2}{n_2}$ represent the same rational number, i.e., if $m_1 n_2 = m_2 n_1$, then also

$$(\xi^{m_1})^{\frac{1}{n_1}} = (\xi^{m_2})^{\frac{1}{n_2}},$$

for $\eta = (\xi^{m_1})^{\frac{1}{n_1}}$ is equivalent to $\eta^{n_1} = \xi^{m_1}$, which is equivalent to $\xi^{m_1 m_2} = \eta^{m_2 n_1}$, and this to $\xi^{m_2} = \eta^{n_2}$ or $\eta = (\xi^{m_2})^{\frac{1}{n_2}}$. Thus we may define for all rational numbers r

$$\xi^r = (\xi^m)^{\frac{1}{n}} \quad \text{for} \quad r = \frac{m}{n}. \tag{9.2}$$

Since power functions with exponents $\frac{1}{n}$ $(n > 0)$ are properly increasing, *power functions with rational exponents* $r > 0$ are properly increasing, and those with rational exponents $r < 0$ properly decreasing. 85

These power functions are furthermore (by Theorem 8.8) continuous in their own domains. Our definitions immediately yield the arithmetical rules for powers:

$$\xi^{r+s} = \xi^r \cdot \xi^s \qquad (\xi \cdot \eta)^r = \xi^r \cdot \eta^r$$
$$\xi^{r \cdot s} = (\xi^r)^s$$

The formation of inverse functions produces nothing new, since

$$\eta = \xi^r \leftrightarrow \xi = \eta^{\frac{1}{r}} \quad \text{where} \quad r \neq 0.$$

Let us therefore consider now the function of r represented by the term α^r for some $\alpha > 0$. This *exponential function* (to the base α) has only the rational numbers as arguments up to now. Its values are positive real numbers.

The exponential function is uniformly continuous in any closed interval $[\beta \mid \gamma]$; i.e.,

$$\bigwedge_\varepsilon \bigvee_\delta \bigwedge_{\substack{r_1, r_2 \\ [\beta|\gamma]}} |r_1 - r_2| < \delta \rightarrow |\alpha^{r_1} - \alpha^{r_2}| < \varepsilon.$$

To prove this, we set $s = r_1 - r_2$. It follows that

$$\alpha^{r_1} - \alpha^{r_2} = (\alpha^s - 1)\alpha^{r_2}.$$

Since α^r with $r \in [\beta|\gamma]$ is bounded, all that remains to be shown is that $\lim_{s \to 0} \alpha^s = 1$.

Let $\alpha > 1$. Then the exponential function is properly increasing:

$$r < s \to \alpha^r < \alpha^s,$$

for we have $\alpha^s = \alpha^r \cdot \alpha^{s-r}$ and $\alpha^{s-r} > 1$ because $s - r > 0$.

Thus all we have left to prove is

$$\wedge_\varepsilon \vee_n \alpha^{\frac{1}{n}} - 1 < \varepsilon,$$

which follows immediately from the following formula:

$$n(\alpha^{\frac{1}{n}} - 1) < \alpha - 1.$$

For the proof (without recourse to the binomial formula), we consider the function ϕ such that

$$\phi r = \alpha^r - 1.$$

86 We have $\phi 0 = 0$. Furthermore, ϕ is *superadditive*, i.e.,

$$\phi(r + s) > \phi r + \phi s \quad \text{where} \quad r, s > 0, \tag{9.3}$$

for

$$\alpha^{r+s} - 1 > \alpha^r - 1 + \alpha^s - 1$$

in view of the fact that

$$\alpha^{r+s} - \alpha^r = \alpha^r(\alpha^s - 1) > \alpha^s - 1.$$

Superadditivity entails that

$$\phi\left(\frac{m+1}{n}\right) - \phi\left(\frac{m}{n}\right) > \phi\left(\frac{1}{n}\right) \qquad (m = 0, \ldots, n - 1),$$

and hence by summation

$$\phi 1 - \phi 0 > n\phi\left(\frac{1}{n}\right)$$

which is the desired inequality.

Now if $\alpha < 1$, then $\frac{1}{\alpha} > 1$. Thus

$$\wedge_\varepsilon \vee_n \left(\frac{1}{\alpha}\right)^{\frac{1}{n}} < 1 + \varepsilon.$$

From this it follows that

$$\wedge_\varepsilon \vee_n \alpha^{\frac{1}{n}} > \frac{1}{1 + \varepsilon}$$

$$\wedge_\varepsilon \vee_n 1 - \alpha^{\frac{1}{n}} < 1 - \frac{1}{1+\varepsilon} = \frac{\varepsilon}{1+\varepsilon} < \varepsilon,$$

whence again

$$\lim_{s \to 0} \alpha^s = 1.$$

By Theorem 8.1, exponential functions are therefore uniquely definable by

$$\alpha^\xi = \lim_{r \to \xi} \alpha^r \tag{9.4}$$

for every real ξ.

These real functions are again monotonic; more specifically, they are

properly increasing for $\alpha > 1$
constant (= 1) for $\alpha = 1$
properly decreasing for $\alpha < 1$.

At the same time, we have succeeded in defining the *power functions* I^β *for real exponents.*

The arithmetical laws are preserved for these new functions.

For $\alpha \neq 1$, there is a uniquely determined inverse function 87 belonging to α^I, the *logarithm function.* We write

$$\xi = \log_\alpha \eta \Leftrightarrow \eta = \alpha^\xi. \tag{9.5}$$

The logarithmic functions are likewise continuous throughout and properly increasing for $\alpha > 1$. Their domain is the open interval $(0 | \infty)$, since $\alpha^r \to \infty$ if $r \to \infty$ and $\alpha^r \to 0$ if $r \to -\infty$. Otherwise, if α^* were bounded, there would be a β such that $\alpha^* \to \beta$. That would imply $\alpha^{*+1} \to \alpha\beta$ and hence $\alpha\beta = \beta$, in contradiction to $\alpha \neq 1$.

The arithmetical laws governing power functions give rise to the following for logarithmic functions (among others):

$$\log_\alpha(\xi \cdot \eta) = \log_\alpha \xi + \log_\alpha \eta$$

$$\log_\alpha(\xi^\eta) = \eta \cdot \log_\alpha \xi$$

and

$$\log_\beta \xi = \log_\alpha \xi \cdot \log_\beta \alpha$$

(in particular

$$1 = \log_\alpha \beta \cdot \log_\beta \alpha).$$

These laws are the basis for the utility of logarithms in practical computation: multiplication is reduced to addition. Logarithms to different bases differ only by a constant factor.

The last of the elementary functions we have to discuss are the *circle functions*, in particular the *trigonometric functions*. His-

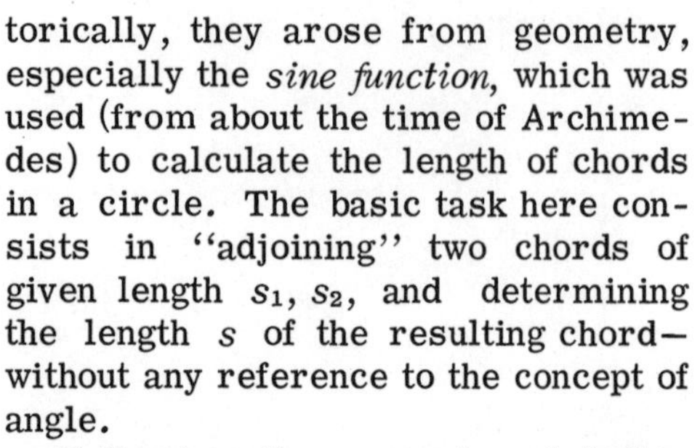

torically, they arose from geometry, especially the *sine function*, which was used (from about the time of Archimedes) to calculate the length of chords in a circle. The basic task here consists in "adjoining" two chords of given length s_1, s_2, and determining the length s of the resulting chord—without any reference to the concept of angle.

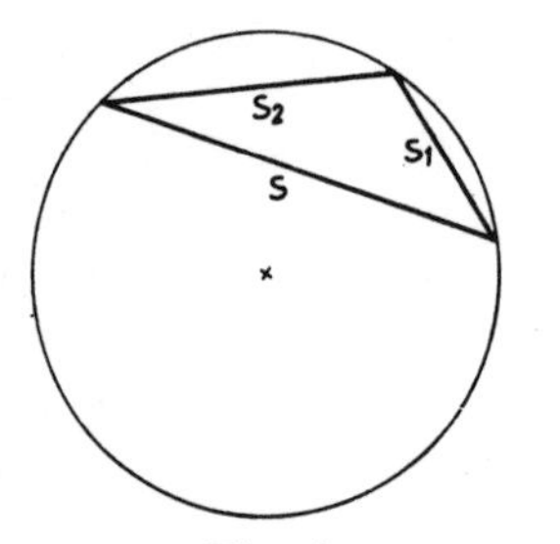

Fig. 4

Following the example of Indian mathematicians (since ca. A.D. 500), we consider not the chords themselves but rather the corresponding semichords. For a unit circle (radius 1) we get Figure 5.

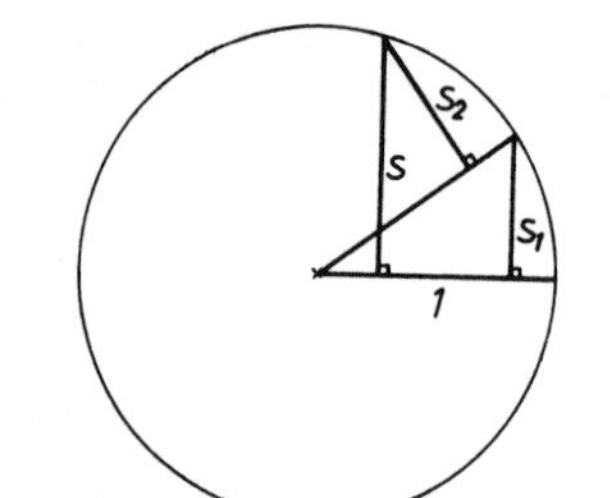

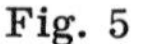

Fig. 5

88

Utilizing the auxiliary lines of lengths

$$c_1 = \sqrt{1 - s_1^2}$$
$$c_2 = \sqrt{1 - s_2^2} \qquad (9.6)$$

we then have

$$s = s_1 c_2 + s_2 c_1. \qquad (9.7)$$

Fig. 6

For a circle with diameter 1 this follows from the Ptolemaic diagonal theorem: the product of the diagonals of an inscribed quadrilateral is equal to the sum of the products of opposite sides.

Proof. Since the shaded angles at A and B (in Fig. 6) are both circumferential angles comprehending CD, they are equal. Thus there must be a point E on AC such that the shaded triangles are similar.

It follows that

$$AE{:}c_2 = s_1{:}s,$$

whence

$$AE \cdot s = s_1 c_2 .$$

In the same way we get $EC \cdot s = s_2 c_1$, from which the theorem follows in view of $AE + EC = AC$ [= 1 in the case of (9.7)].

For the special case $s_1 = s_2$, (9.7) yields

$$s = 2s_1c_1.$$

The equation $s = 2s_1\sqrt{(1 - s_1^2)}$ can be solved for s_1. Using

$$c = \sqrt{1 - s^2} \tag{9.8}$$

we immediately get

$$s_1 = \sqrt{\frac{1 \pm c}{2}}.$$

For $s = 1$, this uniquely yields $s_1 = \frac{1}{2}\sqrt{2}$. The stipulation that $s_1 < \frac{1}{2}\sqrt{2}$ for $s < 1$ therefore implies the unique result that

$$s_1 = \sqrt{\frac{1 - c}{2}}. \tag{9.9}$$

The formulas (9.6)–(9.9) now permit a solution to the following problem. Divide a right angle into two equal parts; i.e., find an s_1 for Figure 7. *89*

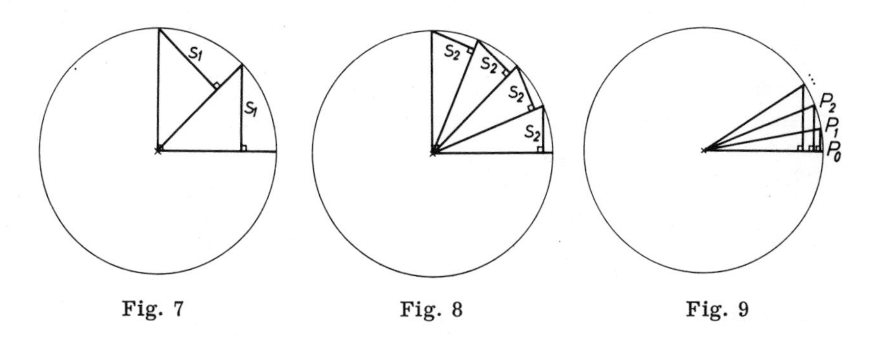

Fig. 7 Fig. 8 Fig. 9

Next, bisect each of the resulting angles, and so on (Fig. 8).

Having divided the right angle into 2^n equal parts, compute the length of the semichord corresponding to each of the bisecting points $P_0^{(n)}, P_1^{(n)}, \ldots, P_{2^n}^{(n)}$ (Fig. 9). We designate this length as $\operatorname{Sin} \frac{m}{2^n}$. This length does in fact depend solely upon the rational number $\frac{m}{2^n}$: if $\frac{m}{2^n} = \frac{m'}{2^{n'}}$, then by construction $P_m^{(n)} = P_{m'}^{(n')}$.

With that we have defined a *function* Sin for all rational numbers $r = \frac{m}{2^n}$ in the interval $[0|1]$.

This function is evidently properly increasing, and its (algebraic) values likewise fall in the interval $[0|1]$. Notice that the definition of this function presupposes no unit of measure for angles in general (nor, in particular, radial measure). All we do is to keep dividing the right angle into 2, 4, 8, ... equal parts —where *angular equality* is defined by identity of chord length.

Proceeding from this function Sin, we can get the usual trigonometric functions by introducing radial measure. Before going on to that, we shall provide an arithmetical foundation for the function Sin, so as to find a path to the trigonometric functions that is independent of elementary geometry. At the same time, the arithmetical account will point up the objective connection that justifies including the trigonometric functions along with the other elementary functions, especially the power functions. This connection is purely arithmetical and requires no geometric knowledge. To be sure, the close relation between potentiation and trigonometric functions only arises after we extend the real-

90 number field to the *complex-number field*. Since this book is confined to real analysis, we give here only a brief sketch of the salient features of complex numbers.

So as to render such equations as $\xi^2 + 1 = 0$ soluble, one introduces *complex numbers* $\alpha + i\beta$, which (as will be seen) is only a handy notation for the real-number pair α, β.

Addition and *multiplication* are defined so as to preserve the arithmetical laws (associativity, commutativity, distributivity) and so as to set $i^2 = -1$.

Thus one must posit

$$(\alpha_1 + i\beta_1) + (\alpha_2 + i\beta_2) = (\alpha_1 + \alpha_2) + i(\beta_1 + \beta_2)$$

$$(\alpha_1 + i\beta_1) \cdot (\alpha_2 + i\beta_2) = (\alpha_1\alpha_2 - \beta_1\beta_2) + i(\alpha_1\beta_2 + \alpha_2\beta_1).$$

Subtraction and division (except by zero, i.e., $0 + i0$) are generally applicable. In particular, one gets

$$\frac{1}{\alpha + i\beta} = \frac{\alpha - i\beta}{\alpha^2 + \beta^2}.$$

$\sqrt{(\alpha^2 + \beta^2)}$ is called the *absolute value* of $\alpha + i\beta$, written as $|\alpha + i\beta| = \sqrt{(\alpha^2 + \beta^2)}$.

From $|\alpha + i\beta|^2 = (\alpha + i\beta)(\alpha - i\beta)$, it easily follows by multiplication that $|(\alpha_1 + i\beta_1) \cdot (\alpha_2 + i\beta_2)| = |\alpha_1 + i\beta_1| \cdot |\alpha_2 + i\beta_2|$, and in particular that

$$|(\alpha + i\beta)^n| = |\alpha + i\beta|^n.$$

This brings us to the task of defining powers of $\alpha + i\beta$ for other exponents besides whole ones. We restrict ourselves to *real exponents*; and in view of $\alpha + i\beta = |\alpha + i\beta| \dfrac{\alpha + i\beta}{|\alpha + i\beta|}$, we can confine our attention to bases $\alpha + i\beta$ with an absolute value of 1. $|(\alpha + i\beta)^n| = 1$ always follows from $|\alpha + i\beta| = 1$ and vice versa.

From $(\alpha_1 + i\beta_1)^2 = \alpha + i\beta$, where $\alpha, \beta \geqslant 0$, we obtain the equation

$$\beta = 2\alpha_1\beta_1$$

and hence for $\alpha_1 \geqslant 0$ $\quad \beta = 2\beta_1\sqrt{(1 - \beta_1^2)}$.

That is nothing but the equation for the chord of a double angle. Its solution yields as in (9.9)

$$\beta_1 = \sqrt{\frac{1 - \alpha}{2}} \quad \text{where} \quad \alpha = \sqrt{1 - \beta^2}.$$

We write

$$\alpha_1 + i\beta_1 = (\alpha + i\beta)^{\frac{1}{2}}.$$

In this way $(\alpha + i\beta)^{\frac{m}{2^n}}$ can be computed for arbitrary m, n.

In the special case of i, we stipulate for $\alpha = 0$, $\beta = 1$ that 91

$$i^{\frac{m}{2^n}} = \text{Cos}\,\frac{m}{2^n} + i\,\text{Sin}\,\frac{m}{2^n}.$$

The function Sin defined over $[0|1]$ by this formula coincides exactly with the function Sin previously introduced geometrically. For the *function* Cos we have

$$\text{Cos}\,\gamma = \sqrt{1 - \text{Sin}^2\,\gamma}. \qquad (9.10)$$

It is not surprising that the same functions are produced by the potentiation of i as by the division of a right angle if we take note of the translatability of geometric theory into analysis (by way of so-called analytic geometry).

If a system of orthogonal coordinates x, y is introduced into a Euclidean plane, every *vector* $\mathfrak{a}$ can be correlated one-to-one with a complex number, viz., $\alpha + i\beta$, where α, β are the coordinate values of the vector.

Furthermore, if O is the origin of the coordinate system, every *point* P can be correlated one-to-one with a complex number, viz., $\alpha + i\beta$, where α, β are the coordinate values of the vector $\vec{OP}$.

In this translation, the sum of two vectors corresponds to the sum of the corresponding complex numbers. If $\alpha + i\beta$ corresponds to the vector $\mathfrak{a}$, then the product $\gamma(\alpha + i\beta)$ corresponds to the vector $\gamma\mathfrak{a}$. The product $i(\alpha + i\beta)$ corresponds to the vector $\mathfrak{a}$ after it has been rotated counterclockwise through a right angle. (The coordinates of the rotated vector are $-\beta$, α.)

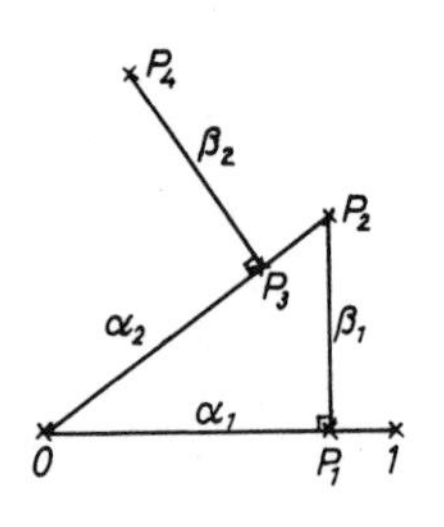

Fig. 10

Two vectors $\alpha_1 + i\beta_1$ and $\alpha_2 + i\beta_2$ with the absolute value 1 therefore give rise to this geometric construction of the product

$$(\alpha_1 + i\beta_1) \cdot (\alpha_2 + i\beta_2).$$

The coordinate values of the points P_1 – P_4 are given successively by the correlated complex numbers:

$$P_1\colon\ \alpha_1$$

$$P_2\colon\ \alpha_1 + i\beta_1$$

$$P_3\colon\ \alpha_2(\alpha_1 + i\beta_1)$$

$$P_4\colon\ \alpha_2(\alpha_1 + i\beta_1) + i\beta_2(\alpha_1 + i\beta_1).$$

The construction for P_4 is precisely the construction required to "adjoin" the semichords β_1 and β_2.

92 With that we have defined the functions Sin and Cos for the dyadic rational numbers in the interval $[0|1]$.

In that interval, Sin is properly increasing and Cos properly decreasing. Furthermore,

$$\operatorname{Sin} 0 = 0 \qquad \operatorname{Sin} 1 = 1$$

$$\operatorname{Cos} 0 = 1 \qquad \operatorname{Cos} 1 = 0.$$

On the basis of multiplication of complex numbers, the following *addition theorems* hold:

$$\begin{aligned} \operatorname{Sin}(r_1 + r_2) &= \operatorname{Sin} r_1 \operatorname{Cos} r_2 + \operatorname{Cos} r_1 \operatorname{Sin} r_2 \\ \operatorname{Cos}(r_1 + r_2) &= \operatorname{Cos} r_1 \operatorname{Cos} r_2 - \operatorname{Sin} r_1 \operatorname{Sin} r_2 . \end{aligned} \tag{9.11}$$

It is also advisable to introduce the following *function* Tan:

$$\operatorname{Tan} = \frac{\operatorname{Sin}}{\operatorname{Cos}}.$$

This function is properly increasing. It satisfies

$$\operatorname{Tan} 0 = 0, \quad \operatorname{Tan} \tfrac{1}{2} = 1, \quad (\operatorname{Tan} 1 = \infty)$$

and the *addition theorem*,

$$\operatorname{Tan}(r + s) = \frac{\operatorname{Tan} r + \operatorname{Tan} s}{1 - \operatorname{Tan} r \operatorname{Tan} s}. \tag{9.12}$$

Now let us define these functions for arbitrary rational arguments. To this end we make use of the uniform continuity of the function Sin in $[0|1]$. Since the values of the function Cos are always $\leqslant 1$, the addition theorem immediately yields

$$\text{Sin}\,(r+\delta) \leqslant \text{Sin}\, r + \text{Sin}\,\delta,$$

whence

$$\text{Sin}\,(r+\delta) - \text{Sin}\, r \leqslant \text{Sin}\,\delta.$$

All that remains to be shown is that

$$\lim_{\delta \rightsquigarrow 0} \text{Sin}\,\delta = 0. \tag{9.13}$$

The proof takes as its point of departure the superadditivity of the function Tan [cf. (9.3)]. Since $0 < 1 - \text{Tan}\, r\ \text{Tan}\, s < 1$, the addition theorem indeed directly yields

$$\text{Tan}\,(r+s) > \text{Tan}\, r + \text{Tan}\, s \quad \text{for} \quad r, s > 0.$$

Exactly as in the case of the exponential function, superadditivity here gives us

$$2^{n-1}\ \text{Tan}\,\frac{1}{2^n} < \text{Tan}\,\frac{1}{2} - \text{Tan}\, 0$$

93

or

$$2^{n-1}\ \text{Tan}\,\frac{1}{2^n} < 1.$$

Since $\text{Sin}\, r \leqslant \text{Tan}\, r$, we get *a fortiori*

$$2^n\, \text{Sin}\,\frac{1}{2^n} < 2. \tag{9.14}$$

This inequality yields the desired limit relationship (9.13).

The function Cos is likewise uniformly continuous in view of (9.10). We have

$$\lim_{\delta \rightsquigarrow 0} \text{Cos}\,\delta = \text{Cos}\, 0 = 1.$$

The functions Sin and Cos are now uniquely definable for all real numbers in the interval $[0|1]$ as

$$\text{Sin}\,\xi = \lim_{r \rightsquigarrow \xi} \text{Sin}\, r$$

$$\text{Cos}\,\xi = \lim_{r \rightsquigarrow \xi} \text{Cos}\, r.$$

Monotony and uniform continuity are preserved, as well as the addition theorems and the equation (9.10).

The transition from the functions Sin and Cos to the functions sin and cos as applied to the usual angular measure in degrees (according to which the right angle is assigned a numerical measure of 90—or more recently 100) would require nothing more than the definitions

$$\sin \xi^\circ = \mathrm{Sin}\ \frac{\xi}{100}, \qquad \cos \xi^\circ = \mathrm{Cos}\ \frac{\xi}{100}.$$

This notation has the disadvantage of making the functions sin and cos appear as functions of the size of the angles rather than as functions of real numbers. We therefore avoid the use of such angular measure. We shall define sin and cos after π has been introduced—but we shall not get to that until later. For the time being, we extend the domain of the functions Sin and Cos beyond the interval $[0|1]$.

The formula

$$\mathrm{i}^{\frac{m}{2^n}} = \mathrm{Cos}\ \frac{m}{2^n} + \mathrm{i}\ \mathrm{Sin}\ \frac{m}{2^n}$$

defines Cos $\frac{m}{2^n}$ and Sin $\frac{m}{2^n}$ for all integers m, whether inside or outside of $[0|2^n]$. We therefore let $\mathrm{i}^\xi = \mathrm{Cos}\ \xi + \mathrm{i}\ \mathrm{Sin}\ \xi$ generally.

94 It follows from the addition theorems that

$$\mathrm{Sin}\ (\xi + 1) = \mathrm{Cos}\ \xi$$

$$\mathrm{Cos}\ (\xi + 1) = -\mathrm{Sin}\ \xi$$

and hence also that

$$\mathrm{Sin}\ (\xi + 2) = -\mathrm{Sin}\ \xi$$

$$\mathrm{Cos}\ (\xi + 2) = -\mathrm{Cos}\ \xi$$

and

$$\mathrm{Sin}\ (\xi + 4) = \mathrm{Sin}\ \xi$$

$$\mathrm{Cos}\ (\xi + 4) = \mathrm{Cos}\ \xi$$

(in keeping with $\mathrm{i}^2 = -1$, $\mathrm{i}^4 = 1$).

The functions Sin and Cos are thus uniformly continuous *throughout* and have a *period* of 4.

If we set $\mathrm{i}^{-\xi} = \frac{1}{\mathrm{i}^\xi}$, then we also get

$$\mathrm{i}^{-\xi} = \mathrm{Cos}\ \xi - \mathrm{i}\ \mathrm{Sin}\ \xi,$$

so that

$$\mathrm{Sin}\ (-\xi) = -\mathrm{Sin}\ \xi$$

$$\mathrm{Cos}\ (-\xi) = \mathrm{Cos}\ \xi,$$

whence

$$\text{Sin}\,(1-\xi) = \text{Cos}\,\xi$$

$$\text{Cos}\,(1-\xi) = \text{Sin}\,\xi.$$

Since Cos $\xi = 0$ only for $\xi = 1 \pm 2n$ $(n = 0, 1, 2, \ldots\ldots)$, Tan is a continuous function throughout—except at those points where Cos $\xi = 0$. Whereas the range of Sin and Cos is the interval $[-1 \mid +1]$, the range of Tan is the class of all real numbers. Tan is periodic, with a *period* of 2.

The *trigonometric functions* Sin, Cos, Tan are not monotonic over their entire domain, but every bounded interval thereof breaks down into finitely many monotonic intervals. Sin and Tan have the monotonic intervals $[n-1 \mid n+1]$ for n even; Cos has the monotonic intervals $[n-1 \mid n+1]$ for n odd.

In every monotonic interval the trigonometric functions are invertible. Their inverses are called "*arc functions*." We denote the inverse functions over the interval $[n-1 \mid n+1]$ as Arc Sin_n and Arc Tan_n (n even), and Arc Cos_n (n odd).

Arc Sin_0, Arc Tan_0, and Arc Cos_1 are also referred to as the *principal functions*, and the others as the *subsidiary functions*.

For example, it is the case that

95

$$\text{Arc Sin}_0\eta = \xi \dot{\leftrightarrow} \text{Sin}\,\xi = \eta \wedge |\xi| \leqslant 1$$

$$\text{Arc Sin}_n\eta = \text{Arc Sin}_0\eta + n \quad (n \text{ even}).$$

The function Arc Cos is reducible to Arc Sin by way of the formula

$$\text{Arc Cos}_1\eta = 1 - \text{Arc Sin}_0\eta.$$

For the reduction of Arc Tan, we use

$$\text{Sin}\,\xi = \frac{\text{Sin}\,\xi}{\sqrt{\text{Cos}^2\xi + \text{Sin}^2\xi}} = \frac{\text{Tan}\,\xi}{\sqrt{1 + \text{Tan}^2\xi}},$$

whence

$$\xi = \text{Arc Sin}_n \frac{\text{Tan}\,\xi}{\sqrt{1 + \text{Tan}^2\xi}} \quad (\text{for an appropriate } n).$$

For $\eta = \text{Tan}\,\xi$ we get

$$\text{Arc Tan}_n\,\eta = \text{Arc Sin}_n \frac{\eta}{\sqrt{1+\eta^2}}.$$

The arc functions are uniformly continuous in every closed interval of their domain, as well as properly monotonic.

§10. Sequences of Functions

From the elementary functions, we can easily arrive at further functions, in particular further continuous functions, by considering function sequences. If $T(n, \xi)$ is a definite term representing a function

$$\phi_n \xi = T(n, \xi)$$

for every n ($n = 1, 2, 3, \ldots\ldots$), then the term $T(n, \xi)$ with variable n represents a *function sequence*:

$$\phi_* = \phi_1, \phi_2, \phi_3, \ldots\ldots$$

A couple of simple examples:
(1) the term ξ^n represents a function sequence ϕ_* such that

$$\phi_1 \xi = \xi, \quad \phi_2 \xi = \xi^2, \quad \phi_3 \xi = \xi^3, \ldots\ldots;$$

(2) the term $\sum_1^n {}_i \dfrac{\text{Sin}\, i\xi}{i^2}$ represents a function sequence ϕ_* such that $\phi_1 \xi = \text{Sin}\ \xi$, $\phi_2 \xi = \text{Sin}\ \xi + \dfrac{\text{Sin}\ 2\xi}{4}$, $\phi_3 \xi = \text{Sin}\ \xi + \dfrac{\text{Sin}\ 2\xi}{4} + \dfrac{\text{Sin}\ 3\xi}{9}, \ldots\ldots$

96 If all members ϕ_n of a sequence ϕ_* have a common domain K, and if the numerical sequence $\phi_* \xi$ is convergent for every $\xi \in K$, then the term $\lim \phi_* \xi$ denotes a real number for every $\xi \in K$. The result is the representation of a new function ϕ, the *limit function* for ϕ_*. In such a case the sequence ϕ_* is said to be *convergent* in K, and we write

$$\phi = \lim \phi_*. \tag{10.1}$$

If the members ϕ_n do not all have the same domain, or if the sequences $\phi_* \xi$ are not all convergent for every ξ of a common domain, then the domain for which (10.1) holds must be expressly specified.

Even if all the members of a convergent function sequence are continuous, the limit function still need not necessarily be continuous. For example, in the closed interval $[0|1]$ the function sequence ϕ_* such that $\phi_n = I^n$ is convergent. If $0 \leqslant \xi < 1$ then

$$\lim \phi_* \xi = \lim_{n \to \infty} \xi^n = 0.$$

For $\xi = 1$, on the other hand,

$$\lim \phi_* \xi = 1.$$

Thus the function lim ϕ_* is discontinuous for the argument 1.

The *convergence* of a function sequence ϕ_* to ϕ in a domain K is defined as

$$\bigwedge_{\substack{\xi \\ K}} \wedge_\varepsilon \vee_N \wedge_{\substack{n \\ >N}} |\phi_n \xi - \phi \xi| < \varepsilon. \tag{10.2}$$

Condition (10.2) has the same "form" as the condition for continuity of a function in K: it begins with 2 universal quantifiers, 1 existential quantifier, and 1 universal quantifier.

Just as we made use of uniform continuity in K (as a stronger condition) along with continuity in K, so we now introduce the corresponding property of *uniform convergence* in K.

DEFINITION 10.1 The function sequence ϕ_* is *uniformly convergent* to ϕ in K (written $\phi_* \simeq \phi$) if

$$\wedge_\varepsilon \vee_N \bigwedge_{\substack{\xi \\ K}} \wedge_{\substack{n \\ >N}} |\phi_n \xi - \phi \xi| < \varepsilon. \tag{10.3}$$

As in the definition of uniform continuity compared to that of continuity, the quantifiers here in (10.3) are permuted in comparison with (10.2). The existential quantifier $\vee_N$ now stands ahead of the universal quantifier $\bigwedge_\xi$.

Since

$$\vee_y \wedge_x A(x, y) \rightarrow \wedge_x \vee_y A(x, y)$$

is universally valid, convergence likewise always follows from uniform convergence.

The convergence of the sequence ϕ_* such that $\phi_n \xi = \xi^n$ is non-uniform in $(0|1)$, for 97

$$\vee_\varepsilon \wedge_N \bigvee_{\substack{\xi \\ (0|1)}} \vee_{\substack{n \\ >N}} |\xi^n - \phi \xi| \geqslant \varepsilon \quad \text{where} \quad \phi = \lim \phi_*.$$

Proof. If $\varepsilon = \frac{1}{2}$, then for any N there is a ξ such that $0 < \xi < 1$ (whence $\phi \xi = 0$) and $\xi^{N+1} \geqslant \varepsilon$, for $\lim\limits_{\xi \to 1} \xi^N = 1$.

The uniform convergence of a sequence can also be formulated without recourse to limit functions.

THEOREM 10.1 A function sequence ϕ_* is uniformly convergent in K if and only if

$$\wedge_\varepsilon \vee_N \bigwedge_{\substack{\xi \\ K}} \wedge_{\substack{m,n \\ >N}} |\phi_m \xi - \phi_n \xi| < \varepsilon.$$

Proof issues directly from our definitions and the Cauchy convergence criterion (§7).

Regarding the uniform convergence of sequences of continuous functions we have

THEOREM 10.2 If a function sequence ϕ_* is uniformly convergent in a domain K, and if every member ϕ_n is continuous in K, then $\phi = \lim \phi_*$ is also continuous in K.

Proof. We must show for every $\xi_0 \in K$ and every $\varepsilon > 0$ that

$$\bigvee_\delta \bigwedge_{\substack{\xi \\ K}} . |\xi - \xi_0| < \delta \rightarrow |\phi\xi - \phi\xi_0| < \varepsilon .$$

Because of the uniform convergence of ϕ_* to ϕ, there is an n such that

$$\bigwedge_{\substack{\xi \\ K}} |\phi_n\xi - \phi\xi| < \frac{\varepsilon}{3}.$$

Since ϕ_n is continuous at ξ_0, it further holds of some δ that

$$\bigwedge_{\substack{\xi \\ K}} . |\xi - \xi_0| < \delta \rightarrow |\phi_n\xi - \phi_n\xi_0| < \frac{\varepsilon}{3}.$$

Thus we obtain for that δ and for all $\xi \in K$ such that $|\xi - \xi_0| < \delta$ the result

$$|\phi\xi - \phi\xi_0| \leqslant |\phi\xi - \phi_n\xi| + |\phi_n\xi - \phi_n\xi_0| + |\phi_n\xi_0 - \phi\xi_0| < \varepsilon .$$

Uniformly convergent sequences of functions not only lend themselves to the construction of new functions from those already available; they can also be used to reduce familiar functions to simpler ones. For reasons that will not become clear until we get to the differential calculus (§12, Taylor series), the most important function sequences are those representable by *power series*.

98 By a *power series* we mean a term $\sum_{0}{}_i a_i \xi^i$ with a definite *sequence of coefficients* a_*. Such a term represents a sequence with the members

$$a_0, a_0 + a_1\xi, a_0 + a_1\xi + a_2\xi^2, \ldots\ldots$$

If this sequence, i.e., the sequence $\sum_{0}^{*}{}_i a_i \xi^i$ is convergent, then its limit is denoted by $\sum_{0}^{\infty}{}_i a_i \xi^i$ or (incompletely) by

$$a_0 + a_1\xi + a_2\xi^2 + \ldots\ldots$$

This limit is called the *sum* of the members of the sequence

$$a_0, a_1\xi, a_2\xi^2, \ldots\ldots$$

or the sum of the series. The members of $\sum_{0}^{*}{}_i a_i \xi^i$ are therefore referred to as *partial sums*.

Usually no distinction is drawn between a series (which is a term) and its sum (a number). Thus one sometimes speaks of a power series

$$a_0 + a_1\xi + a_2\xi^2 + \ldots\ldots$$

even when it is not convergent. Since it is often convenient to give only the initial members in referring to a series (or its sum), we shall resort to the convention of writing 5 dots at the end when we mean the sum, 4 dots for the series (and 3 dots for a finite sum).

Not all continuous functions can be represented by power series, as we shall see in §12. If $\sum_{0}^{*}{}_i a_i(\xi - \xi_0)^i$ is convergent in an interval around ξ_0, then the function ϕ such that

$$\phi\xi = \sum_{0}^{\infty}{}_i a_i(\xi - \xi_0)^i$$

is said to be *analytic*.

Not every power series represents an analytic function; e.g., $\sum_{0}^{*}{}_i i!\,\xi^i$ is not convergent for any $\xi \neq 0$. Because

$$(i+1)!\,\frac{1}{n^{i+1}} \geqslant i!\,\frac{1}{n^i} \quad \text{where} \quad i \geqslant n,$$

it is in fact true of every $\varepsilon > 0$ that

$$\sum_{0}^{\infty}{}_i i!\,\varepsilon^i = \infty.$$

We shall now prove that all analytic functions are continuous functions. This will follow from the uniform convergence of power series. We preface the proof with a lemma about arbitrary function sequences ϕ_*. 99

If we set

$$\psi_0 = \phi_0$$

$$\psi_i = \phi_i - \phi_{i-1} \quad (i \geqslant 1),$$

we get

$$\phi_0 = \psi_0$$

$$\phi_1 = \psi_0 + \psi_1$$

$$\phi_2 = \psi_0 + \psi_1 + \psi_2$$

$$\vdots$$

$$\phi_n = \sum_{0}^{n}{}_i \psi_i\,.$$

The sequence ϕ_* is thus representable by the series $\underset{0}{\overset{*}{\sum}}{}_i\,\psi_i$.

THEOREM 10.3 If a sequence $\underset{0}{\overset{*}{\sum}}{}_i\psi_i$ satisfies the estimate $|\psi_i\xi| \leqslant a_i$ for all ξ in an interval J, and if the sequence $\underset{0}{\overset{*}{\sum}}{}_i a_i$ is convergent, then $\underset{0}{\overset{*}{\sum}}{}_i\psi_i$ is uniformly convergent in J, and it holds in J that

$$\left|\sum_0^\infty{}_i\psi_i\,\xi\right| \leqslant \sum_0^\infty{}_i a_i .$$

Proof. By Theorem 10.1, we must show for every ε that there is an N such that

$$\left|\sum_0^m{}_i\psi_i\,\xi - \sum_0^n{}_i\psi_i\,\xi\right| < \varepsilon$$

for all $m > n > N$ and for all $\xi \in J$. Now,

$$\left|\sum_0^m{}_i\psi_i\,\xi - \sum_0^n{}_i\psi_i\,\xi\right| = \left|\sum_{n+1}^m{}_i\psi_i\,\xi\right| \leqslant \sum_{n+1}^m{}_i a_i .$$

Since $\underset{0}{\overset{*}{\sum}}{}_i a_i$ is by hypothesis convergent, there is in fact for every ε an N such that

$$\sum_{n+1}^m{}_i a_i < \varepsilon$$

for all $m > n > N$.

100 The series $\underset{0}{\sum}{}_i a_i$ is called a *majorant* of the series $\underset{0}{\sum}{}_i\psi_i\,\xi$. As an example we prove the convergence of $\underset{1}{\overset{*}{\sum}}{}_i i\theta^{i-1}$ for $0 < \theta < 1$. For sufficiently large n,

$$\theta < \frac{n}{n+1} < \frac{n+1}{n+2} < \ldots\ldots$$

Hence, if $q = \dfrac{n+1}{n}\,\theta < 1$, then

$$(n+1)\theta^n = \frac{n+1}{n}\cdot n\theta^{n-1}\cdot\theta = qn\theta^{n-1},$$

$$(n+2)\theta^{n-1} = \frac{n+2}{n+1}\,(n+1)\theta^n\cdot\theta < q(n+1)\theta^n = q^2 n\theta^{n-1},$$

$$\vdots$$

i.e., from a certain n on, $\sum_{1} iq^i n\theta^{n-1}$ is a majorant of $\sum_{n+1} i\theta^{i-1}$. But since $q < 1$, $\sum_{0}^{\infty} iq^i = \frac{1}{1-q}$.

Now we easily obtain

THEOREM 10.4 If the sequence $\sum_{0}^{*} {}_i a_i \xi^i$ is convergent for $\xi = \xi_1 \neq 0$, then it is uniformly convergent in every interval $[-\xi_0 | \xi_0]$ such that $0 < \xi_0 < |\xi_1|$.

Proof. From the convergence of $\sum_{0}^{*} {}_i a_i \xi_1^i$ it follows, e.g., that $|a_i \xi_1^i| < 1$ for all sufficiently large i. If $0 < \xi_0 < |\xi_1|$, i.e., $\xi_0 = q|\xi_1|$ where $0 < q < 1$, then we get for $|\xi| \leqslant \xi_0$

$$|a_i \xi^i| \leqslant |a_i| \xi_0^i = |a_i \xi_1^i| \cdot q^i < q^i$$

for all sufficiently large i. Thus, since $\sum_{0}^{\infty} {}_i q^i = \frac{1}{1-q}$, the sequence $\sum_{0}^{*} {}_i a_i \xi^i$ is uniformly convergent in $[-\xi_0 | \xi_0]$ by Theorem 10.3.

For every power series $\sum_{0} {}_i a_i \xi^i$ that converges for some $\xi \neq 0$, there is thus always a maximum open interval, if only the interval $(-\infty | \infty)$, in which the power series is convergent. This maximum interval is known as the *convergence interval* of the power series. By Theorem 10.2, the sum of a power series is a continuous function in its convergence interval. This establishes the continuity of analytic functions.

In §12 we shall see that all elementary functions are analytic. The analytic functions thus form a class of which the elementary functions are a subclass. While it is only an (historically explainable) accident that certain functions are designated as *101* elementary, the distinction accorded to analytic functions is objectively justified.

The distinctiveness of analytic functions is already evident from the fact that two analytic functions ϕ and ψ which coincide in a finite interval for infinitely many arguments also coincide throughout the entire interval. For there will then be a convergent sequence $\xi_* \to \xi_0$ such that $\phi\xi_n = \psi\xi_n$ for all n. If it holds around ξ_0 that

$$\phi\xi = \sum_{0}^{\infty} {}_i a_i (\xi - \xi_0)^i \quad \text{and} \quad \psi\xi = \sum_{0}^{\infty} {}_i b_i (\xi - \xi_0)^i,$$

then in view of continuity it follows that $\phi\xi_0 = \psi\xi_0$, i.e., $a_0 = b_0$.

If $a_i = b_i$ has already been proved for all $i < n$, then it follows that

$$\phi_n \xi \leftrightharpoons \frac{\phi\xi - \sum_{0}^{n-1}{}_i a_i(\xi - \xi_0)^i}{(\xi - \xi_0)^n} = \sum_{n}^{\infty}{}_i a_i(\xi - \xi_0)^{i-n}$$

and similarly that

$$\psi_n \xi \leftrightharpoons \frac{\psi\xi - \sum_{0}^{n-1}{}_i b_i(\xi - \xi_0)^i}{(\xi - \xi_0)^n} = \sum_{n}^{\infty}{}_i b_i(\xi - \xi_0)^{i-n} .$$

As the right-hand sides show, both functions ϕ_n and ψ_n are continuous at ξ_0. Thus $\phi_n \xi_0 = \psi_n \xi_0$ implies that $a_n = b_n$. ϕ and ψ thus coincide around ξ_0 in any event. Analytic induction then yields the conclusion that ϕ and ψ coincide around every point of the interval.

We have yet to demonstrate that the class of analytic functions —like the class of elementary functions—is closed under addition, multiplication, and composition.

In the case of addition, e.g., we must prove that if two power series $\sum_{0}{}_i a_i \xi^i$ and $\sum_{0}{}_i b_i \xi^i$ are convergent for some $\xi \neq 0$, then the power series $\sum_{0}{}_i (a_i + b_i)\xi^i$ is likewise convergent, and furthermore

$$\sum_{0}^{\infty} (a_i + b_i)\xi^i = \sum_{0}^{\infty}{}_i a_i \xi^i + \sum_{0}^{\infty}{}_i b_i \xi^i .$$

Similar propositions must be proved for multiplication, starting with the formation of a *double series*

$$\sum_{0,0}{}_{i,j} a_i b_j \xi^{i+j} .$$

For composition, we have to construct

$$\sum_{0}{}_i a_i \Big(\sum_{0}{}_j b_j \xi^j\Big)^i .$$

102 Since we only have to consider the power series here at a particular point ξ, the only feature of them that we shall use is essentially what is established by the proof of Theorem 10.4, viz., their *absolute convergence* (in the interior of the convergence interval), i.e., the convergence not just of $\sum_{0}^{*}{}_i a_i \xi^i$ but also of $\sum_{0}^{*}{}_i |a_i \xi^i|$.

In general, a series $\sum_{0}{}_i a_i$ is called *absolutely convergent* if $\sum_{0}{}_i |a_i|$ is convergent. $\sum_{0}{}_i |a_i|$ is then a majorant of $\sum_{0}{}_i a_i$. Thus every absolutely convergent series is convergent.

That the converse does not hold, that not every convergent series is absolutely convergent, is shown, e.g., by

$$1 - \frac{1}{2} + \frac{1}{3} - \frac{1}{4} + \frac{1}{5} - \dots\dots$$

The desired properties of power series (and hence of analytic functions) now follow from the fact that in computation we can handle the sums of absolutely convergent series very much as we do finite sums. In particular, comutativity and associativity are preserved. Distributivity,

$$b \sum_0^\infty{}_i a_i = \sum_0^\infty{}_i ba_i,$$

obviously holds of all convergent series.

The commutativity of finite sums $\sum_0^n{}_i a_i$ means that for every invertible mapping $k_0, \dots, k_n$ of the system of numbers $0, \dots, n$ onto itself we shall have

$$\sum_0^n{}_i a_{k_i} = \sum_0^n{}_i a_i.$$

Extended to infinite sums, this gives

THEOREM 10.5 If $\sum_0{}_i a_i$ is absolutely convergent and k_* is a one-to-one mapping of the set of natural numbers with 0 onto itself, then the reordered series $\sum_0{}_i a_{k_i}$ is likewise absolutely convergent, and we get

$$\sum_0^\infty{}_i a_{k_i} = \sum_0^\infty{}_i a_i.$$

Proof. (1) The partial sums $\sum_0^n{}_i a_{k_i}$ are bounded, for where $m = \max(k_0, \dots, k_n)$ we have (because of invertibility)

$$\sum_0^n{}_i |a_{k_i}| \leqslant \sum_0^m{}_i |a_i| \leqslant \sum_0^\infty{}_i |a_i|.$$

Hence the series $\sum_0{}_i a_{k_i}$ is absolutely convergent. 103

(2) For every $\varepsilon > 0$ there is an m such that $\sum_m^\infty{}_i |a_i| < \varepsilon$. Now, if

$$\{0, 1, \dots, m\} = \{k_{j_0}, \dots, k_{j_m}\}$$

and $M = \max\{j_0, \dots, j_m\}$, then $M \geqslant m$ and

$$\{a_0, \dots, a_m\} \subseteq \{a_{k_0}, \dots, a_{k_M}\}.$$

Thus $\sum_{0}^{n}{}_i a_{k_i} - \sum_{0}^{n}{}_i a_i$, where $n \geqslant M$, can be represented as a sum of elements a_j such that $j > m$; i.e.,

$$\left| \sum_{0}^{n}{}_i a_{k_i} - \sum_{0}^{n}{}_i a_i \right| \leqslant \sum_{m}^{\infty}{}_i |a_i| < \varepsilon$$

and hence

$$\sum_{0}^{\infty}{}_i a_{k_i} = \sum_{0}^{\infty}{}_i a_i .$$

In order to formulate associativity, we start with a double sequence $a_{*\dagger}$. Let the finite sums

$$\sum_{0,0}^{m,n}{}_{i,j} |a_{ij}|$$

be bounded. By commutativity, then, which we just proved, all series $\sum_{0}{}_l b_l$ generated by reordering the double sequence $a_{*\dagger}$ into a sequence b_* are absolutely convergent—and they have the same limit, which we denote by

$$\sum_{0,0}^{\infty,\infty}{}_{i,j} a_{ij} .$$

THEOREM 10.6 If the double series $\sum_{0,0}{}_{i,j} a_{ij}$ is absolutely convergent, then the simple series $\sum_{0}{}_j a_{ij}$ are absolutely convergent; and if $s_i = \sum_{0}^{\infty}{}_j a_{ij}$, then $\sum_{0}{}_i s_i$ is absolutely convergent too. Furthermore, we have

$$\sum_{0}^{\infty}{}_i s_i = \sum_{0,0}^{\infty,\infty}{}_{i,j} a_{ij} .$$

Proof. The absolute convergence of the several series $\sum_{0}{}_j a_{ij}$ is trivial, seeing that

$$\sum_{0}^{n}{}_j |a_{ij}| \leqslant \sum_{0,0}^{\infty,\infty}{}_{i,j} |a_{ij}| .$$

104 Since the double series is absolutely convergent, there will be for every ε an M and an N such that $\sum_{M,N}^{\infty,\infty}{}_{i,j} |a_{ij}| < \varepsilon$. If $m > M$, then $\left(\sum_{0}{}_j a_{ij} \text{ being convergent}\right)$ there will be for every $i \leqslant m$ an $n_i > N$ such that

$$\left| s_i - \sum_{0}^{n_i}{}_j a_{ij} \right| < \frac{1}{m}\varepsilon .$$

Thus, if $n = \max(n_0, \ldots, n_m)$, then

$$\left| \sum_{0}^{m}{}_i s_i - \sum_{0}^{m}{}_i \sum_{0}^{n}{}_j a_{ij} \right| < \varepsilon .$$

Furthermore (abbreviating),

$$\left| \sum_{0,0}^{\infty,\infty}{}_{i,j} a_{ij} - \sum_{0,0}^{m,n}{}_{i,j} a_{ij} \right| \leqslant \left| \sum_{0,0}^{\infty,\infty} - \sum_{0,0}^{M,N} \right| + \left| \sum_{0,0}^{m,n} - \sum_{0,0}^{M,N} \right|$$

$$\leqslant \sum_{M,N}^{\infty,\infty}{}_{i,j} |a_{ij}| + \sum_{M,N}^{m,n}{}_{i,j} |a_{ij}| < 2\varepsilon .$$

Adding the two inequalities together, we obtain

$$\left| \sum_{0}^{m}{}_i s_i - \sum_{0,0}^{\infty,\infty}{}_{i,j} a_{ij} \right| < 3\varepsilon, \quad \text{whence} \quad \sum_{0}^{\infty}{}_i s_i = \sum_{0,0}^{\infty,\infty}{}_{i,j} a_{ij} .$$

For an understanding of the connection between Theorem 10.6 and ordinary associativity, the following remark may be helpful. By Theorem 10.5, the sum of an absolutely convergent series is invariant under any reordering of the sequence composed of the series' members. The reorderability of one sequence into another is clearly an equivalence relation. We may therefore regard two such sequences as identical from an abstract point of view: we shall say that they represent the same *infinite complex.* The complex composed of the members of a sequence (or, as we shall say, the *associated complex*) is to be distinguished from the associated set. In a complex, each member occurs with a certain *multiplicity*, which is lost by abstraction in the formation of the associated set.

Theorem 10.5 then says that the sum of an absolutely convergent series depends only on the associated complex. Thus we could speak of *summable complexes* and their *sums.* Theorem 10.6 admits of formulation in such an idiom: if a summable complex is decomposed into (finitely or infinitely many) subcomplexes, the sum of the sums of the subcomplexes equals the sum of the entire complex.

For a finite sum, e.g., a 3-member one $a + b + c$, the possible decompositions into (nontrivial) subcomplexes are precisely $(a + b) + c$, $(b + c) + a$, and $(c + a) + b$. The equality of these sums is a consequence of commutativity and ordinary associativity. *105*

We now proceed to apply these results to power series. There is nothing to prove about the addition of power series, since it is already the case for partial sums that

$$\sum_{0}^{n}{}_i(a_i + b_i)\xi^i = \sum_{0}^{n}{}_i a_i\,\xi^i + \sum_{0}^{n}{}_i b_i\xi^i.$$

For the multiplication of two power series $\sum_{0}{}_i a_i\,\xi^i$ and $\sum_{0}{}_j b_j\xi^j$, assume that each is absolutely convergent. We must then show that $\sum_{0,0}{}_{i,j} a_i b_j\,\xi^{i+j}$ too is absolutely convergent and that

$$\sum_{0,0}^{\infty,\infty}{}_{i,j}\, a_i b_j\,\xi^{i+j} = \sum_{0}^{\infty}{}_i a_i\,\xi^i \cdot \sum_{0}^{\infty}{}_j\, b_j\xi^j.$$

This follows immediately from the more general

THEOREM 10.7 If $\sum_i a_i$ and $\sum_j b_j$ are absolutely convergent, then so is $\sum_{i,j} a_i b_j$, and

$$\sum_{0,0}^{\infty,\infty}{}_{i,j}\, a_i b_j = \sum_{0}^{\infty}{}_i a_i \cdot \sum_{0}^{\infty}{}_j\, b_j.$$

Proof. (1) The partial sums of $\sum_{i,j} |a_i b_j|$ are bounded, because

$$\sum_{0,0}^{p,p}{}_{i,j}\, |a_i\, b_j| = \sum_{0}^{p}{}_i\, |a_i| \cdot \sum_{0}^{p}{}_j\, |b_j| \leqslant \sum_{0}^{\infty}{}_i\, |a_i| \cdot \sum_{0}^{\infty}{}_j\, |b_j|.$$

(2) By Theorem 10.6 (and distributivity, which is trivial),

$$\sum_{0,0}^{\infty,\infty}{}_{i,j}\, a_i b_j = \sum_i \left(\sum_j a_i b_j\right) = \sum_i \left(a_i \sum_j b_j\right) = \sum_{0}^{\infty}{}_i a_i \cdot \sum_{0}^{\infty}{}_j\, b_j.$$

Theorem 10.7 permits the multiplication of absolutely convergent series by one another. In particular, powers $\left(\sum_i a_i\right)^m$ are possible.

The only remaining case is that of the composition of power series.

THEOREM 10.8 Let $\phi\eta = \sum_{0}^{\infty}{}_i a_i\,\eta^i$ and $\psi\xi = \sum_{0}^{\infty}{}_j\, b_j\,\xi^j$ be the sums of absolutely convergent series, where $|\eta| < r$, $|\xi| < s$. Provided that $\sum_{0}^{\infty}{}_j\, |b_j| s^j < r$, the double series

$$\sum_{0}{}_i a_i \left(\sum_{0}{}_j\, b_j\,\xi^j\right)^i$$

is absolutely convergent for $|\xi| < s$ and has the sum $\phi\psi\xi$.

106 *Proof.* Let $\left(\sum_{0}{}_j\, b_j\,\xi^j\right)^i = \sum_j b_{ij}\,\xi^j$. The complex with members $a_i b_{ij}\,\xi^j$ is absolutely summable, because the sums of all its finite subcomplexes are bounded: for every i,

$$\sum_{0}^{n}{}_{j} |a_i b_{ij}\, \xi^j| \leqslant |a_i| \left(\sum_{0}^{n}{}_{j} |b_j| |s^j| \right)^i < |a_i| r^i.$$

Thus by Theorems 10.6 and 10.7 we get

$$\sum_{0,0}^{\infty,\infty}{}_{i,j} a_i b_{ij}\, \xi^j = \sum_{0}^{\infty}{}_{i} a_i \left(\sum_{0}^{\infty}{}_{j} b_j \xi^j \right)^i = \sum_{0}^{\infty}{}_{i} a_i (\psi\xi)^i = \phi\urcorner\psi\xi.$$

Once we have proved in §12 that the basic elementary functions are analytic, it will follow from the above results that all elementary functions are analytic (since they are, after all, built up out of basic elementary functions by addition, multiplication, and composition).

To represent nonanalytic functions as limit functions, polynomials are not especially suited as approximation functions, but rather the still simpler *step functions*.

By a *step function* we mean a function that is piecewise constant; i.e., let $[\alpha \,|\, \beta]$ be partitioned into a *finite* number of pieces

$$\alpha = \alpha_0 < \alpha_1 < \alpha_2 < \ldots < \alpha_n = \beta,$$

and let

$$\phi\xi = \gamma_i,$$

where $\xi \in (\alpha_{i-1} \,|\, \alpha_i)$, with $\phi\alpha_i$ chosen arbitrarily.

The class of functions representable as the limits of uniformly convergent sequences of step functions includes not only (as we shall see) all continuous functions, but also discontinuous functions. An example of one such function, in fact one whose discontinuity points are dense in the interval $[0 \,|\, 1]$, can be constructed in the following way. Divide the interval $[0 \,|\, 1]$ into three equal parts; then divide each of the resulting subintervals into three equal parts; etc. After each trisection, raise the middle third by a distance of 1/2, 1/4, 1/8, ... successively. Call the step function arising from the nth step in this construction ϕ_n. Then we obviously have

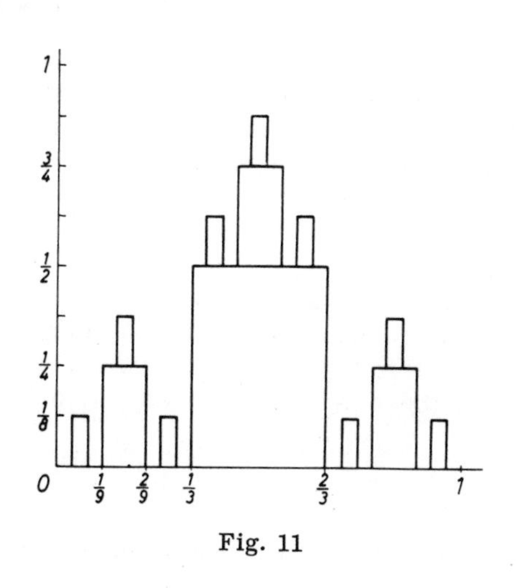

Fig. 11

107

$$m > n \to |\phi_m - \phi_n| < \frac{1}{2^n},$$

which directly entails the uniform convergence of the sequence ϕ_* of step functions.

As in Theorem 10.2, it turns out that in general every uniform limit of a sequence of step functions has left- and right-hand limits throughout, just as do the step functions themselves; i.e., the left-hand limit $\lim_{\xi \underset{<}{\to} \xi_0} \phi\xi$ and the right-hand limit $\lim_{\xi \underset{>}{\to} \xi_0} \phi\xi$ ($\lim_{<\xi_0} \phi$ and $\lim_{>\xi_0} \phi$ for short) exist everywhere. However, these limits are not necessarily equal either to each other or to the function value $\phi\xi_0$ (as they are in the case of continuous functions).

If it is the case that

$$\phi\xi_0 \neq \lim_{<\xi_0} \phi = \lim_{>\xi_0} \phi,$$

then ξ_0 is called a *removable* point of discontinuity. It is reasonable to confine our discussion to functions having no removable points of discontinuity. Functions with left- and right-hand limits throughout and no removable discontinuities may be called "*jump-continuous*." For jump-continuous functions, every point of discontinuity is a *jump* such that

$$\lim_{<\xi_0} \phi \neq \lim_{>\xi_0} \phi.$$

The set of jumps in a function ϕ is definite if its rational subfunction ϕ_0 is definite. This follows from the fact that the jumps ξ_0 are already characterized by

$$\lim_{r \underset{<}{\to} \xi_0} \phi r \neq \lim_{r \underset{>}{\to} \xi_0} \phi r.$$

In fact, we can even give a term for a sequence whose members are none other than the jumps: the set of jumps is *definitely denumerable*. [For an example of an indefinite term, cf. (7.3).] For there is an n for every jump ξ_0 such that

$$\left| \lim_{r \underset{<}{\to} \xi_0} \phi r - \lim_{r \underset{>}{\to} \xi_0} \phi r \right| \geq \frac{1}{n}.$$

If ξ_0 is such a jump $\left(\text{with a "}\textit{saltus}\text{" or jump height} \geq \frac{1}{n}\right)$, then, since $\lim_{r \underset{<}{\to} \xi_0} \phi r$ and $\lim_{r \underset{>}{\to} \xi_0} \phi r$ exist, there is an m such that

$$\begin{cases} \bigwedge_s . \xi_0 - \dfrac{1}{m} < s < \xi_0 \rightarrow \left| \phi s - \lim\limits_{r \gtrsim \xi_0} \phi r \right| < \dfrac{1}{2n}. \\ \bigwedge_s . \xi_0 < s < \xi_0 + \dfrac{1}{m} \rightarrow \left| \phi s - \lim\limits_{r \gtrsim \xi_0} \phi r \right| < \dfrac{1}{2n}. \end{cases} \qquad 108$$

For each m, there can be at most finitely many ξ_0s in a finite interval that satisfy these conditions. The open interval

$$\left(\xi_0 - \frac{1}{m} \middle| \xi_0 + \frac{1}{m} \right)$$

evidently contains only the one point ξ_0, so no open subinterval of the length $\frac{1}{m}$ can contain more than one jump satisfying the above conditions.

For all n, therefore, the number of jumps of height $\geqslant \frac{1}{n}$ is definitely denumerable. Furthermore, the set of all jumps whatsoever turns out to be definitely denumerable.

If we accordingly call a jump-continuous function "*left-normalized*" when it is left-continuous, then corresponding to every jump-continuous function there will be a uniquely determined left-normalized jump-continuous function which differs from the original function at no more than a definitely denumerable number of points.

We shall now show that every left-normalized jump-continuous function can be represented as the uniform limit of a definite sequence of left-normalized step functions.

THEOREM 10.9 Every left-normalized jump-continuous function in $[\alpha | \beta]$ is the uniform limit of (left-normalized) step functions.

Proof. We wish to demonstrate for arbitrary ε the existence of a step function τ such that

$$\bigwedge_{\substack{\xi \\ [\alpha|\beta]}} |\phi\xi - \tau\xi| < \varepsilon.$$

To that end we consider the statement form $A(\xi)$:

$$\bigvee_\tau \bigwedge_{\substack{r \\ [\alpha|\xi]}} |\phi r - \tau r| < \varepsilon.$$

We have (1) $A(\alpha)$ (trivially), and

$$(2)\ \bigwedge_\xi . \xi < \xi_0 \rightarrow A(\xi). \rightarrow \bigvee_\delta A(\xi_0 + \delta),$$

for left-normalized jump-continuity implies the existence of some δ such that

$$\bigwedge_{r \atop [\xi_0 - \delta | \xi_0]} |\phi r - \phi \xi_0| < \varepsilon \quad \text{and} \quad \bigwedge_{r \atop [\xi_0 | \xi_0 + \delta]} |\phi r - \lim_{>\xi_0} \phi| < \varepsilon.$$

109 Thus, if τ_- is an ε-approximation in $[\alpha \mid \xi_0 - \delta]$, then

$$\tau_+ \xi = \begin{cases} \phi \xi_0 \text{ in } [\xi_0 - \delta \mid \xi_0] \\ \lim\limits_{>\xi_0} \phi \text{ in } [\xi_0 \mid \xi_0 + \delta] \end{cases}$$

extends τ_- to an ε-approximation in $[\alpha \mid \xi_0 + \delta]$.

Since $A(\xi)$ contains the indefinite quantifier $\mathbf{V}_\tau$, we cannot apply the analytic induction principle. However, the extension of τ_- to τ_+ just given shows that we can confine ourselves to step functions τ which satisfy the following conditions:

(1) The only jumps in τ are points that are jumps in ϕ too.

(2) The values of the function τ are rational.

Since the set of jumps in ϕ is definitely denumerable, so is the set of such step functions τ. They form a sequence τ_1, τ_2, τ_3, ...

Thus, if we restrict the range of the variable "τ" to step functions satisfying (1)–(2), we can replace the indefinite quantifier $\mathbf{V}_\tau$ in $A(\xi)$ by the definite quantifier $\vee_\tau$.

Now it follows by the analytic induction principle that there exists a step function τ such that

$$\bigwedge_{r \atop [\alpha | \beta]} |\phi r - \tau r| < \varepsilon.$$

In view of the left-continuity of ϕ and τ, this implies in turn that

$$\bigwedge_{\xi \atop [\alpha | \beta]} |\phi \xi - \tau \xi| < \varepsilon.$$

We choose $\varepsilon = \dfrac{1}{n}$, and we let τ_{k_n} be the ε-approximation with the smallest subscript. This gives us a definite sequence τ_{k_*} of step functions that uniformly converges to ϕ.

COROLLARY. If we note that the values of no jump-continuous function deviate from left-normalization at more than a definitely denumerable number of points, then the assumption of left-normalization turns out to be superfluous. For if the sequence ξ_* includes all the jumps, then all we have to do is to alter the member τ_{k_n} of a sequence $\tau_{k_*} \succeq \phi$ at the points $\xi_1, \ldots,$ ξ_n in such a way that

$$\tau_{k_n} \xi_i = \phi \xi_i \quad (i = 1, \ldots, n).$$

Since monotonic functions have no removable discontinuities, they are included among the jump-continuous functions.

If we limit our attention to functions without removable discontinuities, then the jump-continuous functions will also include the functions of *bounded variation.* A function ϕ is of bounded variation in the interval $[\alpha \mid \beta]$ if the set of numbers *110*

$$\sum_{0}^{n-1}{}_i \, |\phi\alpha_i - \phi\alpha_{i+1}|$$

is bounded for any partitioning $\alpha = \alpha_0 < \alpha_1 < \ldots < \alpha_n = \beta$ of the interval $[\alpha \mid \beta]$.

All monotonic functions are of bounded variation, for any monotonic function ϕ will satisfy

$$\sum_{0}^{n-1} |\phi\alpha_i - \phi\alpha_{i+1}| = |\phi\alpha - \phi\beta|.$$

As to the jump-continuity of functions of bounded variation, we now show

THEOREM 10.10 If ϕ is of bounded variation in $[\alpha \mid \beta]$, then ϕ has both a left-hand and a right-hand limit everywhere in $[\alpha \mid \beta]$.

Proof. Let $\xi_0 \in [\alpha \mid \beta]$. We proceed by indirect proof; i.e., in order to prove the existence of a left-hand limit,

$$\wedge_\varepsilon \vee_\delta \quad \bigwedge_{\substack{\xi_1,\ \xi_2 \\ \xi_1 < \xi_2 < \xi_0}} . \xi_0 - \xi_1 < \delta \wedge \xi_0 - \xi_2 < \delta \dot{\rightarrow} |\phi\xi_1 - \phi\xi_2| < \varepsilon.,$$

we try assuming the existence for some ε and for all δ of $\xi_1 < \xi_2 < \xi_0$ such that

$$\xi_0 - \xi_1 < \delta \wedge \xi_0 - \xi_2 < \delta \wedge |\phi\xi_1 - \phi\xi_2| \geqslant \varepsilon.$$

For $\delta' = \xi_0 - \xi_2$ there would then also be $\xi_1' < \xi_2'$ such that

$$\xi_1 < \xi_2 < \xi_1' < \xi_2' < \xi_0 \wedge |\phi\xi_1' - \phi\xi_2'| \geqslant \varepsilon.$$

In this way there follows for all n the existence of numbers

$$\xi_1 < \xi_2 < \xi_1' < \xi_2' < \ldots < \xi_1^{(n)} < \xi_2^{(n)} < \xi_0$$

such that

$$|\phi\xi_1^{(i)} - \phi\xi_2^{(i)}| \geqslant \varepsilon \quad \text{where} \quad i = 0, \ldots, n.$$

Thus we get

$$\sum_{0}^{n}{}_i \, |\phi\xi_1^{(i)} - \phi\xi_2^{(i)}| \geqslant (n+1)\varepsilon,$$

which is to say that ϕ is not of bounded variation. Contradiction!

In the same way we get a right-hand limit. (As far as the indefinite quantifiers are concerned, all we infer here is $\bigwedge\neg\ldots$ from $\neg\bigvee\ldots$, *not* $\bigvee\neg\ldots$ from $\neg\bigwedge\ldots$)

111 Every function of bounded variation can furthermore be represented as a difference of monotonic functions. This is usually done by defining a function σ as follows:

$$\sigma(\xi) = \text{lub}\ \epsilon_\sigma \underset{\alpha=\alpha_0<\ldots<\alpha_n=\xi}{\bigvee_{\alpha_0,\ldots,\alpha_n}} \sigma = \sum_{0}^{n-1}{}_i |\phi\alpha_i - \phi\alpha_{i+1}|.$$

However, since this involves an indefinite quantifier on the right, we first have to show that the left class of

$$\epsilon_\sigma \underset{\alpha=\alpha_0<\ldots<\alpha_n=\xi}{\bigvee_{\alpha_0,\ldots,\alpha_n}} \sigma = \sum_{0}^{n-1}{}_i |\phi\alpha_i - \phi\alpha_{i+1}|$$

is definite. It is in fact sufficient in dealing with the partitionings $\alpha_0 < \ldots < \alpha_n$ to restrict our attention to the rational arguments along with the jumps of ϕ. (Together these form a definitely denumerable set.) From

$$\sigma < \sum_{0}^{n-1}{}_i |\phi\alpha_i - \phi\alpha_{i+1}|$$

it follows that

$$\sigma < \sum_{0}^{n-1}{}_i |\phi\alpha_i' - \phi\alpha_{i+1}'|,$$

where $\alpha_i' = \alpha_i$ if α_i is a jump, and α_i' is a suitable rational number if ϕ is continuous at α_i.

Thus the lubs $\sigma(\xi)$ exist for $\xi \in [\alpha|\beta]$.

Clearly it holds of $\xi_1 < \xi_2$ that

$$\sigma(\xi_1) + |\phi\xi_1 - \phi\xi_2| \leqslant \sigma(\xi_2).$$

The function σ is therefore increasing; and since

$$\phi\xi_2 - \phi\xi_1 \leqslant \sigma(\xi_2) - \sigma(\xi_1),$$

$\sigma - \phi$ is also increasing.

$\phi = \sigma - (\sigma - \phi)$ is the desired representation of ϕ as a difference of monotonic functions.

§11. Differentiation

§10 established in particular that every continuous function can be uniformly approximated by step functions. The graphical representation of such an approximation will look something like Figure 12.

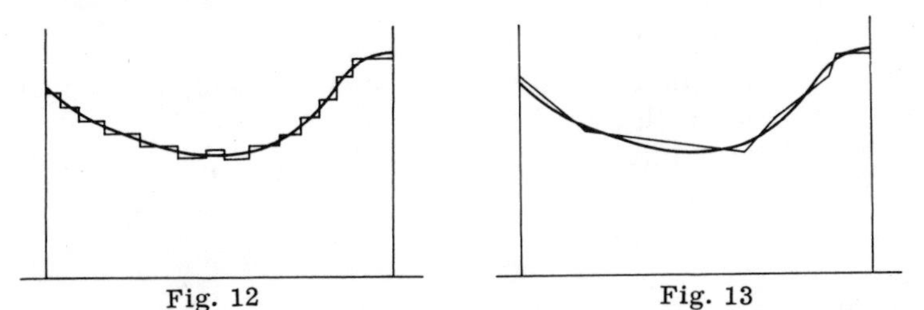

Fig. 12 Fig. 13

We can obviously do a better job of approximation by working with general linear functions as in Figure 13 rather than approximating piecewise with constant functions.

The question arises as to what general linear function L can "best" be used to approximate a given function ϕ at a point ξ_0. Put geometrically, this amounts to asking for the *tangent* of a curve.

We place the following requirement on the approximation of ϕ by L:

$$\bigwedge_\varepsilon \bigvee_\delta \bigwedge_\xi . |\xi - \xi_0| < \delta \rightarrow |\phi\xi - L\xi| < \varepsilon |\xi - \xi_0|. \qquad (11.1)$$

If there is any general linear function L at all satisfying (11.1), then $L\xi_0 = \phi\xi_0$; i.e.,

$$L\xi = \phi\xi_0 + q(\xi - \xi_0).$$

The number q must furthermore be the limit of the difference quotient $\frac{\phi\xi - \phi\xi_0}{\xi - \xi_0}$ (which is not defined for $\xi = \xi_0$!):

$$\lim_{\xi \to \xi_0} \frac{\phi\xi - \phi\xi_0}{\xi - \xi_0} = q, \qquad (11.2)$$

for

$$\bigwedge_\varepsilon \bigvee_\delta \bigwedge_\xi . 0 < |\xi - \xi_0| < \delta \rightarrow \left| \frac{\phi\xi - \phi\xi_0 - q(\xi - \xi_0)}{\xi - \xi_0} \right| < \varepsilon.$$

There is thus at most one general linear function L.

DEFINITION 11.1 A function ϕ is *differentiable* at the point ξ_0 if there is a general linear function L satisfying (11.1).

THEOREM 11.1 A function ϕ is differentiable at the point ξ_0 if and only if the limit (11.2) exists.

If ϕ is differentiable at the point ξ_0, then

$$q = \lim_{\xi \to \xi_0} \frac{\phi\xi - \phi\xi_0}{\xi - \xi_0}$$

is called the *differential quotient* of ϕ at ξ_0. The function $L = \lambda_\xi . \phi\xi_0 + q(\xi - \xi_0)$ is called the *tangential function* of ϕ at the point ξ_0.

113 We denote the differential quotient of ϕ at ξ_0 by $\phi'\xi_0$. The function ϕ' is called the *derivative* of ϕ: it takes as arguments all numbers at which ϕ is differentiable.

If ϕ is differentiable in an interval, then it will hold in that interval that

$$\phi'\xi_0 = \lim_{\gamma \rightharpoondown \xi_0} \frac{\phi\gamma - \phi\xi_0}{\gamma - \xi_0}.$$

Since ϕ is rationally definite (as we always presuppose in keeping with §8), the derivative of ϕ is then a definite function also.

Instead of ϕ' we sometimes also write $\mathscr{D}\phi$, and for $\phi'\xi$ we then write ${}_\xi\mathscr{D}\phi$ in case $\mathscr{D}\phi\xi$ is ambiguous. If ϕ is represented by a term $T(\xi)$, so that $\phi\xi = T(\xi)$, then we write $\mathscr{D}\iota_\xi T(\xi)$ for short as $\mathscr{D}_\xi T(\xi)$.

The derivative functions serve primarily to provide estimates of the differences. On the basis of our definition it is of course the case that

$$\phi\xi - \phi\xi_0 = (\phi'\xi_0 + h)(\xi - \xi_0),$$

where $h \rightharpoondown 0$ as $\xi \rightharpoondown \xi_0$. A more precise statement about this h is possible, however. We now prove it as the

BOUND-THEOREM. If ϕ is differentiable in a closed interval $[\alpha | \beta]$, then every bound of ϕ' in $[\alpha | \beta]$ is also a bound of the difference quotients $\dfrac{\phi\xi_1 - \phi\xi_2}{\xi_1 - \xi_2}$ of ϕ in $[\alpha | \beta]$.

The proof makes use of the following trivial estimate:

LEMMA:

$$\min\left(\frac{\alpha_1 - \alpha_2}{\beta_1 - \beta_2}, \frac{\alpha_2 - \alpha_3}{\beta_2 - \beta_3}\right) \leqslant \frac{\alpha_1 - \alpha_3}{\beta_1 - \beta_3} \leqslant \max\left(\frac{\alpha_1 - \alpha_2}{\beta_1 - \beta_2}, \frac{\alpha_2 - \alpha_3}{\beta_2 - \beta_3}\right)$$

where $\beta_1 > \beta_2 > \beta_3$.

We have only to consider here that, e.g., it follows directly from $\alpha_1 - \alpha_2 \geqslant \gamma(\beta_1 - \beta_2)$ and $\alpha_2 - \alpha_3 \geqslant \gamma(\beta_2 - \beta_3)$ that $\alpha_1 - \alpha_3 \geqslant \gamma(\beta_1 - \beta_3)$, since

$$\alpha_1 - \alpha_3 = \alpha_1 - \alpha_2 + \alpha_2 - \alpha_3 \geqslant \gamma(\beta_1 - \beta_2 + \beta_2 - \beta_3) = \gamma(\beta_1 - \beta_2).$$

We now demonstrate that $\phi' \geqslant \gamma$ in $[\alpha | \beta]$ always entails

$$\frac{\phi\beta - \phi\alpha}{\beta - \alpha} \geqslant \gamma.$$

We employ analytic induction and consider for some $\gamma_0 < \gamma$ the definite statement form $A(\xi)$,

$$\bigwedge_{r \atop (\alpha | \xi]} \frac{\phi r - \phi \alpha}{r - \alpha} \geqslant \gamma_0.$$

That $A(\alpha)$ is obvious, since $(\alpha | \alpha]$ is empty; and if 114

$$\bigwedge_{\xi} . \xi < \xi_0 \rightarrow A(\xi).,$$

it follows that $\dfrac{\phi \xi_0 - \phi \alpha}{\xi_0 - \alpha} \geqslant \gamma_0$. On the other hand, since $\phi' \xi_0 > \gamma_0$, there is a δ such that for all r, if $\xi_0 < r < \xi_0 + \delta$, then

$$\frac{\phi r - \phi \xi_0}{r - \xi_0} \geqslant \gamma_0.$$

By the lemma, these inequalities together yield

$$\frac{\phi r - \phi \alpha}{r - \alpha} \geqslant \gamma_0.$$

Hence $A(\xi_0 + \delta)$. By the analytic induction principle $A(\beta)$ follows, whence immediately

$$\frac{\phi \beta - \phi \alpha}{\beta - \alpha} \geqslant \gamma_0.$$

According to this proof,

$$\frac{\phi \xi_1 - \phi \xi_2}{\xi_1 - \xi_2} \geqslant \gamma_0$$

holds for the differential quotients throughout the entire interval, provided that $\phi' > \gamma_0$ throughout. Accordingly, we then also have

$$\frac{\phi \xi_1 - \phi \xi_2}{\xi_1 - \xi_2} \geqslant \gamma.$$

The bound-theorem is proved exactly analogously for upper bounds.

As a simplified notation for difference quotients we write

$$\overset{\xi_2}{\underset{\xi_1}{\Delta}} \phi \quad \text{for} \quad \phi \xi_2 - \phi \xi_1.$$

For the identity function I this gives

$$\overset{\xi_2}{\underset{\xi_1}{\Delta}} I = \xi_2 - \xi_1.$$

The difference quotient of ϕ (a 2-place function) is then denoted by $\dfrac{\Delta \phi}{\Delta I}$, and we can formulate the bound-theorem as

$$\gamma_1 \leqslant \phi' \leqslant \gamma_2 \rightarrow \gamma_1 \frac{\Delta \phi}{\Delta I} \leqslant \gamma_2 .$$

As every differential quotient is the limit of difference quotients, it will even be the case that

$$\gamma_1 \leqslant \phi' \leqslant \gamma_2 \leftrightarrow \gamma_1 \leqslant \frac{\Delta\phi}{\Delta I} \leqslant \gamma_2 .$$

115 If the derivative function ϕ' is continuous, then it follows from the bound-theorem together with the intermediate-value theorem that

$$\frac{\phi\xi_2 - \phi\xi_1}{\xi_2 - \xi_1} = \phi'\xi_0$$

for a suitable "*mean value*" $\xi_0 \in [\xi_1 | \xi_2]$.

It is well known that this *mean-value theorem* holds even without the assumption of continuity for ϕ'. We could prove it in the usual manner via Rolle's theorem:

$$\phi\xi_1 = \phi\xi_2 \rightarrow \bigvee_{\substack{\xi_0 \\ (\xi_1 | \xi_2)}} \phi'\xi_0 = 0 .$$

We omit the proof, however, because we can always use the bound-theorem in place of the mean-value theorem.

The bound-theorem has a decisive advantage over the mean-value theorem in that its estimates remain valid even when the function is not differentiable everywhere in the interval, so long as it is everywhere continuous and differentiable except for at most denumerably many points. Thus we have the

STRENGHTHENED BOUND-THEOREM. If ϕ is continuous in $[\alpha | \beta]$ and fails to be differentiable at not more than a definitely denumerable number of points in $[\alpha | \beta]$, then every bound of the values of ϕ' is also a bound of the difference quotients $\frac{\Delta\phi}{\Delta I}$ in $[\alpha | \beta]$.

Proof. Let ζ_* be a definite sequence of the points of non-differentiability. We set

$$\sigma(\xi) = \sum_{\substack{i \\ \zeta_i < \xi}} \frac{1}{2^i} ,$$

for which it will be the case that $0 \leqslant \sigma(\xi) \leqslant 1$. Let γ be, say, an upper bound, and $\phi' \leqslant \gamma < \gamma_0$. For arbitrary ε we consider the following definite statement form:

$$A(\xi) \leftrightharpoons \bigwedge_{\substack{r \\ (\alpha | \xi]}} \frac{\phi r - \phi\alpha}{r - \alpha} \leqslant \gamma_0 + \varepsilon\sigma(\xi).$$

We clearly have (1) $A(\alpha)$

(2) $\xi_1 < \xi_2 \dot{\rightarrow} A(\xi_2) \rightarrow A(\xi_1)$.

But it is also true that

$$(3)\ \bigwedge_{\xi} . \xi < \xi_0 \to A(\xi) . \to \bigvee_{\delta} A(\xi_0 + \delta),$$

for if ϕ is differentiable at ξ_0, then for an appropriate δ

$$\frac{\phi r - \phi \xi_0}{r - \xi_0} \leqslant \gamma_0 \quad \text{where} \quad r \in [\xi_0 \mid \xi_0 + \delta].$$

It follows by the earlier lemma that 116

$$\frac{\phi r - \phi \alpha}{r - \alpha} \leqslant \gamma_0 + \varepsilon\sigma(\xi_0) \leqslant \gamma_0 + \varepsilon\sigma(\xi_0 + \delta).$$

If, on the other hand, ξ_0 is one of the nondifferentiable points ζ_n, then—ϕ being continuous at ξ_0—we obtain for suitable δ

$$\frac{\phi r - \phi \alpha}{r - \alpha} \leqslant \gamma_0 + \varepsilon\sigma(\xi_0) + \frac{\varepsilon}{2^n} \leqslant \gamma_0 + \varepsilon\sigma(\xi_0 + \delta)$$

$$\text{where} \quad r \in [\xi_0 \mid \xi_0 + \delta].$$

It must be stressed that $A(\xi)$ has to be a definite formula for this proof. It is thus necessary for ζ_* to be a *definite* sequence of nondifferentiable points. The proof proceeds similarly for lower bounds.

As a briefer formulation, we call a function *ω-everywhere differentiable* if it is differentiable everywhere in an interval except for at most a definitely denumerable number of points.

A direct consequence of the bound-theorem is then

THEOREM 11.2 If a function ϕ is everywhere continuous and ω-everywhere differentiable in $[\alpha \mid \beta]$, then

$\phi' \geqslant 0 \leftrightarrow \phi$ increasing in $[\alpha \mid \beta]$
$\phi' > 0 \leftrightarrow \phi$ properly increasing in $[\alpha \mid \beta]$
$\phi' \leqslant 0 \leftrightarrow \phi$ decreasing in $[\alpha \mid \beta]$
$\phi' < 0 \leftrightarrow \phi$ properly decreasing in $[\alpha \mid \beta]$
$\phi' = 0 \leftrightarrow \phi$ constant in $[\alpha \mid \beta]$.

Proof. $\phi' \geqslant 0$ implies $\frac{\phi\beta_0 - \phi\alpha_0}{\beta_0 - \alpha_0} \geqslant 0$, i.e., $\phi\beta_0 \geqslant \phi\alpha_0$ for all $\beta_0 > \alpha_0$ in $[\alpha \mid \beta]$; and similarly for $\phi' \leqslant 0$, *mutatis mutandis*. If $\phi' > 0$ $\{< 0\}$ then ϕ is properly increasing {properly decreasing}, for otherwise ϕ would be constant in a subinterval and $\phi' = 0$ there. A function that is at once increasing and decreasing is constant. The converses are in each case trivial.

The case $\phi' = 0$ is the most important, since it directly yields

THEOREM 11.3 Any function that is everywhere continuous and ω-everywhere differentiable in a closed interval is determined by its derivative up to an additive constant.

Proof. If $\phi' = \psi'$ ω-everywhere, then the derivative of $\phi - \psi$ is 0 throughout, so that $\phi - \psi$ is constant.

The derivative ϕ' of a function ϕ need not be continuous. But if ϕ' is continuous (at a point or in an interval), then ϕ is said to
117 be *continuously differentiable* there. If ϕ is differentiable in an interval and continuously differentiable at ξ_0, then not only is

$$\lim_{\xi \to \xi_0} \frac{\phi\xi - \phi\xi_0}{\xi - \xi_0} = \phi'\xi_0 ,$$

but in fact (cf. §15) even

$$\lim_{\substack{\xi_1 \to \xi_0, \xi_2 \to \xi_0 \\ \xi_1 \neq \xi_2}} \frac{\phi\xi_1 - \phi\xi_2}{\xi_1 - \xi_2} = \phi'\xi_0 .$$

Proof. By the bound-theorem, $\frac{\phi\xi_1 - \phi\xi_2}{\xi_1 - \xi_2}$ has the same bounds as ϕ' in $[\xi_0 - \varepsilon \mid \xi_0 + \varepsilon]$ for any $\xi_1 \neq \xi_2$ in that interval. As ϕ' is continuous at ξ_0, $\frac{\phi\xi_1 - \phi\xi_2}{\xi_1 - \xi_2}$ thus differs arbitrarily little from $\phi'\xi_0$ for sufficiently small ε.

If we strengthen the requirement of differentiability by stipulating the existence of $\lim_{\substack{\xi_1 \to \xi_0, \xi_2 \to \xi_0 \\ \xi_1 \neq \xi_2}} \frac{\phi\xi_1 - \phi\xi_2}{\xi_1 - \xi_2}$, i.e., if we require *free differentiability*, then the following simple proof of the bound-theorem is possible for functions which are freely differentiable throughout. $\phi' > 0$ immediately entails local and therefore global monotony; i.e., ϕ is properly increasing. Since $(\phi - \gamma I)' = \phi' - \gamma$, $\phi' > \gamma$ accordingly entails that $\phi - \gamma I$ is properly increasing; i.e., for $\beta > \alpha$ we have

$$\phi\beta - \gamma\beta > \phi\alpha - \gamma\alpha$$

$$\phi\beta - \phi\alpha > \gamma(\beta - \alpha).$$

Hence $\frac{\Delta\phi}{\Delta I} > \gamma$ holds generally.

If the derivative function ϕ' is not just continuous but differentiable as well, we can form *higher-order derivatives* ϕ'', ϕ''', ... ϕ'' is called the second derivative, ϕ''' the third derivative, and so on. We denote the *nth derivative* by $\phi^{(n)}$.

We shall not go into applications of these higher-order derivatives here. We shall rather investigate how—and whether—the elementary functions can be differentiated. The differentiability of a function does not follow from the fact that it is continuous. It is trivially true that every differentiable function is continuous (if

$\xi \rightarrow \xi_0$ then necessarily $\phi\xi \rightarrow \phi\xi_0$; otherwise $\frac{\phi\xi - \phi\xi_0}{\xi - \xi_0}$ could have no limit as $\xi \rightarrow \xi_0$); but, to give an example, the continuous but only piecewise linear functions are not differentiable. A function such as that given by $\phi\xi = |\xi|$ is not differentiable at $\xi_0 = 0$. For if $\xi < \xi_0$ the difference quotient is -1, whereas if $\xi > \xi_0$ it is +1. This function is, as we say, only *left-differentiable* and *right-differentiable*.

118

It is even possible to construct functions that are continuous throughout but not differentiable at any point. The first example was given by Bolzano (cf. von Mangoldt and Knopp, II, 100).

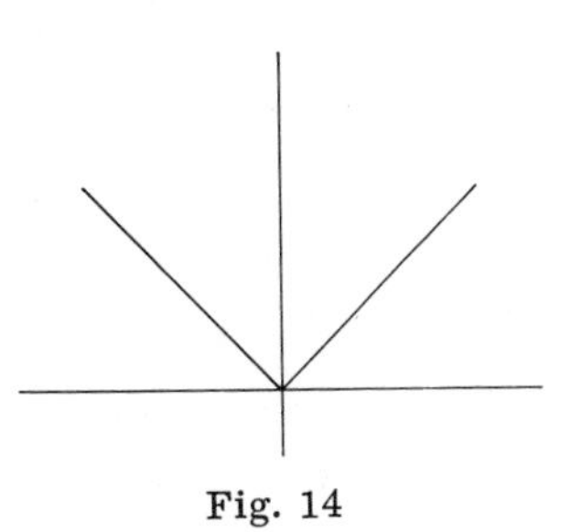

Fig. 14

We begin with the differentiability of general linear functions, which is trivial. Every general linear function is its own tangential function. If $\phi\xi = \alpha\xi + \beta$, then $\phi'\xi = \alpha$.

Turning next to the differentiation of rational functions, we must compute the derivatives of $\phi \pm \psi$, $\phi \cdot \psi$, and $\frac{\phi}{\psi}$ from the derivatives of ϕ and ψ. The following rules apply:

THEOREM 11.4

$$(1)\quad (\phi \pm \psi)' = \phi' \pm \psi'$$

$$(2)\quad (\phi \cdot \psi)' = \phi' \cdot \psi + \phi \cdot \psi'$$

$$(3)\quad \left(\frac{\phi}{\psi}\right)' = \frac{\phi' \cdot \psi - \phi \cdot \psi'}{\psi^2}$$

in particular

$$(4)\quad \left(\frac{1}{\psi}\right)' = -\frac{\psi'}{\psi^2}.$$

Proof. The proof of (1) is trivial, since

$$(\phi\xi \pm \psi\xi) - (\phi\xi_0 \pm \psi\xi_0) = (\phi\xi - \phi\xi_0) \pm (\psi\xi - \psi\xi_0).$$

For (2) we form the difference as follows:

$$\phi\xi \cdot \psi\xi - \phi\xi_0 \cdot \psi\xi_0 = (\phi\xi - \phi\xi_0) \cdot \psi\xi + (\psi\xi - \psi\xi_0) \cdot \phi\xi_0.$$

Since $\psi\xi \rightarrow \psi\xi_0$ as $\xi \rightarrow \xi_0$, division by $\xi - \xi_0$ yields the desired proposition.

For (3) it is enough to prove the special case (4). It must of course be presupposed that $\psi\xi_0 \neq 0$, whence too $\psi\xi \neq 0$ around ξ_0. The difference can be taken thus:

$$\frac{1}{\psi\xi} - \frac{1}{\psi\xi_0} = \frac{\psi\xi_0 - \psi\xi}{\psi\xi \cdot \psi\xi_0} = - \frac{\psi\xi - \psi\xi_0}{\psi\xi \cdot \psi\xi_0}.$$

As $\xi \to \xi_0$ the denominator tends to $(\psi\xi_0)^2$. The general case then follows with the help of (2):

$$\left(\frac{\phi}{\psi}\right)' = \left(\phi \cdot \frac{1}{\psi}\right)' = \phi' \cdot \frac{1}{\psi} - \phi \frac{\psi'}{\psi^2}.$$

119 These rules permit in particular the differentiation of power functions with whole-number exponents:

If $\phi\xi = \xi^n$, then $\phi'\xi = n\xi^{n-1}$.

Proof. Let $\phi_n\xi = \xi^n$. ϕ_1 is a linear function, so

(1) $\phi_1' = 1$.

(2) If $\phi_n'\xi = n\xi^{n-1}$, then, since
$\phi_{n+1}' = (\phi_n \cdot \phi_1)' = \phi_n' \cdot \phi_1 + \phi_n \cdot \phi_1'$,

it follows directly that

$$\phi_{n+1}'\xi = n\xi^{n-1} \cdot \xi + \xi^n \cdot 1 = (n+1)\xi^n.$$

With that the differentiation formula is proved for $n > 0$ by induction. The case $n = 0$ is trivial, and for $n < 0$ the formula follows from

$$\left(\frac{1}{\phi_n}\right)'\xi = - \frac{\phi_n'}{\phi_n^2}(\xi) = - \frac{n\xi^{n-1}}{\xi^{2n}} = -n\xi^{-n-1}.$$

To differentiate power functions with rational exponents, we make use of the fact that the function $I^{\frac{1}{n}}$ is the inverse of I^n.

In general, the inverse ψ of an (invertible) function ϕ is characterized by the fact that $\phi\urcorner\psi$ is the identity function I. Therefore, in order to compute ψ' from ϕ', we first consider the differentiation of the composition $\phi\urcorner\psi$ for arbitrary functions ϕ and ψ.

THEOREM 11.5 (CHAIN RULE). Let ψ be differentiable at ξ_0 and ϕ at $\psi\xi_0$. Then $\phi\urcorner\psi$ is differentiable at ξ_0, and

$$(\phi\urcorner\psi)' = (\phi'\urcorner\psi) \cdot \psi'.$$

Proof. We have

(1) $$\frac{\phi\urcorner\psi\xi - \phi\urcorner\psi\xi_0}{\xi - \xi_0} = \frac{\phi\urcorner\psi\xi - \phi\urcorner\psi\xi_0}{\psi\xi - \psi\xi_0} \cdot \frac{\psi\xi - \psi\xi_0}{\xi - \xi_0} \quad \text{where } \psi\xi \neq \psi\xi_0.$$

If $\psi\xi = \psi\xi_0$, the equation

(2) $$\frac{\phi\urcorner\psi\xi - \phi\urcorner\psi\xi_0}{\xi - \xi_0} = \phi'\urcorner\psi\xi_0 \frac{\psi\xi - \psi\xi_0}{\xi - \xi_0}$$

is trivial: both sides are 0. The first factor on the right—in both (1) and (2)—tends to the limit $\phi'\urcorner\psi\xi_0$ as $\xi \to \xi_0$, since $\lim\limits_{\xi \to \xi_0} \psi\xi = \psi\xi_0$.

Accordingly, we obtain from (1) and (2) by passage to the limit the equation we had to prove:

$$(\phi\urcorner\psi)'\xi_0 = (\phi'\urcorner\psi\xi_0) \cdot \psi'\xi_0.$$

If now ψ is the inverse of an (invertible) function ϕ, and if ϕ and ψ are differentiable, then it follows from $(\phi\urcorner\psi)' = 1$ in accordance with the above that *120*

$$(\phi'\urcorner\psi) \cdot \psi' = 1,$$

i.e.,

$$\psi' = \frac{1}{\phi'\urcorner\psi}. \tag{11.3}$$

Thus, if ψ is differentiable at η_0, then in particular

$$\phi'\urcorner\psi\eta_0 \neq 0.$$

Conversely, we have

THEOREM 11.6 If ϕ is invertible and differentiable, if $\phi'\xi_0 \neq 0$, and if its inverse ψ is continuous at $\phi\xi_0$, then ψ is differentiable at $\phi\xi_0$.

Proof. We set $\eta_0 = \phi\xi_0$. If $\eta \neq \eta_0$, then $\psi\eta \neq \psi\eta_0$ and

$$\frac{\psi\eta - \psi\eta_0}{\eta - \eta_0} = \frac{1}{\dfrac{\phi\urcorner\psi\eta - \phi\urcorner\psi\eta_0}{\psi\eta - \psi\eta_0}}.$$

Furthermore, since ψ is continuous at η_0, we have $\lim\limits_{\eta \to \eta_0} \psi\eta = \psi\eta_0$. Together these give (since $\phi'\urcorner\psi\eta_0 \neq 0$)

$$\psi'\eta_0 = \frac{1}{\phi'\urcorner\psi\eta_0}.$$

Applying this now to the inverse ψ of the function $\phi = I^n (n > 0)$, we get, in view of $\psi\eta = \eta^{\frac{1}{n}}$ for $\eta \neq 0$, the result that

$$\psi'\eta = \frac{1}{n\left(\eta^{\frac{1}{n}}\right)^{n-1}} = \frac{1}{n}\,\eta^{\frac{1}{n}-1}. \tag{11.4}$$

The differentiation formula of Theorem 11.4 thus extends to exponents of the form $\frac{1}{n}$. In fact, it holds for all rational exponents, for (with the help of Theorem 11.5) from $\phi\xi = \xi^{\frac{m}{n}} = \left(\xi^{\frac{1}{n}}\right)^m$ it follows immediately that

$$\phi'\xi = m\left(\xi^{\frac{1}{n}}\right)^{m-1} \cdot \frac{1}{n}\,\xi^{\frac{1}{n}-1} = \frac{m}{n}\,\xi^{\frac{m}{n}-1}. \tag{11.5}$$

Differentiation of power functions with arbitrary real exponents $\alpha = \lim r_*$ can be obtained from the above by passage to the

121 limit. However, we can avoid the discussion that would be necessary regarding the interchangeability of the two limits

$$\lim_{\xi \to \xi_0} \lim \frac{\xi^{r_*} - \xi_0{}^{r_*}}{\xi - \xi_0} = \lim \lim_{\xi \to \xi_0} \frac{\xi^{r_*} - \xi_0{}^{r_*}}{\xi - \xi_0}$$

$$[= \lim r_* \, \xi_0{}^{r_* - 1} = \alpha\, \xi_0{}^{\alpha - 1}]$$

by first treating the differentiation of the exponential function.

Let $\alpha > 1$. For the function $\phi = \alpha^{I}$ we must form the difference quotient

$$\frac{\alpha^{\xi} - \alpha^{\xi_0}}{\xi - \xi_0} = \alpha^{\xi_0}\,\frac{\alpha^{\xi - \xi_0} - 1}{\xi - \xi_0}.$$

Thus, in order to determine the limit as $\xi \to \xi_0$, we must determine $\lim_{\rho \to 0} \frac{\alpha^{\rho} - 1}{\rho}$.

If we assume here that $\rho > 0$, then in view of $\alpha^{\rho} > 1$ it will always be the case that $\frac{\alpha^{\rho} - 1}{\rho} > 0$. For $\rho < 0$ it likewise follows that $\frac{\alpha^{\rho} - 1}{\rho} > 0$, since $\frac{\alpha^{-\rho} - 1}{-\rho} = \alpha^{-\rho}\,\frac{\alpha^{\rho} - 1}{\rho}$.

To demonstrate, therefore, that the right-hand limit exists, we have only to prove that $\frac{\alpha^{\rho} - 1}{\rho}$ is increasing as a function of ρ. To do so, we must show that

$$\frac{\alpha^{\rho_1} - 1}{\rho_1} > \frac{\alpha^{\rho_2} - 1}{\rho_2} \quad \text{when} \quad \rho_1 > \rho_2 .$$

A function ϕ is said to be "*convex*" (in an interval) if its difference quotients are increasing; i.e., if

$$\xi_1 \leqslant \eta_1 \wedge \xi_2 \leqslant \eta_2 \,\dot{\rightarrow}\, \frac{\phi\xi_2 - \phi\xi_1}{\xi_2 - \xi_1} \leqslant \frac{\phi\eta_2 - \phi\eta_1}{\eta_2 - \eta_1}. \tag{11.6}$$

To prove the convexity of a function, the special case $\xi_1 = \eta_1$ is clearly sufficient:

$$\xi_1 \leqslant \xi_2 \rightarrow \frac{\phi\xi_1 - \phi\xi_0}{\xi_1 - \xi_0} \leqslant \frac{\phi\xi_2 - \phi\xi_0}{\xi_2 - \xi_0}. \tag{11.7}$$

In fact, even the following special case suffices for convexity:

$$\xi_1 = \xi, \quad \eta_2 = \eta, \quad \eta_1 = \xi_2 = \frac{\xi + \eta}{2} \quad \text{where} \quad \xi \leqslant \eta,$$

i.e.,

$$\phi\left(\frac{\xi + \eta}{2}\right) - \phi\xi \leqslant \phi\eta - \phi\left(\frac{\xi + \eta}{2}\right)$$

or 122

$$\phi\left(\frac{\xi + \eta}{2}\right) \leqslant \frac{\phi\xi + \phi\eta}{2}, \tag{11.8}$$

provided that ϕ is continuous.

For $\phi = \alpha^{I}$ the validity of (11.8) is obtained immediately:

$$\alpha^{\frac{\xi+\eta}{2}} \leqslant \frac{\alpha^{\xi} + \alpha^{\eta}}{2},$$

for when $\alpha_1 = \alpha^{\xi}$ and $\alpha_2 = \alpha^{\eta}$ it is well known that

$$\sqrt{\alpha_1\alpha_2} \leqslant \frac{\alpha_1 + \alpha_2}{2}.$$

To prove (11.7) from (11.8) we confine ourselves to the case

$$m \leqslant n \rightarrow \frac{\phi(\xi + m\theta) - \phi\xi}{m} \leqslant \frac{\phi(\xi + n\theta) - \phi\xi}{n} \tag{11.9}$$

for integers $m, n \geqslant 0$ and for $\theta > 0$. For continuous functions, (11.7) follows trivially from (11.9). For the proof of (11.9) we set

$$\eta_i = \phi(\xi + i\theta) \quad (i = 0, 1, 2, \ldots\ldots).$$

(11.8) entails the inequalities

$$\eta_1 - \eta_0 \leqslant \eta_2 - \eta_1 \leqslant \ldots \leqslant \eta_m - \eta_{m-1} \leqslant \ldots \leqslant \eta_n - \eta_{n-1}.$$

By the lemma for the bound-theorem, these yield

$$\frac{\eta_m - \eta_0}{m} \leqslant \eta_m - \eta_{m-1} \leqslant \frac{\eta_m - \eta_n}{n - m}.$$

A further application of the lemma produces the desired inequality:

$$\frac{\eta_m - \eta_0}{m} \leqslant \frac{\eta_n - \eta_0}{n}.$$

We denote the right-hand limit of $\dfrac{\alpha^{\rho} - 1}{\rho}$ as $\rho \rightarrow 0$ by $l(\alpha)$. The left-hand limit exists too: it is the same as $l(\alpha)$, for

$$\frac{\alpha^{-\rho} - 1}{-\rho} = \alpha^{-\rho} \frac{\alpha^{\rho} - 1}{\rho}$$

and $\lim_{\rho \to 0} \alpha^{-\rho} = 1$. Taking these results together, we have proved that

$$\lim_{\rho \to 0} \frac{\alpha^{\rho} - 1}{\rho} = l(\alpha)$$

and furthermore that $l(\alpha) > 0$.

123 The function l is now easily shown to be a logarithmic function, for we have

$$\alpha^{\frac{1}{l(\alpha)}} = \lim_{\rho \to 0} \alpha^{\frac{\rho}{\alpha^{\rho}-1}} = \lim_{\beta \to 1} \beta^{\frac{1}{\beta-1}} = \lim_{\gamma \to 0} (1+\gamma)^{\frac{1}{\gamma}} = \lim_{n \to \infty} \left(1 + \frac{1}{n}\right)^{n}.$$

If we let $e = \lim \left(1 + \frac{1}{n}\right)^{n}$, then it follows that $e^{l(\alpha)} = \alpha$; i.e., $l(\alpha) = \log_e \alpha$. e is called the *Euler number.*

From now on we write $\log_e$ for short as ln. The function ln is called the *natural logarithm*.

With that we have differentiated the exponential function as follows:

THEOREM 11.7 If $\phi\xi = \alpha^{\xi}$, then $\phi'\xi = \alpha^{\xi} \ln \alpha$.

The validity of Theorem 11.7 for $\alpha < 1$ follows from

$$\frac{\left(\frac{1}{\alpha}\right)^{\rho} - 1}{-\rho} = -\alpha^{-\rho} \frac{\alpha^{\rho} - 1}{\rho} \to -\ln \alpha = \ln \frac{1}{\alpha}.$$

As an estimate we get (for $\alpha > 0, \rho > 0$)

$$\frac{\alpha^{-\rho} - 1}{-\rho} \leqslant \ln \alpha \leqslant \frac{\alpha^{\rho} - 1}{\rho},$$

and hence for $\rho = 1$

$$1 - \frac{1}{\alpha} \leqslant \ln \alpha \leqslant \alpha - 1;$$

in particular,

$$\ln 2 \geqslant \frac{1}{2}, \quad \text{whence} \quad e = 2^{\frac{1}{\ln 2}} \leqslant 4.$$

Later we shall learn convenient methods for computing e. They lead to the decimal expansion

$$e = 2.71828 \cdot \cdot \cdot \cdot$$

The function e^I is denoted by exp. It then holds that

$$\exp' = \exp.$$

Since $\exp' \xi$ is never 0, the inverse, ln, of the function exp is by Theorems 11.5–6 differentiable in $(0|\infty)$, and we have

$$\ln' = \frac{1}{\exp \ln},$$

i.e.,

$$\ln' = \frac{1}{I}.$$

Since $\log_\alpha = \frac{\ln}{\ln \alpha}$, this entails the general result that 124

THEOREM 11.8 $\qquad \log_\alpha{}' \xi = \dfrac{1}{\xi \ln \alpha}.$

Since $I^\alpha = \exp(\alpha \ln I)$ for any power function I^α, we can differentiate by the chain rule:

$$\begin{aligned} \mathcal{D} I^\alpha &= \exp'(\alpha \ln I) \cdot \alpha \ln' I \\ &= \exp(\alpha \ln I) \cdot \frac{\alpha}{I} \\ &= I^\alpha \cdot \frac{\alpha}{I} \\ &= \alpha I^{\alpha-1}. \end{aligned}$$

The only elementary functions still left are the circle functions. Just as differentiation of the exponential function led us to the number e, so we shall now be led of necessity to the number π.

To find the derivative of the function Sin, we must consider the limit

$$\lim_{\xi \to \xi_0} \frac{\operatorname{Sin} \xi - \operatorname{Sin} \xi_0}{\xi - \xi_0}.$$

From the addition theorems

$$\begin{aligned} \operatorname{Sin} \xi &= \operatorname{Sin}\left(\frac{\xi + \xi_0}{2} + \frac{\xi - \xi_0}{2}\right) \\ &= \operatorname{Sin} \frac{\xi + \xi_0}{2} \operatorname{Cos} \frac{\xi - \xi_0}{2} + \operatorname{Cos} \frac{\xi + \xi_0}{2} \operatorname{Sin} \frac{\xi - \xi_0}{2} \end{aligned}$$

$$\text{Sin } \xi_0 = \text{Sin}\left(\frac{\xi + \xi_0}{2} - \frac{\xi - \xi_0}{2}\right)$$

$$= \text{Sin } \frac{\xi + \xi_0}{2} \text{ Cos } \frac{\xi - \xi_0}{2} - \text{Cos } \frac{\xi + \xi_0}{2} \text{ Sin } \frac{\xi - \xi_0}{2}$$

we get

$$\text{Sin } \xi - \text{Sin } \xi_0 = 2 \text{ Cos } \frac{\xi + \xi_0}{2} \text{ Sin } \frac{\xi - \xi_0}{2},$$

and thus

$$\frac{\text{Sin } \xi - \text{Sin } \xi_0}{\xi - \xi_0} = \text{Cos } \frac{\xi + \xi_0}{2} \frac{\text{Sin } \frac{\xi - \xi_0}{2}}{\frac{\xi - \xi_0}{2}}.$$

125 Since

$$\lim_{\xi \rightarrow \xi_0} \text{Cos } \frac{\xi + \xi_0}{2} = \text{Cos } \xi_0,$$

all that remains to be done is to determine the limit of $\frac{\text{Sin } \rho}{\rho}$ as $\rho \rightarrow 0$ $\left(\text{where } \rho = \frac{\xi - \xi_0}{2}\right)$.

In (9.14) we already proved that

$$\frac{\text{Sin } \frac{1}{2^n}}{\frac{1}{2^n}} < 2.$$

The existence of the limit will thus be assured if we can show that $\frac{\text{Sin } \rho}{\rho}$ is decreasing as a function of ρ. In view of

$$\frac{\text{Sin } (-\rho)}{-\rho} = \frac{\text{Sin } \rho}{\rho}$$

it suffices to consider the right-hand limit.

A function with decreasing difference quotients is called a *concave* function. The concavity of the Sin function dates back all the way to Aristarchos (third century B.C.). The ancient proofs, e.g., that of Ptolemy—like our own school books—made use of circle measurements, especially of the existence of a measure for the area of a circle.

The concavity of a continuous function ϕ is already guaranteed by the following condition:

$$\phi\left(\frac{\xi + \eta}{2}\right) \geqslant \frac{\phi\xi + \phi\eta}{2}.$$

The proof is exactly parallel to that of the convexity of the exponential function.

The formula

$$\text{Sin } \frac{\xi + \eta}{2} \geqslant \frac{\text{Sin } \xi + \text{Sin } \eta}{2}$$

follows directly from the addition theorems, since

$$\text{Sin } \xi + \text{Sin } \eta = 2 \text{ Sin } \frac{\xi + \eta}{2} \text{ Cos } \frac{\xi - \eta}{2}.$$

Geometrically, of course,

$$n \text{ Sin } \frac{1}{n} = \frac{\text{Sin } \frac{1}{n}}{\frac{1}{n}}$$

is an approximation of the circumference of a quarter-circle (with a radius of 1). But without even going into the measurement of circumference, we have established the existence of $\lim\limits_{\rho \to 0} \frac{\text{Sin } \rho}{\rho}$. Geometrically, the circumference of a circle with a diameter of 1 is denoted (since the seventeenth century) by the letter π. We therefore define the number π here as 126

$$\pi \Leftarrow 2 \lim_{\rho \to 0} \frac{\text{Sin } \rho}{\rho}. \qquad (11.10)$$

By (9.14), $\pi < 4$. Methods of calculation to be discussed later yield the decimal expansion

$$\pi = 3.14159 \cdots.$$

Using $\frac{\text{Sin } \rho}{\rho} \to \frac{\pi}{2}$ we now obtain the differentiation formula

$$\text{Sin}' \xi_0 = \frac{\pi}{2} \text{ Cos } \xi_0$$

or

$$\text{Sin}' = \frac{\pi}{2} \text{ Cos}. \qquad (11.11)$$

This formula justifies introducing the familiar trigonometric functions. We set

$$\sin\,\xi = \text{Sin}\,\frac{\xi}{\frac{\pi}{2}}, \qquad \cos\,\xi = \text{Cos}\,\frac{\xi}{\frac{\pi}{2}}. \tag{11.12}$$

We then have, e.g.,

$$\sin 0 = 0, \qquad \sin\frac{\pi}{2} = 1$$

$$\cos 0 = 1, \qquad \cos\frac{\pi}{2} = 0.$$

In particular, it follows that

$$\lim_{\rho \to 0} \frac{\sin \rho}{\rho} = \lim_{\rho \to 0} \frac{2}{\pi} \frac{\text{Sin}\,\frac{\rho}{\frac{\pi}{2}}}{\frac{\rho}{\frac{\pi}{2}}} = \frac{2}{\pi} \cdot \frac{\pi}{2} = 1 \tag{11.13}$$

and (by the chain rule)

$$\sin'\,\xi = \frac{2}{\pi}\,\text{Sin}'\,\frac{\xi}{\frac{\pi}{2}} = \frac{2}{\pi} \cdot \frac{\pi}{2}\,\text{Cos}\,\frac{\xi}{\frac{\pi}{2}} = \cos\,\xi\,;$$

127 i.e.,

THEOREM 11.9 $\sin' = \cos.$

The formulas

$$\text{Cos}\,\xi = \text{Sin}(1 - \xi), \qquad \text{Sin}\,\xi = \text{Cos}(1 - \xi)$$

now yield

$$\cos\,\xi = \sin\left(\frac{\pi}{2} - \xi\right), \qquad \sin\,\xi = \cos\left(\frac{\pi}{2} - \xi\right),$$

and from these we get

$$\begin{aligned} \cos'\,\xi &= -\sin'\left(\frac{\pi}{2} - \xi\right) \\ &= -\cos\left(\frac{\pi}{2} - \xi\right) = -\sin\,\xi. \end{aligned}$$

THEOREM 11.10 $\cos' = -\sin.$

For the trigonometric function defined in analogy to Tan,

$$\tan = \frac{\sin}{\cos},$$

we get the derivative

$$\tan' = \frac{\sin' \cos - \cos' \sin}{\cos^2}$$

$$= \frac{\cos^2 + \sin^2}{\cos^2} = \frac{1}{\cos^2} = 1 + \tan^2.$$

Finally, we consider the arc functions, arc sin, arc cos, and arc tan, that arise as the inverses of the trigonometric functions sin, cos, and tan, respectively. The monotonic interval of sin and tan is, let us say, $\left[-\frac{\pi}{2}\middle|\frac{\pi}{2}\right]$; that of cos, $[0 \mid \pi]$. Since the differential quotients of the trigonometric functions in the interior of the given monotonic intervals are distinct from zero, the arc functions are differentiable in the interior of their domains. The results for the principal functions are

$$\text{arc sin}_0{}' \, \xi = \frac{1}{\cos(\text{arc sin } \xi)} = \frac{1}{\sqrt{1 - \xi^2}}$$

$$\text{arc cos}_1{}' \, \xi = -\frac{1}{\sin(\text{arc cos } \xi)} = -\frac{1}{\sqrt{1 - \xi^2}}$$

$$\text{arc tan}_0{}' \, \xi = \frac{1}{1 + \tan^2(\text{arc tan } \xi)} = \frac{1}{1 + \xi^2}.$$

For arc functions in general we get 128

$$\text{arc sin}_{2n}' = (-1)^n \frac{1}{\sqrt{1 - I^2}}$$

$$\text{arc cos}_{2n+1}' = (-1)^{n+1} \frac{1}{\sqrt{1 - I^2}}$$

$$\text{arc tan}_{2n}' = \frac{1}{1 + I^2}.$$

§12. Taylor Series

The search for a linear approximation of functions has led us to differentiation. A differentiable function ϕ is approximated at the point ξ by

$$L(\eta) = \phi\xi + \phi'\xi(\eta - \xi).$$

We treat the "*remainder*" $\phi\eta - L(\eta)$ as a function R_η of ξ; i.e., we set

$$R_\eta \xi = \phi\eta - \phi\xi - \phi'\xi(\eta - \xi). \tag{12.1}$$

For $\xi = \eta$ we of course get $R_\eta \eta = 0$.

To estimate the remainder, we differentiate (12.1), obtaining

$$R_\eta' \xi = -[\phi'\xi + \phi''\xi(\eta - \xi) - \phi'\xi]$$

$$R_\eta' \xi = -\phi''\xi(\eta - \xi). \tag{12.2}$$

Thus the bound-theorem, in producing an estimate of ϕ'' in the interval $[\xi | \eta]$, also gives an estimate of the remainder.

This treatment can be generalized to the approximation of a function by polynomials:

$$P(\eta) = c_0 + c_1(\eta - \xi) + \ldots + c_n(\eta - \xi)^n.$$

For a given function ϕ and a point ξ we want to specify a polynomial $P(\eta)$ such that at the point ξ the derivatives coincide up to the nth order; i.e.,

$$\begin{aligned} P(\xi) &= \phi\xi \\ P'(\xi) &= \phi'\xi \\ &\vdots \\ P^{(n)}(\xi) &= \phi^{(n)}\xi. \end{aligned} \tag{12.3}$$

For this purpose we must of course presuppose that ϕ is differentiable up to the nth order at ξ. The equations (12.3) successively entail

129

$$\begin{aligned} c_0 &= \phi\xi \\ 1 \cdot c_1 &= \phi'\xi \\ 1 \cdot 2 \cdot c_2 &= \phi''\xi \\ &\vdots \\ n! c_n &= \phi^{(n)}\xi. \end{aligned}$$

The function ϕ may thus be approximated by the polynomial

$$P(\eta) = \sum_0^n {}_i \frac{\phi^{(i)}\xi}{i!}(\eta - \xi)^i = \phi\xi + \phi'\xi(\eta - \xi) + \ldots + \frac{\phi^{(n)}\xi}{n!}(\eta - \xi)^n.$$

This polynomial is called for short the *Taylor polynomial*, and its coefficients, *Taylor coefficients*. For the *remainder*

$$R_\eta \xi = \phi\eta - \sum_0^n {}_i \frac{\phi^{(i)}\xi}{i!}(\eta - \xi)$$

we get, similarly to (12.2),

$$R_\eta' \xi = -\frac{\phi^{(n+1)}\xi}{n!}(\eta - \xi)^n. \tag{12.4}$$

Since $R_\eta \eta = 0$, the difference quotient of R_η is

$$\frac{R_\eta \eta - R_\eta \xi}{\eta - \xi} = - \frac{R_\eta \xi}{\eta - \xi},$$

and by the bound-theorem we obtain the following estimate (*Cauchy's remainder estimate*):

THEOREM 12.1 From

$$\left| \frac{\phi^{(n+1)} \xi}{n!} (\eta - \xi)^n \right| \leqslant K_\eta \quad \text{for all} \quad \xi \in [\xi_0 | \eta]$$

it follows that

$$\left| \phi\eta - \sum_0^n {}_i \frac{\phi^{(i)} \xi_0}{i!} (\eta - \xi_0)^i \right| \leqslant K_n |\eta - \xi_0|.$$

In general, the larger the natural number n, the better we may expect the Taylor polynomial to approximate the function in question.

In the case of functions that are differentiable up to order ∞, i.e., up to arbitrarily high order, the Taylor polynomials constitute the partial sums of a power series, the *Taylor series*:

$$\phi \xi_0 + \frac{\phi' \xi_0}{1!} (\eta - \xi_0) + \frac{\phi'' \xi_0}{2!} (\eta - \xi_0)^2 + \ldots .$$

The first thing we have to show is that the sequence of coefficients is definite. The sequence of differential quotients $\phi^{(*)} \xi_0$ is indeed definite, for 130

$$\phi^{(0)} \xi_0 = \phi \xi_0$$

$$\phi^{(n+1)} \xi_0 = \lim_{r \to \xi_0} \frac{\phi^{(n)} r - \phi^{(n)} \xi_0}{r - \xi_0}$$

provides an inductive definition with no indefinite quantifiers.

The following cases are now possible:

(1) The Taylor series is convergent and has the sum

$$\phi\eta = \sum_0^\infty {}_i \frac{\phi^{(i)} \xi_0}{i!} (\eta - \xi_0)^i;$$

(2) The Taylor series is convergent but has a sum other than $\phi\eta$;

(3) The Taylor series is not convergent.

Only in case (1) is the function ϕ said to be *representable* by its Taylor series at the point η with respect to ξ_0.

From the definition of analytic functions, it follows that if a function is representable by its Taylor series (with respect to ξ_0) in an interval around ξ_0, then the function is analytic at ξ_0.

The converse also holds: every analytic function is representable by its Taylor series. For if, in an interval around ξ_0,

$$\phi\xi = \sum_{0}^{\infty}{}_i a_i (\xi - \xi_0)^i, \tag{12.5}$$

then what we must show is that (12.5) implies the differentiability of ϕ up to order ∞, as well as

$$a_i = \frac{\phi^{(i)}\xi_0}{i!}.$$

The differentiability of analytic functions follows easily on the basis of a general theorem about sequences of differentiable functions. The partial sums of a power series are, after all, polynomials and thus functions differentiable up to arbitrarily high order.

Theorem 12.2 If ϕ_* is a sequence in an interval of differentiable functions and the sequence of the derivative functions ϕ'_* is uniformly convergent in the interval, then, where $\phi = \lim \phi_*$, ϕ too is differentiable in the interval and

$$\phi' = \lim \phi'_*.$$

131 *Proof*. We set

$$\psi_{m,n} = \phi_m - \phi_n.$$

Then

$$\psi'_{m,n} = \phi'_m - \phi'_n,$$

and by our assumption of the uniform convergence of ϕ'_* it also holds in the convergence interval that

$$\bigwedge_\varepsilon \bigvee_n \bigwedge_{m>n} |\psi'_{m,n}| < \varepsilon.$$

It therefore follows directly (by the bound-theorem) from

$$\phi_m\xi - \phi_m\xi_0 = \phi_n\xi - \phi_n\xi_0 + \psi_{m,n}\xi - \psi_{m,n}\xi_0$$

that for all ε and sufficiently large n

$$\bigwedge_{m>n} \left| \frac{\phi_m\xi - \phi_m\xi_0}{\xi - \xi_0} - \frac{\phi_n\xi - \phi_n\xi_0}{\xi - \xi_0} \right| < \frac{\varepsilon}{3},$$

and hence that

$$\left| \frac{\phi\xi - \phi\xi_0}{\xi - \xi_0} - \frac{\phi_n\xi - \phi_n\xi_0}{\xi - \xi_0} \right| \leqslant \frac{\varepsilon}{3},$$

whence furthermore

$$\left|\frac{\phi\xi - \phi\xi_0}{\xi - \xi_0} - \lim \phi'_*\xi_0\right| \leqslant \left|\frac{\phi\xi - \phi\xi_0}{\xi - \xi_0} - \phi'_n\xi_0\right| + |\phi'_n\xi_0 - \lim \phi'_*\xi_0| + \frac{\varepsilon}{3}.$$

For sufficiently large n, the second addend on the right will be $< \frac{\varepsilon}{3}$; and for sufficiently small $|\xi - \xi_0|$, the first addend will then likewise be $< \frac{\varepsilon}{3}$.

Thus, as $\xi \rightarrow \xi_0$,

$$\left|\frac{\phi\xi - \phi\xi_0}{\xi - \xi_0} - \lim \phi'_*\xi_0\right| \rightarrow 0,$$

i.e., ϕ is differentiable at ξ_0, and we get

$$\phi'\xi_0 = \lim \phi'_*\xi_0.$$

A strengthened version of this theorem will be important later on.

CODICIL TO THEOREM 12.2 Even if (under the above conditions on the ϕ_ns and on the sequence ϕ'_*) we assume of ϕ_* only that, for some argument α, the sequence $\phi_*\alpha$ is convergent, it still follows that ϕ_* is uniformly convergent in the interval.

Proof. It will be true of an arbitrary argument ξ in the interval that 132

$$\begin{aligned}\phi_m\xi - \phi_n\xi &= (\phi_m\xi - \phi_m\alpha) + (\phi_m\alpha - \phi_n\alpha) + (\phi_n\alpha - \phi_n\xi)\\ &= \overset{\xi}{\underset{\alpha}{\Delta}}\phi_m - \overset{\xi}{\underset{\alpha}{\Delta}}\phi_n + (\phi_m\alpha - \phi_n\alpha).\end{aligned} \tag{12.6}$$

As $\phi_*\alpha$ is convergent, the last addend on the right will tend to zero as m, n increase.

It holds by hypothesis of $(\phi_m - \phi_n)' = \phi'_m - \phi'_n$ that

$$\bigwedge_\varepsilon \bigvee_N \bigwedge_{\substack{m,n\\ >N}} \bigwedge_\xi |\phi'_m\xi - \phi'_n\xi| < \varepsilon.$$

Hence, by the bound-theorem, we shall get for sufficiently large m, n

$$\overset{\xi}{\underset{\alpha}{\Delta}}\phi_m - \overset{\xi}{\underset{\alpha}{\Delta}}\phi_n = \overset{\xi}{\underset{\alpha}{\Delta}}(\phi_m - \phi_n) < \varepsilon|\xi - \alpha|.$$

Thus the first addend on the right in (12.6) will likewise approach zero for sufficiently large m, n; and it will do so uniformly for all ξ. The sequences $\phi_*\xi$ are concentrated and therefore convergent. The sequence ϕ_* is uniformly convergent.

The differentiability of analytic functions requires by Theorem 12.2 only a demonstration of the uniform convergence of the power series arising from *member-by-member differentiation* of (12.5):

$$\sum_{0} {}_i\, ia_i(\xi - \xi_0)^{i-1}.$$

For present purposes, let the power series (12.5) be convergent for ξ_1, and let $0 < \theta < 1$. We need to show the uniform convergence of the derivative series in the interval $\epsilon_\xi\, |\xi - \xi_0| < \theta |\xi_1 - \xi_0|$. It holds in this interval that

$$|ia_i(\xi - \xi_0)^{i-1}| \leqslant i\, |a_i|\, \theta^{i-1} |\xi_1 - \xi_0|^{i-1}.$$

In view of the absolute convergence of (12.5), the summands

$$a_i(\xi_1 - \xi_0)^{i-1} = \frac{a_i(\xi_1 - \xi_0)^i}{\xi_1 - \xi_0}$$

are uniformly bounded with respect to i. Since furthermore—as shown on page 102—the series $\sum_i i\theta^i$ is convergent, by Theorem 10.3 the derivative power series is likewise convergent. The differentiability of the power series in the interior of their convergence interval is thereby assured.

Thus it follows from (12.5) that

$$\phi'\xi = \sum_{0}^{\infty} {}_i\, ia_i(\xi - \xi_0)^{i-1}. \tag{12.7}$$

133 The derivative function ϕ' is also the sum of a power series and hence likewise differentiable. Taking everything together, we conclude that the analytic functions are differentiable up to order ∞.

From (12.5) it follows as a matter of course for $\xi = \xi_0$ that

$$\phi\xi_0 = a_0.$$

In the same way, (12.7) implies

$$\phi'\xi_0 = 1 \cdot a_1.$$

Iterated member-by-member differentiation in this way yields all the desired equations

$$\phi^{(i)}\xi_0 = i!\,a_i.$$

The analytic functions are thus indeed representable by their Taylor series.

We will now show that all elementary functions are analytic. In specifying power-series representations of the elementary functions, we may confine ourselves here to the basic elementary functions from which all others can be constructed by addition, multiplication, and composition. Thus we need consider only the power functions, exponential functions, logarithmic functions, and circle functions.

The Taylor coefficients for the power functions $\phi\xi = \xi^\alpha$ work out as follows:

$$\phi'\xi = \alpha\,\xi^{\alpha-1}$$

$$\frac{1}{2}\,\phi''\xi = \frac{\alpha(\alpha-1)}{2}\,\xi^{\alpha-1}$$

$$\frac{1}{3!}\,\phi'''\xi = \frac{\alpha(\alpha-1)(\alpha-2)}{3!}\,\xi^{\alpha-3}$$

$$\vdots$$

Thus the *binomial coefficients*

$$\binom{\alpha}{i} = \frac{\alpha(\alpha-1)\ldots(\alpha-i+1)}{i!}$$

make their appearance here, and we get in general

$$\frac{1}{i!}\,\phi^{(i)}\,\xi = \binom{\alpha}{i}\,\xi^{\alpha-i}.$$

For sufficiently large i, the exponents $\alpha - i$ are negative. $\xi = 0$ is then an infinity point of $\phi^{(i)}$. There is thus no power-series representation around 0. For $\xi = 1$ the Taylor coefficients are

$$\frac{1}{i!}\,\phi^{(i)}\,1 = \binom{\alpha}{i}.$$

The Taylor series thus takes the form 134

$$\sum_0 \binom{\alpha}{i}(\xi-1)^i.$$

To prove the adequacy of the representation

$$\xi^\alpha = \sum_0^\infty \binom{\alpha}{i}(\xi-1)^i$$

we use Cauchy's remainder estimate for the nth remainder R_n (Theorem 12.1),

$$R_n \leqslant K\,|\xi - 1|,$$

provided that

$$\frac{1}{n!}\,\alpha(\alpha-1)\ldots(\alpha-n)\,|\eta|^{\alpha-n-1}|\xi-\eta|^n \leqslant K \quad \text{for} \quad \eta \in [1|\xi].$$

To begin with here, $\dfrac{(\alpha-1)\ldots(\alpha-n)}{n!} = \dbinom{\alpha-1}{n} \to 1$ as $n \to \infty$;

and furthermore, $\left|\dfrac{\xi-\eta}{\eta}\right|^n \to 0$ if $|\xi-\eta| < |\eta|$.

Since $\eta \in [1 | \xi]$, this inequality indeed holds for $|\xi - 1| < 1$; for $0 < \xi \leqslant \eta \leqslant 1$ implies $0 \leqslant \eta - \xi < \eta$, and $1 \leqslant \eta \leqslant \xi < 2$ implies $0 \leqslant \xi - \eta < 1$. For sufficiently large n, therefore, the bound K can be chosen to be arbitrarily small; i.e., $R_n \rightsquigarrow 0$ as $n \rightsquigarrow \infty$, where $0 < \xi < 2$. The representation

$$\xi^{\alpha} = \sum_{0}^{\infty}{}_i \binom{\alpha}{i} (\xi - 1)^i$$

holds for the open interval $(0 | 2)$.

For the exponential function $\phi = \exp$, the derivatives are

$$\phi'\xi = e^{\xi}$$
$$\phi''\xi = e^{\xi}$$
$$\vdots$$

Thus for $\xi = 0$ the Taylor coefficients are

$$\phi 0 = 1$$
$$\phi' 0 = 1$$
$$\frac{1}{2}\phi'' 0 = \frac{1}{2}$$
$$\frac{1}{3!}\phi''' 0 = \frac{1}{3!}$$
$$\vdots$$

135 The Taylor series is

$$\sum_{0}{}_i \frac{1}{i!} \xi^i.$$

By Theorem 12.1 the remainder will be such that

$$|R_n| \leqslant K |\xi|, \quad \text{provided} \quad \frac{e^{\eta}}{n!} |\xi - \eta|^n \leqslant K \quad \text{for} \quad \eta \in [0 | \xi].$$

If $\eta \in [0 | \xi]$, then $e^{\eta} < e^{|\xi|}$ and $|\xi - \eta| < |\xi|$. Furthermore, $\lim_{n \rightsquigarrow \infty} \frac{|\xi^n|}{n!} = 0$ for all ξ. Thus here too $R_* \rightsquigarrow 0$, whence

$$e^{\xi} = \sum_{0}^{\infty}{}_i \frac{1}{i!} \xi^i \quad \text{for all } \xi.$$

For the special case $\xi = 1$ we get

$$e = \sum_{0}^{\infty}{}_i \frac{1}{i!}.$$

This series representation yields a way of computing e, for a remainder estimate is now available: $|R_n| < \frac{e}{n!} < \frac{4}{n!}$.

A representation of the exponential function α^I is obtained with the help of

$$\alpha^{\xi} = e^{\xi \ln \alpha}.$$

In the case of the logarithmic functions we can likewise restrict ourselves to natural logarithms. As $\ln \xi$ is not defined for $\xi = 0$, we shall look for a power series for $\xi = 1$.

From $\ln'\xi = \frac{1}{\xi}$, there follow for the higher derivatives

$$\ln''\xi = -\frac{1}{\xi^2}$$

$$\ln'''\xi = \frac{2}{\xi^3}$$

$$\ln''''\xi = -\frac{3!}{\xi^4}$$

$$\vdots$$

The Taylor coefficients for $\xi = 1$ are

$$\ln 1 = 0$$

$$\ln'1 = 1$$

$$\frac{1}{2}\ln''1 = -\frac{1}{2}$$

136

$$\frac{1}{3!}\ln'''1 = \frac{1}{3}$$

$$\vdots$$

The Taylor series is accordingly $\sum_{1}{}_i \frac{(-1)^{i-1}}{i}(\xi - 1)^i$.

For the nth remainder term, Theorem 12.1 gives

$$|R_n| \leq K|\xi - 1|, \quad \text{provided} \quad \frac{1}{|\eta|^{n-1}}|\xi - \eta|^n \leq K \quad \text{for} \quad \eta \in [1|\xi].$$

As with the remainder estimate for the power functions, we have $|\xi - \eta| < |\eta|$ if $|\xi - 1| < 1$. Furthermore, $|\eta| > 1 - |\xi - 1|$, so that for $0 < \xi < 2$, K can again be chosen to be arbitrarily small for sufficiently large n. Thus $R_* \rightsquigarrow 0$ and

$$\ln \xi = \sum_{1}^{\infty}{}_i \frac{(-1)^{i-1}}{i} (\xi - 1)^i \quad \text{where} \quad |\xi - 1| < 1.$$

Incidentally, this representation also holds for $\xi = 2$, as §13 will show:

$$\ln 2 = 1 - \frac{1}{2} + \frac{1}{3} - \frac{1}{4} + \ldots\ldots$$

The only basic elementary functions we still have to take up are the circle functions. We may limit our attention to sin, cos, arc sin, and arc tan, since tan is computable as $\frac{\sin}{\cos}$, and arc cos as $\frac{\pi}{2}$ - arc sin. In view of arc tan I = arc sin $\frac{I}{\sqrt{1+I^2}}$, even arc tan is dispensable.

The derivatives of the sine function are

$$\sin'\xi = \cos \xi$$
$$\sin''\xi = -\sin \xi$$
$$\sin'''\xi = -\cos \xi$$
$$\sin''''\xi = \sin \xi$$
$$\vdots$$

137 Accordingly, the Taylor coefficients (around 0) are as follows:

$$\sin 0 = 0$$
$$\sin'0 = 1$$
$$\frac{1}{2}\sin''0 = 0$$
$$\frac{1}{3!}\sin'''0 = -\frac{1}{3!}$$
$$\vdots$$

The Taylor series is

$$\xi - \frac{1}{3!}\xi^3 + \frac{1}{5!}\xi^5 - + \ldots.$$

To estimate the remainder, we have by Theorem 12.1 to consider

$$\frac{1}{n!}\,|\xi - \eta|^n \quad \text{for} \quad |\eta| < |\xi| \quad \text{and } n \text{ even}.$$

Thus, as in the case of the exponential function, we get for all ξ

$$\sin \xi = \xi - \frac{1}{3!}\,\xi^3 + \frac{1}{5!}\,\xi^5 - + \ldots\ldots$$

In exactly the same way we get

$$\cos \xi = 1 - \frac{1}{2}\,\xi^2 + \frac{1}{4!}\,\xi^4 - + \ldots\ldots$$

To form the remaining Taylor series for arc sin and arc tan, it is not feasible to calculate the Taylor coefficients explicitly: the expressions become too imperspicuous. It is easier to proceed from the fact that

$$\text{arc sin}'\,\xi = \frac{1}{\sqrt{1 - \xi^2}}$$

$$\text{arc tan}'\,\xi = \frac{1}{1 + \xi^2}.$$

On the basis of these equations, the requisite derivatives can be represented without further ado as power series. From 138

$$\begin{aligned}(1 + \eta)^{-\frac{1}{2}} &= 1 + \binom{-\frac{1}{2}}{1}\eta + \binom{-\frac{1}{2}}{2}\eta^2 + \ldots\ldots \\ &= 1 - \frac{1}{2}\,\eta + \frac{1 \cdot 3}{2 \cdot 2}\,\frac{\eta^2}{2!} - \frac{1}{2} \cdot \frac{3}{2} \cdot \frac{5}{2}\,\frac{\eta^3}{3!} + - \ldots\ldots \\ &= 1 - \frac{1}{2}\,\eta + \frac{1 \cdot 3}{2 \cdot 4}\,\eta^2 - \frac{1 \cdot 3 \cdot 5}{2 \cdot 4 \cdot 6}\,\eta^3 + - \ldots\ldots \quad \text{for } |\eta| < 1\end{aligned}$$

it follows that

$$\text{arc sin}'\,\xi = 1 + \frac{1}{2}\,\xi^2 + \frac{1 \cdot 3}{2 \cdot 4}\,\xi^4 + \frac{1 \cdot 3 \cdot 5}{2 \cdot 4 \cdot 6}\,\xi^6 + \ldots\ldots \quad \text{for } |\xi| < 1.$$

Now, it is easy to find a power series that gives rise to the series on the right through member-by-member differentiation:

$$\xi + \frac{1}{2}\,\frac{\xi^3}{3} + \frac{1 \cdot 3}{2 \cdot 4}\,\frac{\xi^5}{5} + \ldots.$$

This series converges for $\xi = 0$ and hence also, by the codicil to Theorem 12.2, for $|\xi| < 1$. If we call the function represented

ϕ, then in addition $\phi' = \text{arc sin}'$, i.e., $(\phi - \text{arc sin})' = 0$. In view of the bound-theorem, ϕ - arc sin is constant. Since $\phi 0 = \text{arc sin } 0$, it therefore follows that

$$\text{arc sin } \xi = \xi + \frac{1}{2}\frac{\xi^3}{3} + \frac{1 \cdot 3}{2 \cdot 4}\frac{\xi^5}{5} + \ldots\ldots \quad \text{for} \quad |\xi| < 1.$$

The same procedure can be used for arc tan. In that case,

$$\text{arc tan}' \xi = \frac{1}{1 + \xi^2} = 1 - \xi^2 + \xi^4 - + \ldots\ldots \quad \text{for} \quad |\xi| < 1.$$

The series

$$\xi - \frac{\xi^3}{3} + \frac{\xi^5}{5} - + \ldots .$$

is transformed through member-by-member differentiation into the series for arc tan′; furthermore, it converges to arc tan 0 when $\xi = 0$. Hence we get

$$\text{arc tan } \xi = \xi - \frac{\xi^3}{3} + \frac{\xi^5}{5} - + \ldots\ldots \quad \text{for} \quad |\xi| < 1.$$

Closer examination of the remainder in the representation of arc tan′ easily discloses that the representation of arc tan holds for $\xi = 1$ too. In this way—since $\tan \frac{\pi}{4} = 1$—we obtain the Leibniz series (which already appeared in the work of J. Gregory):

$$\frac{\pi}{4} = 1 - \frac{1}{3} + \frac{1}{5} - \frac{1}{7} + - \ldots .$$

This series converges very slowly, to be sure, but the arc tan series yields several handy alternatives for computing π. (Cf. von Mangoldt and Knopp, II, No. 56.)

139

§13. Integration

As a consequence of the strengthened bound-theorem we have already concluded (11.3) that if the derivative of a continuous function is 0 ω-everywhere in an interval, then that function will have 0 as difference quotients throughout the interval; i.e., the function will be constant.

Thus two continuous functions Φ, Ψ such that $\Phi' = \Psi'$ ω-everywhere in some interval will have the same difference quotients there. Let us use $\Delta\Phi$ as a 2-place function which takes the value $\Phi\xi_2 - \Phi\xi_1$ for the arguments ξ_1, ξ_2; we write this as

$$\overset{\xi_2}{\underset{\xi_1}{\Delta}}\Phi = \Phi\xi_2 - \Phi\xi_1.$$

It will now hold (in intervals) that

$$\Phi' = \Psi'\ \omega\text{-everywhere} \rightarrow \Delta\Phi = \Delta\Psi \text{ everywhere.}$$

The derivative function Φ' thus uniquely determines the 2-place increment function $\Delta\Phi$.

In contrast to §11, where we started with a function Φ and sought the possible ways of differentiating Φ and computing Φ', we now turn the search around. Beginning with a function ϕ (in an interval), we seek a continuous function Φ, if there be one, for which $\Phi' = \phi$ ω-everywhere.

DEFINITION 13.1 If a definite function Φ is continuous in $[\alpha \mid \beta]$, and if $\Phi' = \phi$ ω-everywhere, then Φ is called a *primitive function* of ϕ in $[\alpha \mid \beta]$.

The primitive functions of a function ϕ are determined only up to an additive constant. However, the 2-place increment functions of the primitive functions Φ are uniquely determined by ϕ. We call $\Delta\Phi$ the *integral function* of ϕ, and we write

$$\mathcal{J}\phi = \Delta\Phi.$$

If ϕ is given by a term $T(\xi)$, then we write

$$\mathcal{J}_{\xi}\, T(\xi)$$

rather than

$$\mathcal{J}\,\imath_{\xi}T(\xi).$$

We avoid the traditional use of differentials for the integral functions; we shall not introduce them until Chapter V. The same goes for the definition of $\int$.

The definition of the integral function is

$$\mathcal{J}\phi \leftrightharpoons \Delta\Phi \quad \text{provided} \quad \Phi' = \phi\ \omega\text{-everywhere.}$$

Since $\mathcal{J}\mathcal{D}\Phi = \Delta\Phi$, we may write in brief *140*

$$\mathcal{J}\mathcal{D} = \Delta. \tag{13.1}$$

The integral function of ϕ of course exists if and only if there is a primitive function of ϕ, in which case ϕ is said to be *integrable* (in the interval in question).

The values of the integral function are called *integrals*. We write

$$\overset{\xi_2}{\underset{\xi_1}{\mathcal{J}}}\phi = \overset{\xi_2}{\underset{\xi_1}{\Delta}}\Phi. \tag{13.2}$$

These integrals are often called "definite integrals," and the primitive functions, "indefinite integrals." As we shall make no use of that terminology, we are free to say "integral" rather than the lengthier "definite integral."

If ϕ is integrable in $[\xi_0 | \xi_1]$, then $\iota_\xi \mathcal{J}_{\xi_0}^{\xi} \phi$ is a primitive function; for a primitive function Φ will, after all, be such that

$$\iota_\xi \mathcal{J}_{\xi_0}^{\xi} \phi = \iota_\xi(\Phi\xi - \Phi\xi_0) = \Phi - \Phi\xi_0.$$

It follows that

$$\mathcal{D}_\xi \mathcal{J}_{\xi_0}^{\xi} \phi = \mathcal{D}\Phi = \phi \quad \omega\text{-everywhere in } [\xi_0 | \xi_1].$$

For short we write

$$\mathcal{D} \mathcal{J}_{\xi_0} \phi = \phi \qquad \omega\text{-everywhere}.$$

Only if ϕ has a primitive function that is differentiable throughout will it be the case everywhere that

$$\mathcal{D} \mathcal{J}_{\xi_0} \phi = \phi.$$

The introduction of integrals so far constitutes nothing more than a new notation for consequences of the bound-theorem, which now (in unstrengthened form) admits of the following formulation:

If ϕ is integrable in $[\alpha | \beta]$ and $\gamma_1 \leqslant \phi \leqslant \gamma_2$ in $[\alpha | \beta]$, then

$$\gamma_1(\xi_2 - \xi_1) \leqslant \mathcal{J}_{\xi_1}^{\xi_2} \phi \leqslant \gamma_2(\xi_2 - \xi_1) \tag{13.3}$$

for all $\xi_1, \xi_2 \in [\alpha | \beta]$.

For a constant function $\phi = \gamma$, γI is a primitive function. Hence we have

$$\mathcal{J}_{\xi_1}^{\xi_2} \gamma = \gamma(\xi_2 - \xi_1).$$

The above formulation of the bound-theorem is thus a special case of

141 THEOREM 13.1 If $\phi \leqslant \psi$ in an interval, it follows that

$$\mathcal{J}\phi \leqslant \mathcal{J}\psi$$

therein.

This property of monotony or, more precisely, isotony that attaches to integration (i.e., the operation leading from ϕ to $\mathcal{J}\phi$) is in turn an immediate consequence of the bound-theorem. From $\phi \leqslant \psi$ it follows that $\psi - \phi \geqslant 0$, so that

$$\mathcal{J}\psi - \mathcal{J}\phi = \mathcal{J}(\psi - \phi) \geqslant 0.$$

The formula

$$\mathcal{J}(\psi - \phi) = \mathcal{J}\psi - \mathcal{J}\phi$$

we are using here is a special case of linearity:

THEOREM 13.2 $\mathcal{J}(\gamma_1\phi_1 + \gamma_2\phi_2) = \gamma_1\mathcal{J}\phi_1 + \gamma_2\mathcal{J}\phi_2.$

Proof $\quad (\gamma_1\Phi_1 + \gamma_2\Phi_2)' = \gamma_1\Phi_1{}' + \gamma_2\Phi_2{}'.$

Admittedly, everything we have done with integrals so far is nothing but a reformulation of the bound-theorem. But even that is already a gain, as may be seen in the estimation of remainders for Taylor polynomials.

You will recall that for the remainder term

$$R_\eta(\xi) = \phi\eta - \left[\phi\xi + \phi'\xi(\eta - \xi) + \ldots + \frac{\phi^{(n)}\xi}{n!}(\eta - \xi)^n\right]$$

we had obtained the derivative

$$R_\eta{}'(\xi) = -\frac{\phi^{(n+1)}\xi}{n!}(\eta - \xi)^n.$$

Since $R_\eta(\eta) = 0$, it now turns out that the remainder can serve as a "*remainder integral,*" as we shall call it:

$$R_\eta(\xi) = \mathop{\mathcal{J}_\zeta}_{\xi}^{\eta} \frac{\phi^{(n+1)}\zeta}{n!}(\eta - \zeta)^n.$$

With that we are in a position to assert—instead of an estimate—an equation, the so-called *Taylor formula* (which in this form goes back to Johann Bernoulli):

$$\phi\eta = \phi\xi + \phi'\xi(\eta - \xi) + \ldots + \frac{\phi^{(n)}\xi}{n!}(\eta - \xi)^n + \mathop{\mathcal{J}_\zeta}_{\xi}^{\eta} \frac{\phi^{(n+1)}\zeta}{n!}(\eta - \zeta)^n. \tag{13.4}$$

The remainder estimate is an estimate of the integral. The monotony theorem (Theorem 13.1) suggests the following estimate: if $|\phi^{(n+1)}| \leqslant K$ in $[\xi|\eta]$, then

$$|\phi^{(n+1)}\zeta(\eta - \zeta)^n| \leqslant K|\eta - \zeta|^n,$$

whence

142

$$|\mathcal{J}_\zeta \phi^{(n+1)}\zeta(\eta - \zeta)^n| \leqslant K|\mathcal{J}_\zeta(\eta - \zeta)^n|.$$

The primitive function of $(\eta - I)^n$ is easily seen to be $-\dfrac{(\eta - I)^{n+1}}{n+1}$, so that

$$\mathscr{T}_{\zeta}{}_{\xi}^{\eta}(\eta - \zeta)^n = \frac{(\eta - \xi)^{n+1}}{n+1}.$$

We thus get the following estimate of the remainder integral of the Taylor formula (*Lagrange's remainder estimate*):

$$\frac{1}{n!}\left|\mathscr{T}_{\zeta}{}_{\xi}^{\eta}\phi^{(n+1)}\zeta(\eta - \zeta)^n\right| \leqslant \frac{K}{(n+1)!}\,|\eta - \xi|^{n+1}$$

$$\text{for} \quad |\phi^{(n+1)}| \leqslant K \quad \text{in } [\xi\,|\,\eta].$$

To show at least one application, we estimate the remainder

$$\ln 2 - \left[1 - \frac{1}{2} + \frac{1}{3} - + \ldots (-1)^{n+1}\frac{1}{n}\right].$$

What we have to estimate is

$$R_n = \frac{1}{n!}\mathscr{T}_{\zeta}{}_{1}^{2} \ln^{(n+1)}\zeta(2 - \zeta).$$

As $\ln^{(n+1)} = (-1)^n \dfrac{n!}{I^n + 1}$, we have $|\ln^{(n+1)}| \leqslant n!$ in $[1\,|\,2]$. The remainder integral will thus be such that

$$|R_n| \leqslant \frac{n!}{(n+1)!}(2 - 1)^{n+1} = \frac{1}{n+1},$$

i.e., $R_* \rightsquigarrow 0$.

As we said, the uses to which we are putting integrals here could be handled just as well without integration by means of the bound-theorem alone.

The justification of integration—other than the gain in perspicuity of formulation—is that the determination of integrals also solves problems that arose quite independently of differentiation. We are brought up against these problems when we consider the integration of step functions. For every step function there are corresponding primitive functions, which are piecewise linear. If the step function τ is defined by partitioning the argument interval $[\alpha\,|\,\beta]$ by means of the numbers

$$\alpha = \alpha_0 < \alpha_1 < \ldots < \alpha_n = \beta$$

and by letting

$$\tau = \gamma_i \quad \text{in} \quad (\alpha_i\,|\,\alpha_{i+1}) \quad \text{where} \quad i = 0, \ldots, n-1,$$

143 then we shall have

$$\mathcal{J}_{\alpha}^{\beta} \tau = \sum_{0}^{n-1}{}_i \gamma_i (\alpha_{i+1} - \alpha_i).$$

Proof. First of all,

$$T\xi = \gamma_i (\xi - \alpha_i) + \sum_{0}^{i-1}{}_j \gamma_j (\alpha_{j+1} - \alpha_j) \quad \text{where} \quad \xi \in [\alpha_i \,|\, \alpha_{i+1}]$$

is a continuous function, for

$$T\alpha_{i+1} = \gamma_i (\alpha_{i+1} - \alpha_i) + \sum_{0}^{i-1}{}_j \gamma_j (\alpha_{j+1} - \alpha_j) = \sum_{0}^{i}{}_j \gamma_j (\alpha_{j+1} - \alpha_j).$$

It holds in particular that

$$T\alpha = 0, \quad T\beta = \sum_{0}^{n-1}{}_i \gamma_i (\alpha_{i+1} - \alpha_i).$$

Furthermore,

$$T'\xi = \gamma_i \quad \text{for} \quad \xi \in (\alpha_i \,|\, \alpha_{i+1}),$$

so that $T' = \tau$ everywhere except for finitely many arguments.

The integrability of the step functions now entails the integrability of all jump-continuous functions, since these are uniformly approximable by step functions (Theorem 10.9). In this connection we prove the following general theorem:

THEOREM 13.3 If ϕ_* is a function sequence which is uniformly convergent (in an interval), and if for every n, Φ_n is a primitive function of ϕ_n such that for an argument α the sequence $\Phi_*\alpha$ is convergent, then the sequence Φ_* is uniformly convergent in the interval, and $\Phi = \lim \Phi_*$ is a primitive function of $\phi = \lim \phi_*$. In particular, Φ is differentiable at all points at which infinitely many ϕ_ns are differentiable.

Proof. The proof of the uniform convergence of Φ_* is the same word for word as that of the codicil to Theorem 12.2, except that the strengthened bound-theorem is substituted for the unstrengthened form.

The function $\Phi = \lim \Phi_*$ is therefore continuous. If $\Phi_n'\xi = \phi_n\xi$ is true for infinitely many n, then there is a subsequence Φ_{k_*} such that $\Phi_{k_n}'\xi = \phi_{k_n}\xi$ for all n. Since $\Phi\xi = \lim \Phi_{k_*}\xi$ and $\phi\xi = \lim \phi_{k_*}\xi$, it follows that Φ is differentiable at ξ and, as in the proof of Theorem 12.2, that $\Phi'\xi = \phi\xi$. Thus $\Phi' = \phi$ holds ω-everywhere; i.e., Φ is a primitive function of ϕ.

In order to derive the integrability of the jump-continuous functions ϕ—in an interval $[\alpha \,|\, \beta]$, let us say—from Theorem 13.3, we have only to approximate ϕ uniformly by means of a definite sequence τ_* of step functions. If T_n is a primitive function of τ_n, then

$$\Phi_n = T_n - T_n\alpha$$

144 too. Since $\Phi_n\alpha = 0$, $\Phi_*\alpha$ is convergent. Therefore Φ_* converges uniformly in the interval to a definite function, and $\Phi = \lim \Phi_*$ is a primitive function of ϕ.

The primitive functions of continuous (not just jump-continuous) functions are everywhere (not just ω-everywhere) differentiable, for every continuous function is trivially approximable by step functions that have no jump at a given arbitrary point and hence are differentiable there. Thus it is universally true of continuous functions ϕ that

$$\mathcal{D}\underset{\xi_0}{\mathcal{J}}\phi = \phi.$$

Using the concept of integral, we immediately get the following from Theorem 13.3:

THEOREM 13.4 If a sequence ϕ_* of integrable functions is uniformly convergent in the interval $[\alpha \mid \beta]$, then $\phi = \lim \phi_*$ is also integrable, and

$$\mathcal{J}\phi = \lim \mathcal{J}\phi_*.$$

Together these results yield for jump-continuous functions the

FUNDAMENTAL THEOREM. A function ϕ which is jump-continuous in $[\alpha \mid \beta]$ is integrable, and the equation

$$\underset{\alpha}{\overset{\beta}{\mathcal{J}}}\phi = \lim \sum_{0}^{m_*}{}_i \gamma_i^{(*)} (\alpha_{i+1}^{(*)} - \alpha_i^{(*)}) \tag{13.5}$$

holds of every sequence τ_* of step functions which uniformly approximates ϕ, where

$$\tau_n = \gamma_i^{(n)} \quad \text{in} \quad (\alpha_i^{(n)} \mid \alpha_{i+1}^{(n)}) \quad \text{for} \quad i = 0, 1, \ldots, m_n.$$

This formation of limits for sums of the form $\sum_0^n{}_i \gamma_i (\alpha_{i+1} - \alpha_i)$ is a task that also arises independently of differentiation. The most familiar example is that of *surface measurement.*

Consider the graph of a function $\phi \geqslant 0$ in $[\alpha \mid \beta]$, and let ϕ be approximated by a step function τ. The corresponding sum $\sum \gamma_i (\alpha_{i+1} - \alpha_i)$ will then be precisely the surface area enclosed by the image of the step function, the ordinates $\xi = \alpha$ and $\xi = \beta$, and the ξ-axis. For that surface is of course composed of rectangles of width $\alpha_{i+1} - \alpha_i$ and height γ_i. If ϕ is jump-continuous, then we know that, for every sequence τ_* of step functions that uniformly

145

converges to ϕ, the surface areas in all cases converge to the same value. Thus it would be natural to take this common limit as a definition of the "*surface area*" enclosed by the ξ-axis, the ordinates $\xi = \alpha$ and $\xi = \beta$, and the image of the function ϕ.

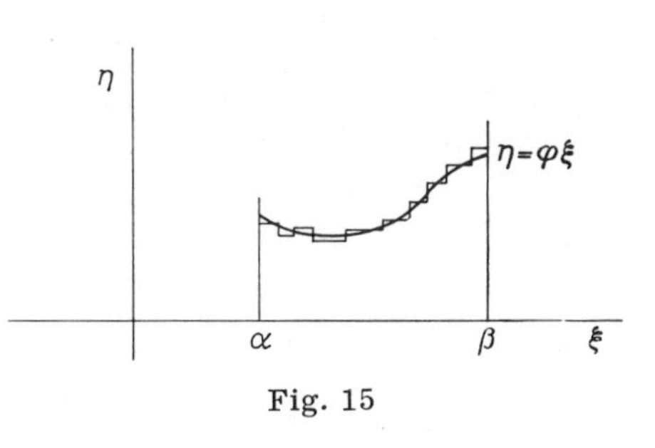

Fig. 15

Taking the limit of sums of the form $\sum \gamma_i (\alpha_{i+1} - \alpha_i)$ is an operation that is by no means confined to geometry. It arises in many areas of physics and similarly mathematicized sciences as well. Suppose, e.g., that the velocity v of a motion is given as a function of the time t, i.e., $v = \phi t$. Then the distance s traversed from t_0 till t_1 will be

$$s = v(t_1 - t_0),$$

provided ϕ is constant.

If ϕ is a step function, then s will be given by

$$s = \sum v_i (t_{i+1} - t_i).$$

For jump-continuous ϕ, s will be defined as the limit of such sums; i.e. (in a notation to be justified later on: cf. Chap. V),

$$s = \int v dt.$$

In the same way, the formula

$$W = F(s_1 - s_0)$$

(work = force · distance) leads to the integral formula

$$W = \int F ds.$$

All that is needed are two measurable physical quantities f and g which are meaningfully combinable in the form $f(g_2 - g_1)$: that is enough to give rise to the integral $\int f dg$, provided f is a nonconstant function of g.

The fundamental theorem provides the insight that in order to calculate the limits of such sums we have only to integrate the approximated function ϕ; i.e., we have only to find a primitive function Φ.

Conversely, the problem of calculating a primitive function can in the case of jump-continuous functions be reduced by means of the fundamental theorem to finding limits of sums. Of particular interest, of course, are those cases in which one can find a primitive function by applying known differentiation rules, thus avoiding the need to take limits of sums. That was the crucial breakthrough that Newton and Leibniz accomplished in comparison with Archimedes.

146 We have already made use of the fact, e.g., in §12, that

$$\mathcal{D}_\xi\, \xi^\alpha = \alpha I^{\alpha-1}$$

entails

$$\mathcal{D}_\xi\, \frac{1}{\alpha+1}\, \xi^{\alpha+1} = I^\alpha \quad (\alpha \neq -1),$$

and hence

$$\mathcal{J}_\xi\, \xi^\alpha = \Delta_\xi \frac{1}{\alpha+1}\, \xi^{\alpha+1} \quad \text{for} \quad \alpha \neq -1. \tag{13.6}$$

For $\mathcal{J}_\xi \xi^{-1}$, $\ln' = \frac{1}{I}$ immediately yields

$$\mathcal{J}\, \frac{1}{I} = \Delta \ln. \tag{13.7}$$

Furthermore, by §11 we get

$$\left.\begin{aligned} \mathcal{J}\exp &= \Delta \exp \\ \mathcal{J}\sin &= -\Delta \cos \\ \mathcal{J}\cos &= \Delta \sin \\ \mathcal{J}\frac{1}{1+I^2} &= \Delta \operatorname{arc\,tan} \\ \mathcal{J}\frac{1}{\sqrt{1-I^2}} &= \Delta \operatorname{arc\,sin} \end{aligned}\right\} \tag{13.8}$$

In contrast to differentiation, there is no general procedure for integration by which, for any term representing an elementary function, we can find an elementary term representing the integral function. In fact, it can even be proved that the primitive functions of several elementary functions, e.g., $\frac{e^I}{I}$, are not elementary. The first proofs of this kind were given by Liouville in 1834 (cf. J. F. Ritt, *Differential Algebra,* 1950).

The basic elementary functions, however, are all elementarily integrable, as the following formulas show:

$$\left.\begin{aligned}
\mathcal{J}\,\ln &= \Delta(I \ln I - I)\\
\mathcal{J}\,\tan &= -\Delta \ln \cos\\
\mathcal{J}\,\text{arc sin} &= \Delta(I \text{ arc sin } I + \sqrt{1 - I^2})\\
\mathcal{J}\,\text{arc cos} &= \Delta(I \text{ arc cos } I - \sqrt{1 - I^2})\\
\mathcal{J}\,\text{arc tan} &= \Delta\left[I \text{ arc tan } I - \frac{1}{2} \ln (1 + I^2)\right]
\end{aligned}\right\} \quad (13.9)$$

These formulas can immediately be verified by differentiation. *147*

While the differentiation formulas for the basic elementary functions enable us to differentiate all elementary functions (by recourse to the differentiation rules for addition, multiplication, and composition), the corresponding strategy does not work for integration. For although

$$\mathcal{J}(\phi + \psi) = \mathcal{J}\phi + \mathcal{J}\psi,$$

there are no integration rules for products and compositions of functions.

By "inverting" the corresponding differentiation rules we can get certain integration rules, to be sure, but not the desired ones. From

$$\mathcal{D}\phi\psi = (\mathcal{D}\phi)\psi + \phi(\mathcal{D}\psi)$$

it follows that

$$\Delta\phi\psi = \mathcal{J}(\phi'\psi) + \mathcal{J}(\phi\psi'I). \quad (13.10)$$

which permits so-called *partial integration.*

Thus we can reduce integrals of the form $\mathcal{J}(\phi'\psi)$ to ones of the form $\mathcal{J}(\phi\psi')$; e.g.,

$$\mathcal{J}\ln = \mathcal{J}(1 \cdot \ln) = \Delta(I \cdot \ln I) - \mathcal{J}(I \cdot \ln' I).$$

As $\ln' I = \frac{1}{I}$, we get the above formula for $\mathcal{J}\ln$.

The chain rule

$$(\Phi \urcorner \psi)' = (\Phi' \urcorner \psi)\psi'$$

yields

$$\Delta(\Phi \urcorner \psi) = \mathcal{J}(\Phi' \urcorner \psi)\psi'.$$

If we let $\phi = \Phi'$, i.e., $\Delta\Phi = \mathcal{J}\phi$, then we get

$$(\mathcal{J}\phi)\urcorner\psi = \mathcal{J}(\phi \urcorner \psi)\psi',$$

i.e.,

$$\mathcal{J}_{\psi\alpha}^{\psi\beta}\phi = \mathcal{J}_{\alpha}^{\beta}(\phi\urcorner\psi)\psi'. \tag{13.11}$$

Thus integrals with integrands of the form $(\phi\urcorner\psi)\psi'$ can be reduced to $\mathcal{J}\phi$: *integration by substitution.*

Example:

$$\mathcal{J}(I\cdot e^{I^2}) = \frac{1}{2}\,[e^{I^2}(I^2)'] = \frac{1}{2}\,(\mathcal{J}e^{I})\urcorner I^2 = \frac{1}{2}\,\Delta e^{I^2}.$$

In the case of an invertible function ψ, the formula can also be used to reduce $\mathcal{J}\phi$ to $\mathcal{J}(\phi\urcorner\psi)\psi'$.

148 *Example*:

$$(\mathcal{J}\sqrt{1-I^2}\,)\urcorner\sin = \mathcal{J}(\sqrt{1-\sin^2}\cdot\sin') = \mathcal{J}\cos^2 = \mathcal{J}\frac{1+\cos 2I}{2}$$

$$= \Delta\left(\frac{I}{2}+\frac{1}{4}\sin 2I\right) = \Delta\left(\frac{I}{2}+\frac{1}{2}\sin I\,\sqrt{1-\sin^2 I}\right)$$

$$\mathcal{J}\sqrt{1-I^2} = \frac{1}{2}\,\Delta\text{ arc sin } I + I\sqrt{1-I^2}\,).$$

The integration of an inverse function is always possible:

$$\phi\urcorner\psi = I \rightarrow \mathcal{J}\psi = \Delta(I\psi) - \mathcal{J}_{\psi}^{\psi}\phi,$$

for $\mathcal{J}(I\phi') = \Delta(I\phi) - \mathcal{J}\phi$ and

$$\mathcal{J}\psi = \mathcal{J}_{\psi}^{\psi}(\psi\urcorner\phi)\cdot\phi' = \mathcal{J}_{\psi}^{\psi}(I\phi') = \Delta(\psi\cdot I) - \mathcal{J}_{\psi}^{\psi}\phi.$$

The most important class of elementary functions which can be integrated elementarily are the rational functions of I and $\sqrt{a+bI+cI^2}$. To be sure, even the elementary integration of all rational functions of I demands some work and some knowledge of algebra. To be specific, we must make use of the fact that every rational function can be represented as the sum of a polynomial and a linear combination of special rational functions of the form

$$\frac{1}{(a+bI+cI^2)^n} \quad\text{and}\quad \frac{I}{(a+bI+cI^2)^n}.$$

By linear substitutions we get as integrands

$$\frac{1}{I^n},\quad \frac{1}{(1+I^2)^n},\quad\text{and}\quad \frac{I}{(1+I^2)^n}.$$

Thus only $\mathcal{J}\frac{1}{(1+I^2)^n}$ and $\mathcal{J}\frac{I}{(1+I^2)^n}$ remain to be figured out. If

$\psi = 1 + I^2$, then

$$\frac{I}{(1+I^2)^n} = \frac{1}{2}\frac{\psi'}{\psi^n};$$

thus if $\phi = \frac{1}{2I^n}$, then

$$\frac{I}{(1+I^2)^n} = (\phi \urcorner \psi) \cdot \psi'.$$

$\mathcal{J}\phi$ is already known, and ψ is invertible in $[0 \mid \infty]$.

For the integral $\mathcal{J}\frac{1}{(1+I^2)^n}$, the case $n = 1$ is already familiar. For $n > 1$ we use

$$\frac{1}{(1+I^2)^n} = \frac{1}{(1+I^2)^{n-1}} - \frac{I^2}{(1+I^2)^n}$$

149

and

$$\mathcal{J}\frac{I}{(1+I^2)^n} = \mathcal{J}\left(\frac{I}{2} \cdot \frac{2I}{(1+I^2)^n}\right)$$

$$= \Delta\left(\frac{I}{2}\,\frac{1}{1-n}\,\frac{1}{(1+I^2)^{n-1}}\right) - \mathcal{J}\left(\frac{1}{2}\,\frac{1}{1-n}\,\frac{1}{(1+I^2)^{n-1}}\right).$$

These formulas permit the reduction of the integral of $\frac{1}{(1+I^2)^n}$ to that of $\frac{1}{(1+I^2)^{n-1}}$. Thus iteration will lead to $\mathcal{J}\frac{1}{1+I^2}$.

The integration of rational functions of I and $\sqrt{a+bI+cI^2}$ can now be reduced by substitution to the integration of rational functions.

In case $c \neq 0$, a linear substitution will first of all yield as integrands rational functions of I and $\sqrt{I^2 \pm 1}$ (if $c > 0$) or $\sqrt{1 - I^2}$ (if $c < 0$). In the first case we let

$$\psi = I - \sqrt{I^2 \pm 1},$$

and we obtain

$$I = \frac{\psi^2 \mp 1}{2\psi},$$

whence

$$\sqrt{I^2 \pm 1} = \frac{-\psi^2 \mp 1}{2\psi}$$

and by differentiation

$$1 = \frac{2\psi^2 - (\psi^2 \mp 1)}{2\psi^2} \cdot \psi' = \frac{\psi^2 \pm 1}{2\psi^2} \cdot \psi'.$$

If the integrand is given by $R(I, \sqrt{I^2 \pm 1})$, where $R(\xi, \eta)$ is a rational term, then we get

$$R(I, \sqrt{I^2 \pm 1}\,) = R\left(\frac{\psi^2 \mp 1}{2\psi}, \frac{-\psi^2 \mp 1}{2\psi}\right) \cdot \frac{\psi^2 \pm 1}{2\psi^2} \cdot \psi' = \Phi(\psi) \cdot \psi',$$

where Φ is a rational function.

If $\sqrt{1 - I^2}$ occurs in the integrand, then we let

$$\psi = \frac{\sqrt{1 - I^2} - 1}{I}.$$

150 This gives us

$$I = \frac{-2\psi}{1 + \psi^2},$$

whence

$$\sqrt{1 - I^2} = \frac{1 - \psi^2}{1 + \psi^2}$$

and by differentiation

$$1 = 2\,\frac{\psi^2 - 1}{(1 + \psi^2)^2} \cdot \psi'.$$

The result is

$$R(I, \sqrt{1 - I^2}\,) = R\left(\frac{-2\psi}{1 + \psi^2}, \frac{1 - \psi^2}{1 + \psi^2}\right) \cdot 2\,\frac{\psi^2 - 1}{(1 + \psi^2)^2} \cdot \psi' = \Phi(\psi) \cdot \psi',$$

again with a rational function Φ.

When $c = 0$, a linear substitution yields an integrand $R(I, \sqrt{I}\,)$. Letting $\psi = \sqrt{I}$, so that $I = \psi^2$ and $1 = 2\psi \cdot \psi'$, we get

$$R(I, \sqrt{I}\,) = R(\psi^2, \psi) 2\psi \cdot \psi' = \Phi(\psi) \cdot \psi',$$

where Φ is a rational function. All other rational functions of I and $\sqrt[n]{a + bI}$ can obviously be handled in the same way.

Elementary integrands for which these methods are inadequate can be taken care of through the use of the Taylor-series representation.

For any analytic function

$$\phi = \sum_{0}^{\infty}{}_i a_i I^i,$$

a primitive function Φ can immediately be found through "*member-by-member integration,*"

$$\Phi = \sum_{0}^{\infty}{}_i \frac{a_i}{i + 1} I^{i+1},$$

as we have already seen in the examples for arc sin and arc tan.

IV. MANY-PLACE FUNCTIONS

§14. The N-Dimensional Number Space

In §3 we introduced functions by abstraction from terms. In the case of real functions we have thus far restricted ourselves to terms containing only one free variable. This restriction to one variable is arbitrary. Even rational terms, such as $\frac{\xi^2 + \eta}{\xi^2 - \zeta}$, provide examples of terms with several variables. Such terms give rise by abstraction to many-place functions.

Many-place functions are indispensable for the kind of "description" of empirical dependencies mentioned in §8; the pressure of a gas, e.g., depends on the temperature *and* the volume.

Following §3, we call two terms $S(\xi_1, \ldots, \xi_n)$ and $T(\xi_1, \ldots, \xi_n)$ equivalent if

$$\bigwedge_{\xi_1, \ldots, \xi_n} S(\xi_1, \ldots, \xi_n) = T(\xi_1, \ldots, \xi_n).$$

We denote the abstract object represented by the term $T(\xi_1, \ldots, \xi_n)$ with respect to $\xi_1, \ldots, \xi_n$ (and with respect to this equivalence) by

$$\daleth_{\xi_1, \ldots, \xi_n} T(\xi_1, \ldots, \xi_n).$$

This object is called a many-place function or, more precisely, in view of the n variables, an n-place function. We employ the letters $\phi, \psi, \ldots$ for many-place functions too, and where

$$\phi = \daleth_{\xi_1, \ldots, \xi_n} T(\xi_1, \ldots, \xi_n)$$

we again stipulate that

$$\phi\daleth(\xi_1, \ldots, \xi_n) = T(\xi_1, \ldots, \xi_n).$$

The system $(\xi_1, \ldots, \xi_n)$ of the n numbers $\xi_1, \ldots, \xi_n$ is now an argument of the function ϕ, and $\phi\daleth(\xi_1, \ldots, \xi_n)$ or for short $\phi(\xi_1, \ldots, \xi_n)$ is the value of the function for that argument. ξ_i is called the ith place in $(\xi_1, \ldots, \xi_n)$.

Before going into many-place functions in more detail in the next section, we shall first consider the "space" of the arguments.

Let the (indefinite) set of all N-place systems $\xi_1, \ldots, \xi_N$ of real numbers be called the *N-dimensional number space* $\mathfrak{Z}^N$. For 152 present purposes, no special significance need be attached to this geometric locution. It is only in Chapter V that connections with geometry will take on importance. Until then, it is only for the sake of brevity that we shall refer to systems of numbers (Ntuples) as "*points*."

Intervals are defined—as in the one-dimensional case—by a lower and an upper "*end point*." If $\Xi = \xi_1, \ldots, \xi_N$ and $H = \eta_1, \ldots, \eta_N$ satisfy for every i ($i = 1, \ldots, N$) the relation

$$\xi_i < \eta_i,$$

then the *open interval* $(\Xi \mid H)$ includes all points $Z = \zeta_1, \ldots, \zeta_N$ such that

$$\xi_i < \zeta_i < \eta_i. \tag{14.1}$$

The *closed interval* $[\Xi \mid H]$ includes all points Z such that

$$\xi_i \leq \zeta_i \leq \eta_i$$

(where again $i = 1, \ldots, N$). The simple term "*interval*" will always be used to refer to a set of points between the open interval $(\Xi \mid H)$ and the closed interval $[\Xi \mid H]$.

In 1-dimensional space we essentially confined ourselves to intervals as the domains of functions. Intervals play an important role in $\mathfrak{Z}^N$ too. But in contrast to $\mathfrak{Z}^1$, the union of two nondisjoint intervals will not in general itself be an interval. Furthermore, some very simply definable sets, such as

$$\epsilon_{\xi,\eta}\ \xi^2 + \eta^2 < 1,$$

do not constitute a finite interval sum.

It is therefore necessary in multidimensional spaces to admit other "domains" besides intervals to serve as the argument-domains of many-place functions.

Customarily, every union of (finitely or infinitely many) open intervals is called an "open set." It must be noted that in our terminology all open intervals, and hence all unions thereof, are indefinite sets (or classes). We must also keep in mind that in a union of infinitely many open intervals, the sets of end points can be either definite or indefinite. The usual theorems about open sets can be proved in constructive analysis only for definite end-point sets.

Since every open interval is a union of open rational intervals, i.e., intervals with end points all places of which are rational numbers, we can confine ourselves to unions of open rational intervals.

If a point Z belongs to an open rational interval, then the interval is called a "*neighborhood*" *of* Z. We therefore call all open rational intervals "*neighborhoods*" for short. We use U, V, . . . as variables for neighborhoods. A neighborhood U is always given by the pair consisting of its rational end points R, S: $U = (R \mid S)$. The set of all pairs of rational points is denumerable. Thus we can limit our attention to the unions of the members of sequences of neighborhoods U_1, U_2, ...

153

For $U_i = (R_i | S_i)$ we have two sequences R_* and S_* of rational points.

Thus every union of neighborhoods can be represented by two such sequences of rational points. If R_* and S_* are definite sequences (and hence representable by terms containing no indefinite quantifiers), then let us call the union

$$\mathsf{U}(R_* | S_*)$$

an *open domain.*

Since the intersection of two open intervals is an open interval, the intersection of two open domains $\mathsf{U}U_*$ and $\mathsf{U}V_\dagger$ is likewise an open domain:

$$\mathsf{U}U_* \cap \mathsf{U}V_\dagger = \mathsf{U}(U_* \cap V_\dagger).$$

A definite sequence of open domains arises from a definite double sequence $U_{*\dagger}$ of neighborhoods by virtue of

$$D_n = \mathsf{U}U_{*n}.$$

The union $\mathsf{U}D_*$ is then likewise an open domain. To use the terms of lattice theory, the open domains form a σ-lattice. We must keep in mind, however, that only in the case of definite sequences is the union itself a domain.

We regard the open domains of $\mathfrak{Z}^N$ as a fitting generalization of the open intervals of $\mathfrak{Z}^1$. However, 1-dimensional intervals do have an important property that is lost in this generalization to domains: 1-dimensional intervals are connected.

We call an open domain D "*connected*" if for any two neighborhoods U, V contained in D there is a finite chain U_0, U_1, . . . , U_n of neighborhoods in D such that

(1) $U_0 = U$ and $U_n = V$

(2) U_{i+1} is not disjoint from U_i but is closedly disjoint from U_j for $j < i$ ($i = 0, \ldots, n-1$).

We call connected open domains "*open regions.*"

Closed regions and closed domains are now defined in $\mathfrak{Z}^N$ analogously to closed intervals.

If $D = \bigcup U_*$ is an open domain, then we call a point Z an "*exterior*" point of D if Z has a neighborhood V which is disjoint from D. Every point Ξ of D, on the other hand, is an "*interior*" point: there is a neighborhood U_i such that $\Xi \in U_i \subseteq D$.

154

Thus no interior point is an exterior point. However (except for the case where D is the entire space $\mathfrak{Z}^N$), there are always points that are neither interior nor exterior points of D. They are called *boundary points* of D.

The set of all interior points and all boundary points of D is called the *closure* of D. We refer to the closure of an open domain D for short as a *closed domain*, denoted by $\overline{D}$. The difference $\overline{D} \llcorner D$, i.e., the set of boundary points, is called the *boundary* of D. By a *domain* (without the qualification "open" or "closed") we mean an *open domain together with an arbitrary part of its boundary.*

The set of all interior points of a domain D is also called the *open core* $\underline{D}$ of D.

The complement of $\overline{D}$ is the set of the exterior points of D and hence again an open domain. If D_1 is open and $\overline{D}_2$ closed, then their difference $D_1 \llcorner \overline{D}_2$ will always be open, since it is an intersection of open sets. If D is an open region, then $\overline{D}$ is called a *closed region.*

If a *closed domain* is also *bounded,* i.e., wholly contained in an appropriate neighborhood (this corresponds to the finite intervals in one dimension), the domain is said to be *compact*—and similarly for *regions.*

These concepts become important now as we proceed to define convergence for point sequences in $\mathfrak{Z}^N$. If Ξ^* is a point sequence with members $\Xi^n = \xi_1^n, \ldots, \xi_N^n$, then Ξ^* is convergent if all the real sequences $\xi_1^*, \ldots, \xi_N^*$ are convergent.

We stipulate that

$$\lim \Xi^* = (\lim \xi_1^*, \ldots, \lim \xi^*). \tag{14.2}$$

For $\lim \Xi^* = \Xi$ we also write $\Xi^* \rightarrow \Xi$.

The definition of convergence is equivalent to the following. $\text{Lim}\ \Xi^* = \Xi$ if and only if for every neighborhood U of Ξ there is an n_0 such that for all $n > n_0$, $\Xi^n \in U$; i.e.,

$$\lim \Xi^* = \Xi \Leftrightarrow \bigwedge_{\substack{U \\ \Xi \in U}} \bigvee_{n_0} \bigwedge_{\substack{n \\ n > n_0}} \Xi^n \in U.$$

This definition is said to be *topological,* since it defines the limit solely in terms of neighborhoods.

If, instead of arbitrary intervals containing Ξ, we specially choose cubes with Ξ as the center [the edges of the interval with end points $(\xi_1, \ldots, \xi_N)$ and $(\eta_1, \ldots, \eta_N)$ have the *length* $\eta_i - \xi_i$; *cubes* are intervals whose edges are all of the same length], then we get

$$\lim \Xi^* = \Xi \leftrightarrow \bigwedge_\varepsilon \bigvee_{n_0} \bigwedge_{\substack{n \\ n>n_0}} \bigwedge_{1}^{N}{}_i \, |\xi_i^n - \xi_i| < \varepsilon.$$

The condition

155

$$\bigwedge_{1}^{N}{}_i \, |\xi_i^n - \xi_i| < \varepsilon$$

can be further transformed as follows: we let

$$|\Xi^n - \Xi| = \max_i |\xi_i^n - \xi_i|. \tag{14.3}$$

The definition of the limit thereby takes on the same form as in one dimension:

$$\lim \Xi^* = \Xi \leftrightarrow \bigwedge_\varepsilon \bigvee_{n_0} \bigwedge_{\substack{n \\ n>n_0}} |\Xi^n - \Xi| < \varepsilon.$$

Notice that (14.3) involves taking the difference of two points. That is a particular application of the following operations on points $\Xi = (\xi_1, \ldots, \xi_N)$, $H = (\eta_1, \ldots, \eta_N)$ and real numbers λ:

$$\begin{aligned} \Xi + H &= (\xi_1 + \eta_1, \ldots, \xi_N + \eta_N) \\ \lambda \Xi &= (\lambda \xi_1, \ldots, \lambda \xi_N) \end{aligned} \tag{14.4}$$

$\Xi - H$ is then defined as $\Xi + (-1)H$. In place of (14.3), we can now define as a so-called *norm*

$$|\Xi| = \max_i |\xi_i|. \tag{14.5}$$

Rather than (14.5) one quite often finds another norm, which however is only motivated by geometric applications; viz.,

$$||\Xi|| = \sqrt{\sum_{1}^{N}{}_i |\xi_i|^2}.$$

It is easily seen that

$$1 \leqslant \frac{||\Xi||}{|\Xi|} \leqslant \sqrt{N}$$

where $\Xi \neq 0$ [i.e., $\neq (0, \ldots, 0)$]. Since the quotient is thus bounded, replacing $|\ldots|$ by $||\ldots||$ in the definition of limit would leave convergence and the limit itself unchanged. The two kinds of norm are said to be "*topologically equivalent*."

A norm $||\ldots||$ is generally expected to satisfy

$$\begin{aligned} &(1) \quad ||\Xi|| > 0 \leftrightarrow \Xi \neq 0 \\ &(2) \quad ||\lambda \Xi|| = |\lambda| \, ||\Xi|| \\ &(3) \quad ||\Xi_1 + \Xi_2|| \leqslant ||\Xi_1|| + ||\Xi_2||. \end{aligned} \tag{14.6}$$

These conditions are clearly fulfilled for (14.5), as well as for the occasionally employed

$$|\Xi|_+ = \sum_1^N {}_i |\xi_i|.$$

156 It already follows from (1) and (2) that topologically equivalent norms must always have bounded quotients, for any two equivalent norms $|\ldots|_1$ and $|\ldots|_2$ must satisfy $\bigvee_r \bigwedge_\Xi . |\Xi|_1 \leqslant r \to |\Xi|_2 \leqslant 1$. Since $\left|\frac{r\,\Xi}{|\Xi|_1}\right|_1 = r$, it therefore follows that $\left|\frac{r\,\Xi}{|\Xi|_1}\right|_2 \leqslant 1$; i.e., $\frac{|\Xi|_2}{|\Xi|_1} \leqslant \frac{1}{r}$ for some $r > 0$. Thus we may confine ourselves to the norm (14.5).

The Cauchy convergence criterion can immediately be carried over from a single dimension:

Ξ^* is convergent if and only if

$$\bigwedge_\varepsilon \bigvee_{n_0} \bigwedge_{\substack{n_1, n_2 \\ > n_0}} |\Xi^{n_1} - \Xi^{n_2}| < \varepsilon. \tag{14.7}$$

Furthermore, we again have

THEOREM 14.1 Every definite bounded sequence (i.e., $\bigvee_r \bigwedge_n |\Xi^n| < r$) has a convergent subsequence.

Proof. If Ξ^* is definite and bounded, then all sequences ξ_i^* are definite and bounded. Thus there is a convergent subsequence of ξ_1^*, say $\xi_1^{k_1^*}$. The sequence $\xi_2^{k_1^*}$ then likewise has a convergent subsequence, say $\xi_2^{k_2^*}$, and so on up to a convergent sequence $\xi_N^{k_N^*}$. The point sequence $\Xi^{k_N^*}$ is the desired convergent subsequence of Ξ^*.

As far as the "completeness" of $\mathfrak{Z}^N$ is concerned, the same considerations apply as for $\mathfrak{Z}^1$, of course. $\mathfrak{Z}^N$, like $\mathfrak{Z}^1$, is not a definite set, but rather an indefinite set or a class. The formation of sequences which lead to new limit points is an indefinite procedure that is never completed. But we need not go into that again.

Instead, we come now to the connection between convergent sequences and closed domains, in particular compact ones.

If K is a set of points and Ξ^* is a convergent sequence of points in K, then $\lim \Xi^*$ is called a *limit point* of K.

THEOREM 14.2 Every limit point of a closed domain $\overline{D}$ belongs to $\overline{D}$.

Proof. Let Ξ^* be a convergent sequence of points in $\overline{D}$ such that $\lim \Xi^* = \Xi$. If $\Xi \notin \overline{D}$, then Ξ would be an exterior point of D. There would be an appropriate neighborhood of Ξ which likewise contained only exterior points of D, and which hence was disjoint from $\overline{D}$, in contradiction to $\lim \Xi^* = \Xi$.

The following converse of Theorem 14.2 also holds:

Every boundary point (and every interior point) of an open domain D is a limit point of D.

Proof. If Z is not an exterior point of D, then every neighborhood U of Z includes at least one point of D. If we now consider, say, a sequence W_* of cubes around Z with edges of lengths $\delta_* \rightsquigarrow 0$, then every W_n will contain a point of D and hence—because D is open—a rational point of D. But, since the set of all rational points is denumerable by the principle of choice, we can then infer from $\bigwedge_n \bigvee_R |R - Z| < \delta_n$ that

157

$$\bigvee_{R^*} \bigwedge_n |R^n - Z| < \delta_n .$$

Thus there are even sequences of rational points of D which converge to Z.

As a consequence of Theorems 14.1 and 14.2, we therefore conclude that a domain K (i.e., an open domain plus some part of its boundary) is compact if and only if for every definite sequence of points in K, K contains a convergent subsequence together with its limit.

Another characterization of compact domains is possible with the help of neighborhoods as coverings. What is involved here is the following. Let $\bar{J}$ be a closed interval and U_* a definite sequence of rational open intervals such that $\bar{J} \subseteq \bigcup U_*$. The sequence U_* is then said to be a definite *neighborhood covering* of $\bar{J}$. We want to show that finitely many members of U_* are already enough to cover the closed interval $\bar{J}$, i.e., that $\bar{J} \subseteq \bigcup_1^n{}_i U_i$ is true of some n.

As a preliminary to the proof, we take up the one-dimensional case. We will show that from $[\alpha | \beta] \subseteq \bigcup U_*$ it always follows that $[\alpha | \beta] \subseteq \bigcup_1^n{}_i U_i$ for some n. First, however, let us prove a weaker claim, viz., that for some n, the union $\bigcup_1^n{}_i U_i$ contains all the *rational* numbers in $[\alpha | \beta]$. With the help of analytic induction, the proof is trivial. If finitely many members of U_* cover the set of rational numbers in $[\alpha | \xi]$, then that (definite!) property of ξ is (1) left-hereditary, and (2) carries over from ξ to some $\xi + \delta$ or other, since ξ is contained in some U_i.

Thus all we have yet to do is to establish the following auxiliary theorem:

THEOREM 14.3 If a finite union of rational intervals $U_1 \cup \ldots \cup U_n$ includes all the rational numbers in $[\alpha | \beta]$, then the system $U_1, \ldots, U_n$ is a covering of the interval $[\alpha | \beta]$.

Proof. Let $\xi \in [\alpha | \beta]$ be irrational, and let $\xi = \lim r_*$, where $r_i \in [\alpha | \beta]$ $(i = 1, 2, \ldots)$. Every r_i lies in one of the neighborhoods $U_1, \ldots, U_n$. Therefore at least one of these neighborhoods contains a subsequence of r_*, and hence also ξ. (Since ξ is irrational, it cannot be an end point of the neighborhood.)

158 That takes care of the one-dimensional case. The general claim follows by arithmetical induction.

THEOREM 14.4 If $\bar{J} \subseteq \bigcup U_*$ is true in $\mathfrak{Z}^N$ of a closed interval $\bar{J}$ and a definite sequence of neighborhoods U_*, then it is also the case that

$$\bar{J} \subseteq U_1 \cup \ldots \cup U_n$$

for some n.

Proof. Assume that the theorem has been proved for N. For $\mathfrak{Z}^{N+1}$ let

$$\bar{J} = [\alpha_1, \ldots, \alpha_N, \alpha_{N+1} | \beta_1, \ldots, \beta_N, \beta_{N+1}].$$

By the hypothesis of induction for N, the interval

$$[\alpha_1, \ldots, \alpha_N, \xi | \beta_1, \ldots, \beta_N, \xi]$$

is covered by finitely many neighborhoods of U_* for every $\xi \in [\alpha_{N+1} | \beta_{N+1}]$. Those neighborhoods thus also cover the interval

$$[\alpha_1, \ldots, \alpha_N, r_1 | \beta_1, \ldots, \beta_N, r_2],$$

where r_1, r_2 are rational numbers such that $r_1 < \xi < r_2$. By the basis case $N = 1$, which we proved above, we already know that a finite number of the intervals $(r_1 | r_2)$ cover the interval $[\alpha_{N+1} | \beta_{N+1}]$. Altogether, then, we get finitely many U_is which cover $\bar{J}$.

Theorem 14.4 is now easily extended to compact domains.

THEOREM 14.5 (COVERING THEOREM). If a compact domain K is covered by a definite sequence of neighborhoods, then it is already covered by finitely many members of the sequence.

Proof. $\mathfrak{Z}^N \llcorner K$ is open and hence is the union of a definite sequence V_* of neighborhoods. From $K \subseteq \bigcup U_*$ it accordingly follows that

$$\mathfrak{Z}^N = \bigcup (U_* \cup V_*).$$

Now, if K is contained in the closed interval $\bar{J}$, then by Theorem 14.4 a finite number of these neighborhoods are enough (the sequences U_* and V_* being definite) to cover $\bar{J}$. Since all V_is are disjoint from K, finitely many of the U_is also suffice to cover $K \subseteq \bar{J}$.

The converse of Theorem 14.5 is easily established, so that we get

THEOREM 14.6 A domain K is compact (i.e., closed and bounded) just in case every definite covering by means of neighborhoods contains a finite covering.

Proof. (1) If every covering of K contains a finite covering, then (since K is coverable by, say, all neighborhoods) K must at any rate be finitely coverable, i.e., bounded.

(2) If every covering of K contains a finite covering, then K has to be closed. For let U be an open interval containing K, and let W_* be a sequence of closed cubes with the interior point Ξ as their sole common point. Then, if Ξ is not in K, the sequence $U \llcorner \overline{W}_*$ provides a covering of K. Since this covering contains a finite covering, Ξ is an exterior point of K. 159

§15. Continuous and Differentiable Functions

Let ϕ be a function in $\mathfrak{Z}^N$ with the domain K, and let Ξ^0 be an interior point of K.

We begin by defining the concept of *limit*, i.e., what we will mean by $\lim\limits_{\Xi \rightharpoonup \Xi^0} \phi\, \Xi$ ($\lim\limits_{\Xi^0} \phi$ for short). For this purpose we have no need of the value $\phi\, \Xi^0$.

DEFINITION 15.1

$$\lim_{\Xi \rightharpoonup \Xi^0} \phi\, \Xi = \eta \leftrightharpoons \bigwedge_\varepsilon \bigvee_\delta \bigwedge_{\substack{\Xi \\ \Xi \neq \Xi^0}} . \bigwedge_{i=1}^{N} |\xi_i - \xi_i^0| < \delta \rightarrow |\phi\, \Xi - \eta| < \varepsilon.$$

Using the norm

$$|\Xi - \Xi^0| = \max_i |\xi_i - \xi_i^0|,$$

we obtain

$$\lim_{\Xi \rightharpoonup \Xi^0} \phi\, \Xi = \eta \leftrightarrow \bigwedge_\varepsilon \bigvee_\delta \bigwedge_\Xi . \, 0 < |\Xi - \Xi^0| < \delta \rightarrow |\phi\, \Xi - \eta| < \varepsilon. \tag{15.1}$$

If instead of the norm $|\ldots|$ we used a topologically equivalent norm in (15.1), it would not alter the limit concept. It follows as a matter of course from this definition (just as for $N = 1$) that there is at most one η such that $\lim\limits_{\Xi^0} \phi = \eta$.

Thus, if $\lim\limits_{\Xi^0} \phi$ exists, it can be given as the *"iterated" limit* of one-place functions; e.g., if $N = 2$,

$$\lim_{(\xi_0, \eta_0)} \phi = \lim_{\xi \rightharpoonup \xi_0} \lim_{\eta \rightharpoonup \eta_0} \phi(\xi, \eta).$$

However, the existence of both *"partial" limits* $\lim\limits_{\eta \rightharpoonup \eta_0} \phi(\xi_0, \eta)$ and $\lim\limits_{\xi \rightharpoonup \xi_0} \phi(\xi, \eta_0)$ does not necessarily mean that the *"total" limit*

$\lim\limits_{(\xi_0,\eta_0)} \phi$ exists; witness $\lim\limits_{(0,0)} \dfrac{\xi\eta}{\xi^2 + \eta^2}$.

For any sequence Ξ^* that tends to Ξ^0, it immediately follows from $\lim\limits_{\Xi^0} \phi = \eta$ that

$$\lim \phi \Xi^* = \eta.$$

DEFINITION 15.2 ϕ is *continuous* at Ξ^0 if

$$\lim_{\Xi \to \Xi^0} \phi \Xi = \phi \Xi^0.$$

160 Similarly, ϕ is said to be continuous in K if ϕ is continuous at every $\Xi^0 \in K$. Unless K is open, no neighborhood of Ξ^0 need belong to K. In that case, the limit of ϕ at Ξ^0 can only be defined with the help of those points of K that lie in a neighborhood of Ξ^0.

If ϕ is continuous in a neighborhood of Ξ^0, we say for short that ϕ is continuous *around* Ξ^0. Similarly we say, "ϕ is applicable *around* Ξ^0" (i.e., applicable in a neighborhood of Ξ^0); "ϕ is differentiable *around* Ξ^0" (i.e., differentiable in a neighborhood of Ξ^0); etc.

Sums and products of continuous functions are themselves continuous. Concerning the *composition* of many-place functions, let ϕ be an N-place function, and let there be N functions $\psi_1, \ldots,$ ψ_N with the same domain (in $\mathfrak{Z}^m$, let us say). If the functions ψ_i are all continuous at Ξ^0, and if ϕ is continuous at $(\psi_1\Xi^0, \ldots,$ $\psi_N\Xi^0)$, then $\phi \urcorner (\psi_1, \ldots, \psi_N)$ is also continuous at Ξ^0.

We of course stipulate that

$$\phi \urcorner (\psi_1, \ldots, \psi_N) \urcorner \Xi = \phi \urcorner (\psi_1 \Xi, \ldots, \psi_N \Xi).$$

In particular, if an N-place function ψ and a one-place function ϕ are both continuous, then $\phi \urcorner \psi$ will be continuous.

As in one dimension, a function which is continuous in a closed domain is always uniquely determined by its rational subfunction.

Special importance again attaches to *uniform continuity* in an argument domain K.

DEFINITION 15.3 ϕ is *uniformly continuous* in K if

$$\bigwedge_\varepsilon \bigvee_\delta \bigwedge_{\substack{\Xi^1, \Xi^2 \\ K}} . \; |\Xi^1 - \Xi^2| < \delta \to |\phi\Xi^1 - \phi\Xi^2| < \varepsilon. \qquad (15.2)$$

THEOREM 15.1 If a function ϕ is continuous in a compact domain $\overline{D}$, then ϕ is uniformly continuous in $\overline{D}$.

Proof. It is sufficient to show for every ε that

$$\bigvee_\delta \bigwedge_{\substack{R^1, R^2 \\ D}} . \; |R^1 - R^2| < \delta \to |\phi R^1 - \phi R^2| < \varepsilon. \qquad (15.3)$$

Since ϕ is to be continuous on the boundary of D too, we shall then have for arbitrary points Ξ^1, Ξ^2 of $\overline{D}$

$$\bigvee_{\delta} \bigwedge_{\substack{\Xi^1, \Xi^2 \\ D}} . \; |\Xi^1 - \Xi^2| < \delta \rightarrow |\phi\Xi^1 - \phi\Xi^2| \leqslant \varepsilon .$$

If, in contradiction to (15.3), there were an ε such that

$$\bigwedge_{\delta} \bigvee_{\substack{R^1, R^2 \\ D}} . \; |R^1 - R^2| < \delta \wedge |\phi R^1 - \phi R^2| \geqslant \varepsilon .,$$

then we would get for some null sequence δ_* 161

$$\bigwedge_n \bigvee_{R^1, R^2} . \; |R^1 - R^2| < \delta_n \wedge |\phi R^1 - \phi R^2| \geqslant \varepsilon .$$

Since the rational points are denumerable, it would follow that there were two sequences of rational points R^1_*, R^2_* such that

$$\bigwedge_n . \; |R^1_n - R^2_n| < \delta_n \wedge |\phi R^1_n - \phi R^2_n| \geqslant \varepsilon .$$

If $R^1_{k_*}$ is a convergent subsequence of R^1_* with the limit Ξ, then $R^2_{k_*}$ also converges to Ξ. From the continuity of ϕ at Ξ it follows that

$$\lim \phi R^1_{k_*} = \lim \phi R^2_{k_*} ,$$

in contradiction to

$$\bigwedge_n |\phi R^1_{k_*} - \phi R^2_{k_*}| \geqslant \varepsilon .$$

That proves the theorem.

It is noteworthy that the path so often taken of proving this theorem with the aid of the covering theorem falls short of the goal here. If we take as our point of departure the continuity of ϕ at every point of $\overline{D}$, then to be sure we do know that for every ε

$$\bigwedge_{\Xi^0} \bigvee_{\delta} \bigwedge_{\Xi} . \; |\Xi - \Xi^0| < \delta \rightarrow |\phi\Xi - \phi\Xi^0| < \varepsilon .$$

We can then define for every Ξ^0 the cubic neighborhood U_{Ξ^0} with edges of the length 2δ, where δ is determined by the definite property

$$\bigwedge_R . \; |R - \Xi^0| < \delta \rightarrow |\phi R - \phi\Xi^0| < \varepsilon .$$

But the union of all U_{Ξ^0}s yields only an indefinite covering of $\overline{D}$. If we limit ourselves to those U_{Ξ^0}s with rational Ξ^0, then in general we get no covering at all. Thus the inference of a covering involving finitely many U_{Ξ^0}s (which would immediately establish uniform continuity) fails. That is why we chose the indirect proof given above.

An immediate corollary of Theorem 15.1 is the *boundedness* of any function that is continuous in a compact domain $\overline{D}$. For suppose it holds of δ that

$$|\Xi^1 - \Xi^2| < \delta \rightarrow |\phi\Xi^1 - \phi\Xi^2| < 1.$$

Then all we have to do is to cover $\overline{D}$ with a finite number of cubes with edges of length δ (which is trivially possible, $\overline{D}$ being bounded), and we get (where n is the number of cubes required)

$$|\phi\Xi^1 - \phi\Xi^2| < n$$

for all Ξ^1, Ξ^2 in $\overline{D}$.

162 The simplest examples of continuous functions are again the *elementary* functions. As a convenient means of representing elementary functions, we utilize the "*projection*" functions I_i given by

$$I_i(\xi_1, \ldots, \xi_N) = \xi_i \quad (i = 1, \ldots, N).$$

All terms built up out of I_is $(i = 1, \ldots, N)$ and constants by means of addition, multiplication, and division denote *rational* functions; e.g.,

$$\frac{I_1^2 + I_2^2}{I_1 \cdot I_2}(\xi, \eta) = \frac{\xi^2 + \eta^2}{\xi \cdot \eta}.$$

The simplest rational functions are the *linear* functions L. It will be convenient here to represent the function value $L(\Xi) = \alpha_1\xi_1 + \ldots + \alpha_N\xi_N$ as the *matrix product* of a row

$$(\alpha_1, \ldots, \alpha_N)$$

and a column:

$$L(\Xi) = (\alpha_1, \ldots, \alpha_N)\begin{pmatrix}\xi_1 \\ \vdots \\ \xi_N\end{pmatrix}.$$

In order to define the class of all N-place elementary functions, we begin with the constants ± 1, utilizing in addition to the projection functions $I_1, \ldots, I_N$ the basic elementary (1-place) functions

exp, ln, sin, cos, tan, arc sin, arc cos, arc tan.

By addition, multiplication, and composition these give rise to all elementary functions.

Now let us turn our attention from continuity to differentiability. In $\mathfrak{Z}^1$ we specified the *tangential function* $\phi\xi_0 + q(I - \xi_0)$ (if any) corresponding to a function ϕ at ξ_0 by means of

$$\bigwedge_\varepsilon \bigvee_\delta \bigwedge_\xi . \; |\xi - \xi_0| < \delta \rightarrow |\phi\xi - \phi\xi_0 - q(\xi - \xi_0)| < \varepsilon|\xi - \xi_0|.$$

With the help of the $\mathfrak{Z}^N$ norm, this definition can immediately be extended to the present situation.

A function ϕ is (*totally*) *differentiable* at Ξ^0 if there is a linear function L such that

$$\bigwedge_\varepsilon \bigvee_\delta \bigwedge_\Xi . |\Xi - \Xi^0| < \delta \rightarrow$$
$$|\phi\,\Xi - \phi\,\Xi^0 - L(\Xi - \Xi^0)| < \varepsilon\,|\Xi - \Xi^0|. \qquad (15.4)$$

In place of the norm $|\ldots|$, this condition could just as well be formulated in terms of any topologically equivalent norm. If one wishes to avoid recourse to norms altogether, one can stipulate equivalently to (15.4) that

$$\bigwedge_\varepsilon \bigvee_\delta \bigwedge_\Xi . \bigwedge_{i\,1}^{N} |\xi_i - \xi_i^0| < \delta \rightarrow$$
$$\bigvee_{i\,1}^{N} |\phi\,\Xi - \phi\,\Xi^0 - L(\Xi - \Xi^0)| < \varepsilon\,|\xi_i - \xi_i^0|.$$

The (total) differentiability of ϕ at $(\xi_1^0, \ldots, \xi_N^0)$ entails the differentiability of the following one-place functions: 163

$$\phi(\xi_1^0, \ldots, \xi_{i-1}^0, I, \xi_{i+1}^0, \ldots, \xi_N^0) \quad \text{at} \quad \xi_i^0 \quad (i = 1, \ldots, N).$$

The function

$$\daleth_{\xi_1, \ldots, \xi_N} . \phi(\xi_1, \ldots, \xi_{i-1}, I, \xi_{i+1}, \ldots, \xi_N)'\daleth\xi.$$

is called the ith *partial derivative* of ϕ, denoted by $\phi'_{(i)}$ or $\mathcal{D}_i\phi$. For $\phi'_{(i)}\,\Xi$ we write more perspicuously ${}_\Xi\mathcal{D}_i\phi$ rather than $\mathcal{D}_i\phi\,\Xi$. From (15.4) it immediately follows that

$$L(\Xi - \Xi^0) = \sum_{i\,1}^{N} \phi'_{(i)}\,\Xi^0(\xi_i - \xi_i^0). \qquad (15.5)$$

Thus, in particular, L is uniquely determined.

If ϕ is *partially differentiable* with respect to all its argument places, then L is definable by (15.5). But in that case (15.4) need not hold.

We prove, however,

THEOREM 15.2 If the partial derivatives $\phi'_{(i)}$ $(i = 1, \ldots, N)$ exist in a neighborhood of Ξ^0, and if they are continuous at Ξ^0, then ϕ is differentiable at Ξ^0.

Proof. We have

$$\phi\,\Xi - \phi\,\Xi^0 = \sum_{i\,1}^{N} . \phi(\xi_1^0, \ldots, \xi_{i-1}^0, \xi_i, \ldots, \xi_N)$$
$$- \phi(\xi_1^0, \ldots, \xi_i^0, \xi_{i+1}, \ldots, \xi_N). \qquad (15.6)$$

Furthermore, for every ε there is a δ such that $|\Xi - \Xi^0| < \delta$ entails

$$\phi'_{(i)}\,\Xi^0 - \varepsilon < \phi'_{(i)}\,\Xi < \phi'_{(i)}\,\Xi^0 + \varepsilon.$$

Thus, by the bound-theorem, the h_i determined by

$$\phi(\xi_1^0, \ldots, \xi_{i-1}^0, \xi_i, \ldots, \xi_N) - \phi(\xi_1^0, \ldots, \xi_i^0, \xi_{i+1}, \ldots, \xi_N) = (\phi'_{(i)}\,\Xi^0 + h_i)(\xi_i - \xi_i^0)$$

will be such that $|h_i| < \varepsilon$.

Summation gives us

$$\phi\,\Xi - \phi\,\Xi^0 = \sum_1^N{}_i (\phi'_{(i)}\,\Xi^0 + h_i)(\xi_i - \xi_i^0). \tag{15.7}$$

It follows that

$$|\phi\,\Xi - \phi\,\Xi^0 - \sum_1^N{}_i \phi'_{(i)}\,\Xi^0(\xi_i - \xi_i^0)| < \varepsilon \sum_1^N{}_i |\xi_i - \xi_i^0|,$$

Q.E.D.

164 As with continuous differentiability in one dimension, (15.4) can be strengthened under the conditions of Theorem 15.2 to

$$\left|\phi\,\Xi^1 - \phi\,\Xi^2 - \sum_1^N{}_i \phi'_{(i)}\,\Xi^0(\xi_i^1 - \xi_i^2)\right| < \varepsilon\,|\Xi^1 - \Xi^2|$$

for all Ξ^1, Ξ^2 with sufficiently small $|\Xi^1 - \Xi^0|$ and $|\Xi^2 - \Xi^0|$. For from

$$\phi'_{(i)}\,\Xi^0 - \varepsilon \leqslant \phi'_{(i)} \leqslant \phi'_{(i)}\,\Xi^0 + \varepsilon$$

it follows by the bound-theorem for $\phi\,\Xi^1 - \phi\,\Xi^2$, which is to be broken down as in (15.6), that

$$\sum_1^N{}_i (\phi'_{(i)}\,\Xi^0 - \varepsilon)(\xi_i^1 - \xi_i^2) \leqslant \phi\,\Xi^1 - \phi\,\Xi^2 \leqslant \sum_1^N{}_i (\phi'_{(i)}\,\Xi^0 + \varepsilon)(\xi_i^1 - \xi_i^2),$$

and hence,

$$|\phi\,\Xi^1 - \phi\,\Xi^2 - \sum_1^N{}_i \phi'_{(i)}\,\Xi^0(\xi_i^1 - \xi_i^2)| < \varepsilon \sum_1^N{}_i |\xi_i^1 - \xi_i^2|.$$

Partial continuous differentiability thus entails "free" total differentiability. A partially continuously differentiable function is therefore characterized more briefly as "*continuously differentiable*." If the partial derivatives of ϕ are themselves continuously differentiable, then ϕ is said to be *twice continuously differentiable* or to exhibit *double continuous differentiability*. The property of being continuously differentiable n times, i.e., up to the nth order, or *n-fold continuous differentiability*, is characterized similarly. Finally, if a function ϕ is continuously differentiable n times for all $n = 1, 2, \ldots\ldots$, then ϕ is said to be *dif-*

ferentiable up to order ∞, or to possess ∞*-fold differentiability.* The higher-order partial derivatives are denoted as follows:

$$\mathscr{D}_i\,\mathscr{D}_j\,\phi = \mathscr{D}^2_{ij}\,\phi = \phi''_{(ji)}$$

$$\mathscr{D}_i\,\mathscr{D}_j\,\mathscr{D}_k\phi = \mathscr{D}^3_{ijk}\,\phi = \phi'''_{(kji)}.$$

If, rather than ϕ, we are given a term $T(\xi_1, \ldots, \xi_N)$ representing ϕ, then in place of $\mathscr{D}_i\phi(\xi_1^0, \ldots, \xi_N^0)$, i.e., for the value $_{\xi_1^0, \ldots, \xi_N^0}\mathscr{D}_i\phi$ of the function $\mathscr{D}_i\,\phi$ for the arguments $\xi_1^0, \ldots, \xi_N^0$, we can write

$$_{\xi_i^0}\mathscr{D}_{\xi_i}T(\xi_1^0, \ldots, \xi_{i-1}^0, \xi_i, \xi_{i+1}^0, \ldots, \xi_N^0).$$

Note that $\mathscr{D}_{\xi_i}T(\xi_1, \ldots, \xi_N)$ is a 1-place function, while $\mathscr{D}_i\,\phi$ is an N-place function. Similarly,

$$\mathscr{D}^2_{\xi_i\xi_j}\,T(\xi_1, \ldots, \xi_N),$$

where $i \neq j$, is a 2-place function.

In cases of multiple continuous differentiability, the order of the subscripts $i, j, k, \ldots$ is immaterial, as we shall see in §17.

We turn now to the task of extending the most important theorems about differentiation from $\mathfrak{Z}^1$ to $\mathfrak{Z}^N$. The elementary rules of differentiation, such as *165*

$$(\phi + \psi)' = \phi' + \psi'$$

$$(\phi \cdot \psi)' = \phi \cdot \psi' + \phi' \cdot \psi,$$

apply as a matter of course to partial derivatives too. Generalizing the chain rule entails differentiating the function $\phi(\psi_1, \ldots, \psi_M)$ for an M-place function ϕ.

THEOREM 15.3 If the functions $\psi_1, \ldots, \psi_M$ are differentiable at Ξ^0 and the M-place function ϕ is differentiable at $\psi_1\Xi^0, \ldots, \psi_M\Xi^0$, then $\phi(\psi_1, \ldots, \psi_M)$ is differentiable at Ξ^0, and its ith partial derivative there is

$$\phi(\psi_1, \ldots, \psi_M)'_{(i)} = \sum_1^M{}_j\,\phi'_{(j)}(\psi_1, \ldots, \psi_M)\cdot\psi'_{j(i)}.$$

Proof. Since we are only concerned here with partial derivatives in $\mathfrak{Z}^N$, we can choose N to be 1 without loss of generality. In that case, $\psi_1, \ldots, \psi_M$ are one-place functions differentiable, let us say, at ξ_0; and ϕ is differentiable at

$$\psi_1\xi_0, \ldots, \psi_M\,\xi_0.$$

We have to prove that (at ξ_0)

$$\phi(\psi_1, \ldots, \psi_M)' = \sum_{1}^{M}{}_j \phi'_{(j)}(\psi_1, \ldots, \psi_M) \cdot \psi_j'.$$

Since ϕ is differentiable, we get for every ε and for sufficiently small $|\xi - \xi_0|$

$$\Big|\phi(\psi_1\xi, \ldots, \psi_M \xi) - \phi(\psi_1\xi_0, \ldots, \psi_M \xi_0) - \sum_{1}^{M}{}_j \phi'_{(j)} (\psi_1\xi_0, \ldots, \psi_M \xi_0)(\psi_j \xi - \psi_j \xi_0)\Big| < \varepsilon \max_j |\psi_j \xi - \psi_j \xi_0|.$$

But for sufficiently small $|\xi - \xi_0|$ we also have

$$\psi_j \xi - \psi_j \xi_0 = (\psi'_{(j)} \xi_0 + h_j)(\xi - \xi_0),$$

where $|\psi'_{(j)} \xi_0 + h_j| \leqslant c$, and hence

$$|\psi_j \xi - \psi_j \xi_0| \leqslant c\, |\xi - \xi_0|.$$

Taking everything together, we get

$$|\phi(\psi_1\xi, \ldots, \psi_M \xi) - \phi(\psi_1\xi_0, \ldots, \psi_M \xi_0) - \sum_{1}^{M}{}_j \phi'_{(j)} (\psi_1\xi_0, \ldots, \psi_M \xi_0)(\psi_j \xi - \psi_j \xi_0)| < \varepsilon c\, |\xi - \xi_0|.$$

166 Thus it follows for $\xi \neq \xi_0$ that

$$\lim_{\xi \to \xi_0} \frac{\phi(\psi_1\xi, \ldots, \psi_M \xi) - \phi(\psi_1\xi_0, \ldots, \psi_M \xi_0)}{\xi - \xi_0}$$

$$= \lim_{\xi \to \xi_0} \sum_{1}^{M}{}_j \phi'_{(j)} (\psi_1\xi_0, \ldots, \psi_M \xi_0) \frac{\psi_j \xi - \psi_j \xi_0}{\xi - \xi_0}$$

$$= \sum_{1}^{M}{}_j \phi'_{(j)} (\psi_1\xi_0, \ldots, \psi_M \xi_0)\psi_j{}'\xi_0 .$$

If the functions ϕ and $\psi_1, \ldots, \psi_M$ are continuously differentiable up to the order n, then the chain rule guarantees the n-fold continuous differentiability of

$$\phi(\psi_1, \ldots, \psi_M).$$

The chain rule also gives us the derivatives of inverse functions—again just as in one dimension. Let ϕ be an $(N + 1)$-place function and let

$$\eta_0 = \phi(\xi_0, \zeta_1{}^0, \ldots, \zeta_N{}^0),$$

which we abbreviate as

$$\eta_0 = \phi(\xi_0, Z^0).$$

If ϕ is applicable to all ξ, Z^0 such that $|\xi - \xi_0| < \delta$, and if $\phi'_{(1)}(\xi_0, Z^0) \neq 0$, then there is a function ψ which is applicable to all η, Z^0 with sufficiently small $|\eta - \eta_0|$, and which satisfies

$$\eta = \phi(\xi, Z^0) \leftrightarrow \xi = \psi(\eta, Z^0).$$

Furthermore, ψ is thereby uniquely determined, and it is differentiable with respect to the first argument place as follows:

$$\psi'_{(1)}(\eta_0, Z^0) = \frac{1}{\phi'_{(1)}(\xi_0, Z^0)}.$$

That all follows directly from the theorems already proved for one-place functions (Theorems 8.8, 11.6). Now, however, let us presuppose that ϕ is continuously differentiable *around* ξ_0, Z^0. Then, since $\phi'_{(1)}(\xi_0, Z^0) \neq 0$, it will also be the case that $\phi'_{(1)} \neq 0$ in a neighborhood of ξ_0, Z^0. There is a uniquely determined function ψ which is applicable around η_0, Z^0 and satisfies

$$\eta = \phi(\xi, Z) \leftrightarrow \xi = \psi(\eta, Z).$$

We can now show that in addition ψ is continuously differentiable. In view of

$$\eta = \phi[\psi(\eta, Z), Z],$$

the partial derivatives can then be computed from

$$0 = \phi'_{(1)}[\psi(\eta, Z), Z] \cdot \psi'_{(i)}(\eta, Z) + \phi'_{(i)}[\psi(\eta, Z), Z],$$

since

$$\mathcal{D}_{\zeta_i} \zeta_j = \begin{cases} 0 & \text{if } i \neq j \\ 1 & \text{if } i = j. \end{cases}$$

We first prove that ψ is continuous. For every δ there is an ε such that $|\phi(\xi, Z^0) - \phi(\xi_0, Z^0)| < \delta$ for $|\xi - \xi_0| < \varepsilon$. Thus it is also the case for sufficiently small $|Z - Z^0|$ that 167

$$|\phi(\xi, Z) - \phi(\xi_0, Z^0)| < \delta.$$

Hence, in view of the one-to-one correspondence (when Z is held constant) between $\xi = \psi(\eta, Z)$ and $\eta = \phi(\xi, Z)$, we in fact get, for sufficiently small $|Z - Z^0|$,

$$|\xi - \xi_0| < \varepsilon \leftrightarrow |\eta - \eta_0| < \delta;$$

i.e., ψ is continuous. ψ is furthermore continuously differentiable, for by (15.7) it follows from $\phi(\xi_0, Z) = \eta_0$ and $\phi[\psi(\eta_0, Z), Z] = 0$—if we set $\xi = \psi(\eta_0, Z)$—that

$$0 = \phi(\xi, Z) - \phi(\xi_0, Z^0)$$
$$= (\xi - \xi_0)[\phi'_{(1)}(\xi_0, Z^0) + h_1] + \sum_{2}^{N+1}{}_i (\zeta_{i-1} - \zeta^0_{i-1})[\phi'_{(i)}(\xi_0, Z^0) + h_i],$$

where $h_i \rightsquigarrow 0$ as $\xi \rightsquigarrow \xi_0$ and $Z \rightsquigarrow Z^0$.

In fact, $\xi \rightsquigarrow \xi_0$ already follows from $Z \rightsquigarrow Z^0$, since ψ is continuous at η_0, Z^0. Thus we get

$$\left| \xi - \xi_0 + \sum_{2}^{N+1}{}_i (\zeta_{i-1} - \zeta^0_{i-1}) \frac{\phi'_{(i)}(\xi_0, Z^0) + h_i}{\phi'_{(1)}(\xi_0, Z) + h_1} \right| = 0$$

and

$$\left| \xi - \xi_0 + \sum_{2}^{N+1}{}_i (\zeta_{i-1} - \zeta^0_{i-1}) \frac{\phi'_{(i)}(\xi_0, Z^0)}{\phi'_{(1)}(\xi_0, Z^0)} \right| < \varepsilon \sum_{1}^{N}{}_i |\zeta_i - \zeta_i^0|,$$

where $\varepsilon \rightsquigarrow 0$ as $Z \rightsquigarrow Z^0$.

All in all, we have thereby proved

THEOREM 15.4 If ϕ is continuously differentiable in a neighborhood of ξ_0, Z^0, and if $\phi'_{(1)} \neq 0$ there, then exactly one function ψ is continuously differentiable in that neighborhood and satisfies

$$\xi = \psi(\eta, Z) \leftrightarrow \eta = \phi(\xi, Z).$$

§16. Mappings

The many-place functions are distinguished from the one-place functions by the fact that their arguments are not numbers but rather systems of numbers or "points" in $\mathfrak{Z}^N$.

We shall now consider systems of functions. Suppose that $\phi_1, \ldots, \phi_M$ are all functions which are applicable in an interval J of $\mathfrak{Z}^N$. Where $\Xi \in J$, the system consisting of the values $\phi_1 \Xi, \ldots, \phi_M \Xi$ of the functions is an M-place system and thus a point in $\mathfrak{Z}^M$.

168 One could call the system $(\phi_1, \ldots, \phi_M)$ a function with points in $\mathfrak{Z}^N$ as arguments and points in $\mathfrak{Z}^M$ as values. For the sake of clarity, however, we shall speak of functions in this book only when the values are real numbers. If the values are systems of numbers, then we shall speak of *mappings*.

The arguments of a system of functions may accordingly also be called "*original points*," and the systems of values, "*image points*." We use the capital letters $\Phi, \Psi, \ldots$ for mappings.

For the mapping $\Phi = (\phi_1, \ldots, \phi_M)$ and the original point Ξ we then write

$$\Phi\,\Xi = (\phi_1 \Xi, \ldots, \phi_M \Xi).$$

$\Phi\,\Xi$ or $\Phi \urcorner \Xi$ is the image point of Ξ under the mapping Φ.

Specifications concerning the domain and the range of a mapping Φ are customarily made in the following terms:

Φ is called a mapping *from* D if its domain is $\subseteq D$.
Φ is called a mapping *of* D if its domain $= D$.
Φ is called a mapping *into* R if its range is $\subseteq R$.
Φ is called a mapping *onto* R if its range $= R$.

The process of generalization that leads from one-place functions via many-place functions to mappings can of course be carried further. For example, we may consider many-place mappings $\Phi(\Xi^1, \ldots, \Xi^n)$; but only seldom shall we have occasion to make use of them.

Among the various types of mappings, certain ones can—as in the case of functions—be singled out as *linear mappings*.

Φ is *linear* if

$$\Phi(\alpha \Xi + \beta H) = \alpha \Phi \Xi + \beta \Phi H$$

holds for all numbers α, β and all arguments Ξ, H. This definition makes use of the fact that the *linear operations* (addition and multiplication by real numbers) are defined both for the original points and for the image points [cf. (14.4)].

With reference to the "*unit points*" of $\mathfrak{Z}^N$, viz.,

$$E^1 = (1, 0, \ldots, 0), \ \ldots, \ E^N = (0, \ldots, 0, 1),$$

it will of course be true of $\Xi = (\xi_1, \ldots, \xi_N)$ that

$$\Xi = \sum_{1}^{N}{}_i \, \xi_i E^i.$$

For a linear mapping Φ it follows from this that

$$\Phi \Xi = \sum_{1}^{N}{}_i \, \xi_i \Phi E^i.$$

The linear mapping Φ is thus uniquely determined by the system of image points $\Phi E^1, \ldots, \Phi E^N$. Each of these image points is itself an M-place system. If we now write 169

$$\Phi E^i = \begin{pmatrix} \alpha_1^i \\ \vdots \\ \alpha_M^i \end{pmatrix} \qquad (i = 1, \ldots, N),$$

we get a rectangular scheme of $M \cdot N$ numbers, a *matrix*:

$$A = \begin{pmatrix} \alpha_1^1 & \dots & \alpha_1^N \\ \vdots & & \vdots \\ \alpha_M^1 & \dots & \alpha_M^N \end{pmatrix}.$$

This matrix uniquely determines the linear mapping Φ. For $\Phi\,\Xi$ we can write, in accordance with the calculating conventions for matrices,

$$\begin{pmatrix} \sum_i \alpha_1^i \xi_i \\ \vdots \\ \sum_i \alpha_M^i \xi_i \end{pmatrix} = \begin{pmatrix} \alpha_1^1 & \dots & \alpha_1^N \\ \vdots & & \vdots \\ \alpha_M^1 & \dots & \alpha_M^N \end{pmatrix} \begin{pmatrix} \xi_1 \\ \vdots \\ \xi_N \end{pmatrix} = A\,\Xi.$$

The composition of two linear mappings Φ and Ψ, the matrices of which are A and B, yields

$$\Phi\Psi\Xi = AB\Xi.$$

For

$$\Xi = \begin{pmatrix} \xi_1 \\ \vdots \\ \xi_N \end{pmatrix},$$

B here must have N columns and the same number of rows as A has columns. As usual, the rows of A are combined with the columns of B. If Φ is a linear mapping of $\mathfrak{Z}^N$ into itself (i.e., $M = N$), then

$$\Phi\Xi = A\Xi,$$

and A will be a *square* matrix of N rows. The system of equations

$$H = A\Xi$$

is soluble for Ξ just in case $|A|$, the *determinant* of A, is not zero. There will then be exactly one matrix $B = A^{-1}$ such that

$$\Xi = BH \leftrightarrow H = A\Xi.$$

170 Linear mappings are a special kind of *rational mapping* $\Phi = (\phi_1, \dots, \phi_M)$, where all of $\phi_1, \dots, \phi_M$ are rational functions. Rational mappings are in turn a special kind of *elementary mapping*, for which $\phi_1, \dots, \phi_M$ are elementary functions.

A mapping is *continuous* at Ξ^0 if

$$\wedge_\varepsilon \vee_\delta \wedge_\Xi \,.\, |\Xi - \Xi^0| < \delta \to |\Phi\Xi - \Phi\Xi^0| < \varepsilon., \qquad (16.1)$$

where

$$|\Phi\Xi - \Phi\Xi^0| = \max_j |\phi_j\Xi - \phi_j\Xi^0|$$

is the norm in $\mathfrak{Z}^M$. It follows immediately from this definition that Φ is continuous at Ξ^0 if and only if all the functions $\phi_1, \ldots,$ ϕ_M are continuous at Ξ^0. This definition of continuity at Ξ^0 is equivalent to the condition that for every neighborhood V of $\Phi\Xi^0$ there be a neighborhood U of Ξ^0 such that $\Phi U \subseteq V$.

We shall write ΦU here for the set of values $\Phi\Xi$ of Φ such that $\Xi \in U$. I.e., ΦU stands for $\bigcup_{\Xi,U}\{\Phi\Xi\}$ or the shorter $\bigcup_U \Phi$. Strictly speaking, a new operation sign ought to be introduced between Φ and U, since it is not ι that is to be supplied. Instead, $\bigcup_U \Phi$ is usually written as $\Phi[U]$ or

$$\{\Phi\Xi \mid \Xi \in U\}.$$

If a mapping Φ is continuous in its domain, and its range includes the neighborhood V, then the set of points Ξ^0 such that $\Phi\Xi^0 \in V$ contains only interior points: it is an open set. In fact, it is an open domain, for it is the union of neighborhoods U such that $\Phi U \subset V$, where, we recall,

$$V = (\eta_1, \ldots, \eta_M \mid \zeta_1, \ldots, \zeta_M)$$

is such that

$$\eta_i < \phi_i U < \zeta_i \quad (i = 1, \ldots, M).$$

In view of continuity, these inequalities are in turn satisfied if and only if, for all rational points $R \in U$,

$$\eta_i < \phi_i R < \zeta_i.$$

That gives us a definite condition for the Us such that $\Phi U \subseteq V$.

More generally, it follows from the foregoing that for any open domain $D = \bigcup V_*$ which is included in the range of a continuous mapping Φ, the set of points Ξ^0 such that $\Phi\Xi^0 \in D$ is again an open domain. The original of which an open domain is the image under a continuous mapping will itself be an open domain.

If a mapping is continuous in a *compact* domain $\overline{D}$, it follows by Theorem 15.1 that it is *uniformly continuous*:

$$\bigwedge_\varepsilon \bigvee_\delta \bigwedge_{\Xi^1,\Xi^2,\overline{D}} . \; |\Xi^1 - \Xi^2| < \delta \rightarrow |\Phi\Xi^1 - \Phi\Xi^2| < \varepsilon. \quad (16.2)$$

Furthermore, where $\Phi = (\phi_1, \ldots, \phi_M)$, $H = (\eta_1, \ldots, \eta_M)$, we have $\Phi\Xi^* \rightsquigarrow H$ if and only if $\phi_j\Xi^* \rightsquigarrow \eta_j$, for $j = 1, \ldots, M$. The *171* situation is similar with *sequences* of mappings Φ^*, where $\Phi^n = (\phi_1^n, \ldots, \phi_M^n)$: $\Phi^* \rightsquigarrow \Phi = (\phi_1, \ldots, \phi_M)$ holds just in case

$$\phi_j^* \rightsquigarrow \phi_j \quad (j = 1, \ldots, M).$$

If and only if all the sequences ϕ_j^* are uniformly convergent is Φ^* likewise *uniformly convergent*.

The various theorems concerning the continuity of mappings are thus directly derivable from the corresponding theorems about the continuity of functions.

This derivability holds also for differentiation.

Φ is *differentiable* at Ξ^0 if there is a linear mapping (i.e., a matrix A) such that

$$\bigwedge_\varepsilon \bigvee_\delta \bigwedge_\Xi . \, |\Xi - \Xi^0| < \delta \rightarrow$$
$$|\Phi\Xi - \Phi\Xi^0 - A(\Xi - \Xi^0)| < \varepsilon|\Xi - \Xi^0|. \quad (16.3)$$

$\Phi = (\phi_1, \ldots, \phi_M)$ is differentiable if and only if all the ϕ_js $(j = 1, \ldots, M)$ are. The matrix A is therefore uniquely determined by (16.3). It turns out to be the value of the so-called "functional matrix" or, as we prefer to say, the "*derivative-matrix*"

$$\begin{pmatrix} \phi'_{1(1)} & \ldots & \phi'_{1(N)} \\ \vdots & & \vdots \\ \phi'_{M(1)} & \ldots & \phi'_{M(N)} \end{pmatrix}$$

for the argument Ξ^0:

$$A = \begin{pmatrix} \phi'_{1(1)}\Xi^0 & \ldots & \phi'_{1(N)}\Xi^0 \\ \vdots & & \vdots \\ \phi'_{M(1)}\Xi^0 & \ldots & \phi'_{M(N)}\Xi^0 \end{pmatrix}.$$

Using the general notation

$$\mathscr{D}\phi = \phi' = (\phi'_{(1)}, \ldots, \phi'_{(N)})$$

and correspondingly

$$\mathscr{D}\Phi = \Phi' = \begin{pmatrix} \phi'_1 \\ \vdots \\ \phi'_M \end{pmatrix},$$

we get $A = \Phi'(\Xi^0)$ and $A = \mathscr{D}\Phi\Xi^0 = {}_{\Xi^0}\mathscr{D}\Phi$.

The *chain rule* (Theorem 15.3) can be formulated as follows with the use of matrix multiplication:

$$(\phi \urcorner \Psi)' = (\phi' \urcorner \Psi) \cdot \Psi'.$$

For mappings Φ we then get 172

$$(\Phi \urcorner \Psi)' = (\Phi' \urcorner \Psi) \cdot \Psi',$$

i.e.,

$$(\Phi \urcorner \Psi)'(\Xi^0) = \Phi'(\Psi \Xi^0) \cdot \Psi'(\Xi^0).$$

For the case $M = N$, we call the *determinant* of the derivative-matrix the *derivative determinant* (functional determinant), denoted by $|\mathscr{D}\Phi| = |\Phi'|$. For mappings in general ($M \neq N$), we must also take account of *partial derivative-matrices* (functional matrices) and their determinants. If $\phi_1, \ldots, \phi_M$ are N-place functions, $\Phi = \begin{pmatrix} \phi_1 \\ \vdots \\ \phi_M \end{pmatrix}$, and $M \leqslant N$ (for purely mnemonic reasons we write the system Φ here as a column instead of a row), then we denote the matrix

$$\begin{pmatrix} \phi'_{1(i_1)} & \ldots & \phi'_{1(i_M)} \\ \vdots & & \vdots \\ \phi'_{M(i_1)} & \ldots & \phi'_{M(i_M)} \end{pmatrix}$$

by $\mathscr{D}_{i_1 \ldots i_M} \Phi$ or $\mathscr{D}_{i_1 \ldots i_M}(\phi_1, \ldots, \phi_M)$. In place of these we can also use $\mathscr{D}_{\xi_{i_1} \ldots \xi_{i_M}} \Phi(\xi_1, \ldots, \xi_N)$. The latter, however, is an M-place function, whereas $\mathscr{D}_{i_1 \ldots i_M} \Phi$ is an N-place one:

$$\mathscr{D}_{\xi_{i_1} \ldots \xi_{i_M}} \Phi(\xi_1, \ldots, \xi_N)$$

applied to $\zeta_{i_1}, \ldots, \zeta_{i_M}$ is the same as $\mathscr{D}_{i_1 \ldots i_M} \Phi$ applied to

$$\xi_1, \ldots, \xi_{i_1 - 1}, \zeta_{i_1}, \xi_{i_1 + 1}, \ldots, \xi_{i_2 - 1}, \zeta_{i_2},$$

$$\xi_{i_2 + 1}, \ldots, \xi_{i_M - 1}, \zeta_{i_M}, \xi_{i_M + 1}, \ldots, \xi_N.$$

In the case where $M = N$, we have to consider the possibility that the mapping Φ has an *inverse*, i.e., a mapping Ψ such that

$$H = \Phi \Xi \leftrightarrow \Xi = \Psi H.$$

If Φ and Ψ are differentiable, then—since $(\Phi \urcorner \Psi)H = H$ for all H—it follows from the chain rule that for the unit matrix $E = \begin{pmatrix} 1 & & 0 \\ & \ddots & \\ 0 & & 1 \end{pmatrix}$,

$$E = \Phi'(\Psi) \cdot \Psi', \quad \text{i.e.,} \quad \Psi' = \Phi'(\Psi)^{-1}.$$

It likewise holds of the derivative determinant that

$$|\Phi'(\Psi H)| \cdot |\Psi'(H)| = 1.$$

173 In particular, therefore, $|\Phi'(\Xi^0)| \neq 0$ if Φ has a differentiable inverse mapping around Ξ^0.

We wish to show—as with 1-place functions—that the converse also holds. In contrast to the 1-place situation, however, only *"local" invertibility* is inferable from $|\Phi'| \neq 0$ in a domain. This can be seen from a simple example in $\mathfrak{Z}^2$.

The introduction of so-called *polar coordinates* ρ, θ by

$$\xi = \rho \cos \theta$$

$$\eta = \rho \sin \theta$$

furnishes a mapping of the infinite interval $(0, -\infty | \infty, \infty)$, i.e., of the open right half of the ρ, θ-plane, onto the ξ, η-plane minus the point 0, 0. The derivative determinant of the mapping at the point ρ, θ is

$$\begin{vmatrix} \cos \theta & -\rho \sin \theta \\ \sin \theta & \rho \cos \theta \end{vmatrix} = \rho > 0.$$

An inverse of this mapping is representable by

$$\rho = \sqrt{\xi^2 + \eta^2}$$

$$\theta = \begin{cases} \operatorname{arc\,tan}_0 \frac{\eta}{\xi} & \text{if} \quad \xi > 0 \\ \frac{\pi}{2} \quad \text{if} \quad \xi = 0, \quad \eta > 0 \\ -\frac{\pi}{2} \quad \text{if} \quad \xi = 0, \quad \eta < 0 \\ \operatorname{arc\,tan}_2 \frac{\eta}{\xi} & \text{if} \quad \xi < 0 \end{cases}$$

The inverse mapping is continuous at all points except where $\xi = 0, \eta < 0$. But the addition of $2k\pi$ to this θ produces for every

$$k = \pm 1, \pm 2, \ldots\ldots$$

another inverse of the original mapping.

A straight line $\rho = r$ in the ρ, θ-half-plane maps onto a circle in the ξ, η-plane. This (one-dimensional) mapping is not invertible, however. The image-circle is traversed infinitely often when the original line is traversed once.

Thus all we can claim of the mapping $(\rho, \theta) \to -(\xi, \eta)$ is that every (ρ, θ) has a neighborhood in which the mapping is invertible. Such a mapping is therefore said to be *locally invertible*.

We now want to show that the same conditions obtain for mappings in general with nonvanishing derivative determinants.

THEOREM 16.1 If a mapping Φ (from $\mathfrak{Z}^N$ into itself) is continuously differentiable in a neighborhood of Ξ^0 and $|\Phi'(\Xi^0)| \neq 0$, then there is a neighborhood of Ξ^0 in which Φ has an inverse Ψ. The domain of Ψ includes a neighborhood of $H^0 = \Phi\Xi^0$, and Ψ is continuously differentiable at H^0.

174

For $N = 1$ this theorem is already familiar. In fact, we even proved a stronger theorem for that case, viz., Theorem 15.4. We can therefore give an inductive proof. As often with such proofs, it is easier to prove a generalization of the stronger theorem straight off.

THEOREM 16.2 If a mapping Φ from $\mathfrak{Z}^{N+M}$ into $\mathfrak{Z}^N$ is continuously differentiable in a neighborhood of Ξ^0, Z^0, and

$$\mathscr{D}_{\xi_1 \ldots \xi_N} \Phi(\xi_1, \ldots, \xi_N, \zeta_1^0, \ldots, \zeta_M^0) | \neq 0$$

at Ξ^0, then there is a mapping Ψ which is continuously differentiable in a neighborhood of Ξ^0, Z^0 such that

$$\Xi = \Psi(H, Z) \leftrightarrow H = \Phi(\Xi, Z).$$

The domain of Ψ includes a neighborhood of H^0, Z^0, where $H^0 = \Phi(\Xi^0, Z^0)$.

Proof. For $N = 1$ and arbitrary M this theorem is already familiar. We now assume the theorem to hold for N and try to prove it for $N + 1$. Let $\begin{pmatrix}\phi_0\\ \Phi\end{pmatrix}$ be a mapping from $\mathfrak{Z}^{N+1+M}$ into $\mathfrak{Z}^{N+1}$ given by $\eta_0 = \phi_0(\xi_0, \Xi, Z)$ for

$$\Xi = (\xi_1, \ldots, \xi_N), \quad Z = (\zeta_1, \ldots, \zeta_M)$$

and by $H = \Phi(\xi_0, \Xi, Z)$ for

$$H = \begin{pmatrix}\eta_1\\ \vdots\\ \eta_N\end{pmatrix}, \quad \Phi = \begin{pmatrix}\phi_1\\ \vdots\\ \phi_N\end{pmatrix}.$$

Let

$$\mathscr{D}_{1, 2, \ldots, N+1} \begin{pmatrix}\phi_0\\ \Phi\end{pmatrix} \Big| \neq 0$$

at ξ_0^0, Ξ^0, Z^0 and hence also in a neighborhood of ξ_0^0, Ξ^0, Z^0. Also, at least one of the N-row subdeterminants must be distinct from zero. We may accordingly assume

$$\mathscr{D}_{\xi_1 \ldots \xi_N} \Phi(\xi_0^0, \Xi, Z^0) | \neq 0$$

to hold at Ξ^0.

The hypothesis of induction implies the existence of an inverse mapping Ψ such that

$$\Xi = \Psi(\xi_0, H, Z) \leftrightarrow H = \Phi(\xi_0, \Xi, Z)$$

175 in a suitable neighborhood of ξ_0^0, H^0, Z^0. Furthermore, by the induction hypothesis Ψ is continuously differentiable, and its domain includes a neighborhood of ξ_0^0, H^0, Z^0, where $H = \Phi(\xi_0^0, \Xi^0, Z^0)$.

Together with

$$\eta_0 = \phi_0[\xi_0, \Psi(\xi_0, H, Z), Z]$$

and

$$H = \Phi(\xi_0, \Xi, Z)$$

the above results yield the continuously differentiable mappings

$$\begin{pmatrix} \xi_0 \\ \Xi \\ Z \end{pmatrix} \rightarrow \begin{pmatrix} \xi_0 \\ H \\ Z \end{pmatrix} \quad \text{and} \quad \begin{pmatrix} \xi_0 \\ H \\ Z \end{pmatrix} \rightarrow \begin{pmatrix} \eta_0 \\ H \\ Z \end{pmatrix}.$$

The composition of these two mappings is the originally given mapping

$$\begin{pmatrix} \phi_0 \\ \Phi \end{pmatrix}.$$

The product of the derivative determinants of these two mappings is not 0, so that in particular

$$\mathscr{D}_{\xi_0} \phi_0[\xi_0, \Psi(\xi_0, H, Z), Z] \neq 0$$

at ξ_0^0. Thus by Theorem 15.4 there is a function ψ_0 such that

$$\xi_0 = \psi_0(\eta_0, H, Z) \leftrightarrow \eta_0 = \phi_0[\xi_0, \Psi(\xi_0, H, Z), Z].$$

ψ_0 is continuously differentiable, and its domain includes a neighborhood of η_0^0, H^0, Z^0, where $\eta_0^0 = \phi_0(\xi_0^0, \Xi^0, Z^0)$. All these results together with

$$\xi_0 = \psi_0(\eta_0, H, Z)$$

$$\Xi = \Psi[\psi_0(\eta_0, H, Z), H, Z]$$

yield the desired inverse mapping

$$\begin{pmatrix} \eta_0 \\ H \\ Z \end{pmatrix} \rightarrow \begin{pmatrix} \xi_0 \\ \Xi \\ Z \end{pmatrix}.$$

It is an immediate corollary of Theorem 16.1 that a mapping Φ is a locally topological mapping (i.e., locally invertible and invertibly continuous) if there is an open domain in which $|\Phi'| \neq 0$.

The characterization of *constant* 1-place functions ϕ by $\phi' = 0$ (in some interval) can likewise be extended word for word to many-place functions: if $\phi'_{(i)} = 0$ ($i = 1, \ldots, N$) in some *region*, then ϕ is constant. For it follows from $\mathscr{D}_{\xi_1}\phi(\xi_1, \ldots, \xi_N) = 0$ that

$$\phi(\xi_1, \xi_2, \ldots, \xi_N) = \phi_1(\xi_2, \ldots, \xi_N)$$

for some ϕ_1 in each interval and hence—on account of connectivity—in the entire region. In the same way it follows from $\mathscr{D}_{\xi_2}\phi_1(\xi_2, \ldots, \xi_N) = 0$ that

$$\phi_1(\xi_2, \ldots, \xi_N) = \phi_2(\xi_3, \ldots, \xi_N),$$

so that $$\phi(\xi_1, \ldots, \xi_N) = \phi_2(\xi_3, \ldots, \xi_N),$$ and so on.

176

Since all mappings

$$\Phi = \begin{pmatrix} \phi_1 \\ \vdots \\ \phi_M \end{pmatrix} \quad \text{are such that} \quad \Phi' = \begin{pmatrix} \phi'_1 \\ \vdots \\ \phi'_M \end{pmatrix},$$

$\Phi' = 0$ (in some region) will always entail the constancy of Φ.

For many-place functions ϕ in regions, we can infer from $\mathscr{D}_j\phi = 0$, where $M < N$ and $j = 1, \ldots, M$, that

$$\phi(\xi_1, \ldots, \xi_N) = \phi_0(\xi_{M+1}, \ldots, \xi_N) \qquad \text{for some } \phi_0;$$

i.e., ϕ depends on the projections $I_{M+1}, \ldots, I_N$ alone.

In general we say that a function ϕ is "*dependent*" on a system of functions $\psi_1, \ldots, \psi_n$ if there is a function ϕ_0 such that

$$\phi = \phi_0(\psi_1, \ldots, \psi_n)$$

in some domain. The function ϕ_0, if any, is uniquely determined, since for every system $\eta_1, \ldots, \eta_n$ such that

$$\eta_1 = \psi_1(\xi_1, \ldots, \xi_N), \ldots, \eta_n = \psi_n(\xi_1, \ldots, \xi_N)$$

it follows that

$$\phi_0(\eta_1, \ldots, \eta_n) = \phi(\xi_1, \ldots, \xi_N).$$

ϕ_0 exists if and only if

$$\psi_i(\xi_1, \ldots, \xi_N) = \psi_i(\xi_1', \ldots, \xi_N') \quad (i = 1, \ldots, n)$$

implies in every case that

$$\phi(\xi_1, \ldots, \xi_N) = \phi(\xi_1', \ldots, \xi_N').$$

By the chain rule (Theorem 15.3) it follows from the dependency of ϕ on $\psi_1, \ldots, \psi_n$—assuming differentiability—that ϕ' is *linearly dependent* on $\psi_1', \ldots, \psi_n'$. The converse relationship also holds.

Let $\Psi = (\psi_1, \ldots, \psi_n)$ be a system of N-place functions and let the derivatives $\psi_1', \ldots, \psi_n'$ be a linearly independent system (for every point of some region). Then $n \leqslant N$, and the partial derivative determinants

$$|\mathscr{D}_{i_1 \ldots i_n}(\psi_1, \ldots, \psi_n)|$$

177 are not all 0. Let us say that

$$|\mathscr{D}_{1 \ldots n}(\psi_1, \ldots, \psi_n)| \neq 0.$$

By the inversion theorem (Theorem 16.2) there is a mapping Φ such that

$$H = \Psi(\Xi, \xi_{n+1}, \ldots, \xi_N) \leftrightarrow \Xi = \Phi(H, \xi_{n+1}, \ldots, \xi_N).$$

The system $\Phi, I_{n+1}, \ldots, I_N$ is an invertible mapping from $\mathfrak{Z}^N$ into itself.

Now let ϕ be an N-place function in the region. In place of ϕ and $\psi_1, \ldots, \psi_n$ we consider $\bar{\phi} = \phi(\Phi, I_{n+1}, \ldots, I_N)$ and the system

$$\bar{\Psi} = \Psi(\Phi, I_{n+1}, \ldots, I_N).$$

Then we have

$$\bar{\Psi} = \begin{pmatrix} I_1 \\ \vdots \\ I_n \end{pmatrix}.$$

If ϕ' is linearly dependent on $\psi_1', \ldots, \psi_n'$ in the region, then $\bar{\phi}'$ is likewise linearly dependent on $I_1', \ldots, I_n'$. In particular, therefore, $\bar{\phi}'_{(i)} = 0$ for $i = n + 1, \ldots, N$; i.e., $\bar{\phi}$ is dependent on the system $\bar{\Psi}$. From this it follows that ϕ is dependent on the system Ψ.

§17. Multiple Differentiation and Integration

Having extended the most important properties of differentiation in the one-place situation (elementary rules of differentiation, inversion for $\phi' \neq 0$, and constancy for $\phi' = 0$) to many-place functions and mappings, we have yet to inquire into the inverse of differentiation, i.e., integration. The following generalization of the question suggests itself: to what extent is a many-place function ϕ determined by the system ϕ' of its partial derivatives—and how can ϕ be computed from ϕ'?

It is, however, useful to extend integration to N-place functions in another manner beforehand, to wit, by inverting multiple differentiation. To that end we first consider the partial derivatives $\phi'_{(1)}$ and $\phi'_{(2)}$ of an N-place function ϕ for $N = 2$. The derivatives are themselves 2-place functions. If both are differentiable, then we can form the higher-order derivatives

$$\phi''_{(11)}, \phi''_{(12)}, \phi''_{(21)}, \phi''_{(22)}.$$

Now—as we will show directly—if continuity obtains, then

$$\phi''_{(12)} = \phi''_{(21)}.$$

Of this "*mixed derivative*" it of course no longer holds true that it determines ϕ up to a constant. For any function that depends on I_1 or I_2 alone, and hence too for any sum of such *partially constant* functions [e.g., $\psi_1(I_1) + \psi_2(I_2)$ with one-place functions ψ_1, ψ_2], the mixed derivative will of course always be zero. The mixed derivative nevertheless determines ϕ up to a sum of *partially constant* functions. In particular, ϕ can be represented by double integration of its mixed derivative. 178

We give a general demonstration of this in what follows. We begin with the formation of multiple differences.

Let ϕ be an N-place function such that

$$\overset{\zeta}{\underset{\zeta_0}{\Delta_i}}\phi = \phi(I_1, \ldots, I_{i-1}, \zeta, I_{i+1}, \ldots, I_N) - \phi(I_1, \ldots, I_{i-1}, \zeta_0, I_{i+1}, \ldots, I_N).$$

Δ_i thus stands for taking the difference at the ith place. $\Delta_i\phi$ is a function with $(N - 1) + 2 = N + 1$ argument places. It will now be true for the special case of 2-place functions that

$$\Delta_1\Delta_2\phi = \Delta_2\Delta_1\phi, \tag{17.1}$$

since

$$\overset{\xi}{\underset{\xi_0}{\Delta_1}}\overset{\eta}{\underset{\eta_0}{\Delta_2}}\phi = \overset{\xi}{\underset{\xi_0}{\Delta_1}}(\phi(I_1, \eta) - \phi(I_1, \eta_0))$$
$$= \phi(\xi, \eta) - \phi(\xi, \eta_0) - \phi(\xi_0, \eta) + \phi(\xi_0, \eta_0)$$

and

$$\overset{\eta}{\underset{\eta_0}{\Delta_2}}\overset{\xi}{\underset{\xi_0}{\Delta_1}}\phi = \overset{\eta}{\underset{\eta_0}{\Delta_2}}(\phi(\xi, I_2) - \phi(\xi_0, I_2))$$
$$= \phi(\xi, \eta) - \phi(\xi_0, \eta) - \phi(\xi, \eta_0) + \phi(\xi_0, \eta_0).$$

For 2-place functions ϕ we abbreviate

$$\overset{\xi}{\underset{\xi_0}{\Delta_1}}\overset{\eta}{\underset{\eta_0}{\Delta_2}} \quad \text{as} \quad \overset{\xi,\eta}{\underset{\xi_0,\eta_0}{\Delta^2}};$$

and correspondingly, for N-place functions,

$$\overset{\xi_1}{\underset{\xi_1^0}{\Delta_1}} \dots \overset{\xi_N}{\underset{\xi_N^0}{\Delta_N}} \quad \text{as} \quad \overset{\xi_1, \dots, \xi_N}{\underset{\xi_1^0, \dots, \xi_N^0}{\Delta^N}}.$$

The order of the successive difference operations here is clearly permutable: we always get the same (total) *mixed difference*. From mixed differences we now form *mixed-difference quotients*.

179 Let ϕ be an N-place function. Then

$$\frac{\Delta^N \phi}{\Delta_1 I_1 \Delta_2 I_2 \dots \Delta_N I_N}$$

is called the (total) *mixed-difference quotient* of ϕ. For $\Xi^0 = \xi_1^0, \dots, \xi_N^0$ and $\Xi = \xi_1, \dots, \xi_N$ its value is

$$\frac{\overset{\Xi}{\underset{\Xi^0}{\Delta^N}}\phi}{(\xi_1 - \xi_1^0) \dots (\xi_N - \xi_N^0)}.$$

The limit of this mixed-difference quotient as $\Xi \to \Xi^0$ is, when it exists, the (total) *mixed derivative* (for the argument Ξ^0): $_{\Xi^0}\mathscr{D}^N\phi$.

The bound-theorem is easily extended to mixed-difference quotients. For the sake of simplicity we formulate it only for $N = 2$.

THEOREM 17.1 (BOUND-THEOREM). If a 2-place function ϕ has a partial mixed derivative $\mathscr{D}^2_{12}$ in an *interval* in which

$$\gamma_1 \leqslant \mathscr{D}^2_{12}\phi \leqslant \gamma_2,$$

then it also holds in the interval that

$$\gamma_1 \leqslant \frac{\Delta^2\phi}{\Delta_1 I_1 \Delta_2 I_2} \leqslant \gamma_2.$$

Proof. We have

$$\frac{1}{(\xi_1 - \xi_0)(\eta_1 - \eta_0)} \overset{\xi_1,\eta_1}{\underset{\xi_0,\eta_0}{\Delta^2}} \phi$$

$$= \frac{1}{\eta_1 - \eta_0} \; \frac{\phi(\xi_1, \eta_1) - \phi(\xi_1, \eta_0) - [\phi(\xi_0, \eta_1) - \phi(\xi_0, \eta_0)]}{\xi_1 - \xi_0}.$$

By the basic bound-theorem as applied to the 1-place function ψ such that

$$\psi(\xi) = \phi(\xi, \eta_1) - \phi(\xi, \eta_0),$$

the difference quotient lies somewhere between the bounds of

$$\frac{1}{\eta_1 - \eta_0} \psi',$$

i.e., of

$$\frac{1}{\eta_1 - \eta_0} [\phi'_{(1)}(\xi, \eta_1) - \phi'_{(1)}(\xi, \eta_0)] \quad \text{where} \quad \xi \in (\xi_0 \,|\, \xi_1).$$

But—again by the bound-theorem—this difference quotient lies between the bounds of $\phi''_{(12)}$ in $(\xi_0, \eta_0 \,|\, \xi_1, \eta_1)$.

It is an immediate corollary of Theorem 17.1 that the mixed-difference quotient converges to $\phi''_{(12)}(\xi_0, \eta_0)$ as $\xi_1, \eta_1 \rightsquigarrow \xi_0, \eta_0$, providing $\phi''_{(12)}$ is continuous at ξ_0, η_0.

If $\phi''_{(21)}$ also exists and is likewise continuous at ξ_0, η_0, then because $\Delta_1\Delta_2 = \Delta_2\Delta_1$ we get exactly the same result, that the mixed-difference quotient converges to $\phi''_{(21)}(\xi_0, \eta_0)$. If both mixed derivatives $\phi''_{(12)}$ and $\phi''_{(21)}$ are continuous, then we say that ϕ is *continuously mixed-differentiable.* With that we get *180*

THEOREM 17.2 If ϕ is continuously mixed-differentiable at a point ξ_0, η_0 then

$$\mathscr{D}^2_{12}\phi = \mathscr{D}^2_{21}\phi \quad \text{at} \quad \xi_0, \eta_0.$$

We write $\mathscr{D}^2\phi$ for short instead of $\mathscr{D}^2_{12}\phi$ or $\mathscr{D}^2_{21}\phi$.

Generalization to N-place functions yields for continuously mixed-differentiable functions ϕ the result that the (continuous) mixed derivatives

$$\mathscr{D}^N_{i_1 \ldots i_N}$$

all coincide, where $i_1, \ldots, i_N$ is any permutation of $1, \ldots, N$. We abbreviate this as $\mathscr{D}^N\phi$. In addition, this mixed derivative satisfies the *bound-theorem*:

$$\gamma_1 \leqslant \frac{\Delta^N \phi}{\Delta_1 I_1 \ldots \Delta_N I_N} \leqslant \gamma_2 \leftrightarrow \gamma_1 \leqslant \mathscr{D}^N \phi \leqslant \gamma_2$$

in every interval in which ϕ is continuously mixed-differentiable.

Just as in the one-place situation, it follows from the bound-theorem that if two functions ϕ, ψ satisfy $\mathscr{D}^N\phi = \mathscr{D}^N\psi$, then also $\Delta^N\phi = \Delta^N\psi$.

$\Delta^N\phi$ is thus uniquely determined by $\mathscr{D}^N\phi$. We again call the operation leading from $\mathscr{D}^N\phi$ to $\Delta^N\phi$ *integration.* It can be reduced to one-place integration.

For an N-place function ψ we define—in analogy to $\Delta_i\psi$—an $(N+1)$-place function $\mathscr{J}_i\psi$ which for the arguments

$$\xi_1, \ldots, \xi_{i-1}, \xi_{i+1}, \ldots, \xi_N$$

and the pair ζ_0, ζ_1 takes the value

$$\mathop{\mathscr{J}}_{\zeta_0}^{\zeta_1}\psi(\xi_1, \ldots, \xi_{i-1}, I, \xi_{i+1}, \ldots, \xi_N) = \mathop{\mathscr{J}_{\zeta}}_{\zeta_0}^{\zeta_1}\psi(\xi_1, \ldots, \xi_{i-1}, \zeta, \xi_{i+1}, \ldots, \xi_N).$$

Thus if ψ is a 2-place function,

$$\mathscr{J}_1\mathscr{J}_2\psi$$

is a 4-place function with the following meaning:

$$\mathop{\mathscr{J}_1}_{\xi_0}^{\xi_1}\mathop{\mathscr{J}_2}_{0}^{\eta_1}\psi = \mathop{\mathscr{J}_{\xi}}_{\xi_0}^{\xi_1}\mathop{\mathscr{J}_{\eta}}_{\eta_0}^{\eta_1}\psi(\xi, \eta).$$

181 If now $\psi = \mathscr{D}_2\mathscr{D}_1\phi$, then the formula for the one-place situation, $\mathscr{J}\mathscr{D} = \Delta$, yields

$$\mathscr{J}_1\mathscr{J}_2\mathscr{D}_2\mathscr{D}_1\phi = \mathscr{J}_1\Delta_2\mathscr{D}_1\phi.$$

Since $\mathscr{D}_1$ is a linear operation, it holds trivially that

$$\Delta_2\mathscr{D}_1 = \mathscr{D}_1\Delta_2.$$

Thus we get

$$\mathscr{J}_1\mathscr{J}_2\mathscr{D}_2\mathscr{D}_1\phi = \mathscr{J}_1\mathscr{D}_1\Delta_2\phi = \Delta_1\Delta_2\phi$$

and hence

$$\mathscr{J}_1\mathscr{J}_2\mathscr{D}^2\phi = \Delta^2\phi.$$

In exactly the same way, of course,

$$\mathscr{J}_2\mathscr{J}_1\mathscr{D}^2\phi = \Delta^2\phi$$

follows. We write $\mathscr{J}^2$ for short in place of $\mathscr{J}_1\mathscr{J}_2$ or $\mathscr{J}_2\mathscr{J}_1$. $\mathscr{J}^2$ is called the "*mixed integral.*"

As applied to a mixed derivative the mixed integral is independent of the order of the one-place integration operations. It is furthermore the case that

$$\mathcal{J}^2\mathcal{D}^2\phi = \Delta^2\phi,$$

or more briefly,

$$\mathcal{J}^2\mathcal{D}^2 = \Delta^2.$$

For N-place functions, where

$$\mathcal{J}^N = \mathcal{J}_1\,\mathcal{J}_2 \cdots \mathcal{J}_N,$$

we correspondingly get

$$\mathcal{J}^N = \mathcal{J}_{i_1} \cdots \mathcal{J}_{i_N}$$

for every permutation $i_1, \ldots, i_N$ of $1, \ldots, N$, and we have

$$\mathcal{J}^N\mathcal{D}^N = \Delta^N. \tag{17.2}$$

The (total) mixed integral $\mathcal{J}^N\psi$ is defined for N-place functions ψ which admit of representation as mixed derivatives $\psi = \mathcal{D}^N\phi$. To write it out completely, for $\Xi^0 = \xi_1^0, \ldots, \xi_N^0$ and $\Xi^1 = \xi_1^1, \ldots, \xi_N^1$,

$$\mathop{\mathcal{J}^N}_{\Xi^0}^{\Xi^1}\psi = \mathop{\mathcal{J}_{\xi_1}}_{\xi_1^0}^{\xi_1^1} \cdots \mathop{\mathcal{J}_{\xi_N}}_{\xi_N^0}^{\xi_N^1} \psi(\xi_1, \ldots, \xi_N). \tag{17.3}$$

$\mathop{\mathcal{J}^N}_{\Xi^0}\psi$ is accordingly itself an N-place function, and of course it holds of $\psi = \mathcal{D}^N\phi$ that

$$\mathop{\mathcal{J}^N}_{\Xi^0}\psi = \mathop{\Delta^N}_{\Xi^0}\phi = \phi + \text{partially constant functions}.$$

It follows that 182

$$\mathcal{D}^N\mathop{\mathcal{J}^N}_{\Xi^0}\psi = \psi. \tag{17.4}$$

By its derivation this formula is valid only for those functions ψ which are representable in the form $\psi = \mathcal{D}^N\phi$, where ϕ is a continuously mixed-differentiable function throughout the interval in question.

To demonstrate that this requirement is satisfied for, say, all continuous functions ψ, we exploit the possibility of strengthening the bound-theorem as we did in the one-dimensional case. We take as our point of departure the subsets of an interval $J = [\Xi | H]$. If $\Xi = (\xi_1, \ldots, \xi_N)$ and $H = (\eta_1, \ldots, \eta_N)$, then we shall call the set of interval points Z such that

$$I_i(Z) = \zeta_i,$$

where the number $\zeta_i \in [\xi_i | \eta_i]$, a "*partition*" of the interval J, to use a brief, suggestive term.

The partitions take the place in N-dimensional intervals of the points by which we partitioned one-dimensional intervals.

For N-dimensional space we then get the following

STRENGTHENED BOUND-THEOREM. If the mixed derivative $\mathcal{D}^N\phi$ of a continuous N-place function ϕ exists everywhere in a given interval except for the points of not more than a definitely denumerable number of partitions of the interval, then every bound of the mixed derivatives in the interval is also a bound of the (total) mixed-difference quotients.

The proof is exactly parallel to that of Theorem 17.1, with the strengthened one-dimensional bound-theorem in place of the unstrengthened version.

Thus, once again, we may call any continuous N-place definite function ψ which has a mixed derivative $\mathcal{D}^N\psi$ *ω-everywhere* (by which we mean here, everywhere except for a definite sequence of partitions) in some interval a *primitive function* of the N-place function ϕ if

$$\mathcal{D}^N\psi = \phi.$$

All primitive functions are additively distinguished from one another by partially constant functions alone. The integral

$$\mathcal{J}^N\phi = \Delta^N\psi$$

is determined solely by ϕ for all primitive functions ψ of ϕ.

Functions having primitive functions are called (mixed-) *integrable.*

The simplest of the integrable functions are again the *step functions.* We confine ourselves to a discussion of the case $N = 2$, since the generalization to arbitrary N involves nothing really new.

183

Let the interval $(\xi_0, \eta_0 | \xi^0, \eta^0)$ be divided into subintervals by

$$\xi_0 < \xi_1 < \ldots < \xi_m = \xi^0$$

$$\eta_0 < \eta_1 < \ldots < \eta_n = \eta^0.$$

The subintervals are $U_{ij} = (\xi_i, \eta_j | \xi_{i+1}, \eta_{j+1})$.

A function τ which is constant in the interior of every subinterval, e.g., $\tau = c_{ij}$ in U_{ij}, is called a *step function.* For every step function τ we can give an explicit primitive function.

If $\tau(\xi, \eta) = c_{ij}$, where $(\xi, \eta) \in U_{ij}$, then the way we find a primitive function T is to integrate

$$T(\xi, \eta) = \overset{\xi}{\underset{\xi_0}{\mathcal{J}_1}} \overset{\eta}{\underset{\eta_0}{\mathcal{J}_2}} \tau.$$

For $\xi \in (\xi_i \mid \xi_{i+1})$ we get

$$\overset{\eta}{\underset{\eta_0}{\mathcal{J}}} \tau(\xi, I_2) = c_{ij}(\eta - \eta_j) + \sum_{0}^{j-1}{}_k c_{ik}(\eta_{k+1} - \eta_k),$$

where $\eta \in (\eta_j \mid \eta_{j+1})$, by the proof of the one-place case. Correspondingly, it follows that

$$\overset{\xi}{\underset{\xi_0}{\mathcal{J}_1}} \overset{\eta}{\underset{\eta_0}{\mathcal{J}_2}} \tau = \left[c_{ij}(\eta - \eta_j) + \sum_{0}^{j-1}{}_k c_{ik}(\eta_{k+1} - \eta_k)\right](\xi - \xi_i)$$
$$+ \sum_{0}^{i-1}{}_l \left[c_{lj}(\eta - \eta_j) + \sum_{0}^{j-1}{}_k c_{lk}(\eta_{k+1} - \eta_k)\right](\xi_{l+1} - \xi_l).$$

We immediately see from this that

$$\mathcal{J}_1\mathcal{J}_2 \tau = \mathcal{J}_2\mathcal{J}_1 \tau.$$

Hence T must be a primitive function of τ. In particular—since $T(\xi_0, \eta) = T(\xi, \eta_0) = 0$—we get as the integral

$$\overset{\xi^0, \eta^0}{\underset{\xi_0, \eta_0}{\mathcal{J}}}{}^2 \phi = \sum_{0,0}^{m-1, n-1}{}_{i,j} c_{ij}(\xi_{i+1} - \xi_i)(\eta_{j+1} - \eta_j). \tag{17.5}$$

As in one dimension, integrability carries over from the step functions to the limiting functions of uniformly convergent sequences of step functions. Again, we call these limiting functions "*jump-continuous*"—without going into the question of whether or how the jump-continuous functions might otherwise be characterized.

It is enough to remark here that all *continuous* functions ϕ are also jump-continuous. Uniform continuity in a closed interval, after all, immediately yields for any ε an approximating step function τ such that *184*

$$|\phi - \tau| < \varepsilon.$$

We have only to partition the interval sufficiently "finely" and to take as values of τ one of those for ϕ from each subinterval.

The integrability of jump-continuous functions (and in particular of continuous ones) now follows from the extension of Theorem 13.3 to the N-dimensional number space.

THEOREM 17.3 If ϕ_* is a uniformly convergent (in some interval) sequence of N-place functions with ϕ as its limit, and ψ_n is a primitive function of ϕ_n, and ψ_* is convergent at least on the partitions through a point Z of the interval, then ψ_* converges uniformly in the entire interval to a primitive function ψ of ϕ. At all points where the derivatives $\mathcal{D}^N \psi_n$ exist for an infinity of ns, $\mathcal{D}^N \psi$ exists too.

Proof. We limit our attention to the case $N = 2$. Let the end points of the interval in question be (ξ_0, η_0) and (ξ_1, η_1). The partitions through the point $Z = (\xi^0, \eta^0)$ are the 1-dimensional intervals $(\xi^0, \eta_0 | \xi^0, \eta_1)$ and $(\xi_0, \eta^0 | \xi_1, \eta^0)$. In analogy to (12.6) it follows for an arbitrary point (ξ, η) in view of

$$\overset{\xi,\eta}{\underset{\xi^0,\eta^0}{\Delta^2}} \psi_i = \psi_i(\xi, \eta) - \psi_i(\xi^0, \eta) - \psi_i(\xi, \eta^0) + \psi_i(\xi^0, \eta^0)$$

that

$$\begin{aligned}\psi_m(\xi, \eta) - \psi_n(\xi, \eta) = \overset{\xi,\eta}{\underset{\xi^0,\eta^0}{\Delta^2}} \psi_m - \overset{\xi,\eta}{\underset{\xi^0,\eta^0}{\Delta^2}} \psi_n &+ \psi_m(\xi^0, \eta) - \psi_n(\xi^0, \eta)\\ &+ \psi_m(\xi, \eta^0) - \psi_n(\xi, \eta^0)\\ &- \psi_m(\xi^0, \eta^0) + \psi_n(\xi^0, \eta^0).\end{aligned}$$

By the strengthened bound-theorem, the first difference on the right lies within the bounds of $\mathcal{D}^2(\psi_m - \psi_n)$, i.e., of $\phi_m - \phi_n$. Its absolute value approaches zero as m, n become sufficiently large. The same goes for the rest of the differences since ψ_* is by stipulation convergent on the partitions through (ξ^0, η^0).

The existence of the (total) mixed derivatives $\mathcal{D}^2\psi$ at points where an infinity of $\mathcal{D}^2\psi_n$s exist follows as in the proof of Theorem 12.2:

For sufficiently large n and all $m > n$, $|\phi_m - \phi_n|$ by hypothesis tends to zero. By the strengthened bound-theorem, $\frac{\Delta^2(\psi_m - \psi_n)}{\Delta_1 I_1 \Delta_2 I_2}$ then likewise approaches zero. Since

$$\Delta^2(\psi_m - \psi_n) = \Delta^2\psi_m - \Delta^2\psi_n,$$

185 it follows for every ε and sufficiently large n that

$$\left|\frac{\Delta^2\psi_m}{\Delta_1 I_1 \Delta_2 I_2} - \frac{\Delta^2\psi_n}{\Delta_1 I_1 \Delta_2 I_2}\right| < \varepsilon \quad \text{for all} \quad m > n,$$

and hence that

$$\left|\frac{\Delta^2\psi_m}{\Delta_1 I_1 \Delta_2 I_2} - \frac{\Delta^2\psi_n}{\Delta_1 I_1 \Delta_2 I_2}\right| \leqslant \varepsilon.$$

If ψ_{k_*} is a subsequence of ψ_* for which $\mathcal{D}^2\psi_{k_*}(\xi, \eta)$ exists [whence $\lim \mathcal{D}^2\psi_{k_*}(\xi, \eta) = \phi(\xi, \eta)$], then it furthermore follows for sufficiently large n that

$$\left|\frac{\overset{\bar\xi,\bar\eta}{\underset{\xi,\eta}{\Delta^2}}\psi}{(\bar\xi-\xi)(\bar\eta-\eta)} - \phi(\xi,\eta)\right| \leq \left|\frac{\overset{\bar\xi,\bar\eta}{\underset{\xi,\eta}{\Delta^2}}\psi_{k_n}}{(\bar\xi-\xi)(\bar\eta-\eta)} - \mathscr{D}^2\psi_{k_n}(\xi,\eta)\right|$$
$$+ |\mathscr{D}^2\psi_{k_n}(\xi,\eta) - \phi(\xi,\eta)| + \varepsilon.$$

As $n \to \infty$ and $(\bar\xi, \bar\eta) \to (\xi, \eta)$ the addends on the right-hand side become arbitrarily small. Hence we have

$$\mathscr{D}^2\psi(\xi,\eta) = \phi(\xi,\eta).$$

Finally, the conclusion that the primitive functions of the continuous functions ϕ are everywhere differentiable also follows from the same consideration as in the one-dimensional case. For any given point Ξ in the interval there is a uniform approximation of ϕ by step functions none of which has a discontinuity at Ξ.

The formula for step functions (17.5) shows that integrals $\mathscr{J}^2\phi$ of jump-continuous functions are representable as limits of sums of the form

$$\sum_{0,0}^{m-1,\,n-1}{}_{i,j}\, c_{ij}(\xi_{i+1}-\xi_i)(\eta_{j+1}-\eta_j).$$

In a similar way, we get for arbitrary N sums of the form

$$\sum_{0,\,\ldots,\,0}^{m_1,\,\ldots,\,m_N}{}_{i_1,\ldots,i_N}\, c_{i_1\ldots i_N}\prod_{1}^{N}{}_j(\xi_j^{i_j+1}-\xi_j^{i_j}).$$

The task of calculating the limits of such sums arises in many areas of geometry (determination of volume) and other exact sciences. This task is reducible to the inverse of mixed differentiation (*generalized fundamental theorem*).

If ϕ is an elementary function, then the natural way to arrive at a value for the integral $\mathscr{J}^N\phi$ is to replace $\mathscr{J}^N$ by the one-place integrals $\mathscr{J}_{i_1}\ldots\mathscr{J}_{i_N}$ in some appropriate order.

The integral $\overset{1,1}{\underset{0,0}{\mathscr{J}}}{}^2_{\xi,\eta}\,\eta^\xi$ provides a simple example. The equation 186

$$\overset{1,1}{\underset{0,0}{\mathscr{J}}}{}^2_{\xi,\eta}\,\eta^\xi = \overset{1}{\underset{0}{\mathscr{J}}}_\eta\overset{1}{\underset{0}{\mathscr{J}}}_\xi\,\eta^\xi$$

leads via

$$\mathscr{J}_\xi\,\eta^\xi = \Delta_\xi\,\frac{\eta^\xi}{\ln\eta},$$

i.e.,

$$\overset{1}{\underset{0}{\mathscr{J}}}_\xi\,\eta^\xi = \frac{\eta-1}{\ln\eta},$$

to the (insoluble) task of giving an elementary determination of

$$\mathscr{J}_\eta \frac{\eta - 1}{\ln \eta}.$$

Reversing the order, on the other hand, we get

$$\mathscr{J}_{\xi\,0}^{\;1}\mathscr{J}_{\eta\,0}^{\;1}\,\eta^\xi = \mathscr{J}_{\xi\,0}^{\;1}\overset{1}{\underset{0}{\Delta}}{}_\eta \frac{\eta^{\xi+1}}{\xi + 1} = \mathscr{J}_{\xi\,0}^{\;1} \frac{1}{\xi + 1} = \overset{1}{\underset{0}{\Delta}}{}_\xi \ln(\xi + 1) = \ln 2.$$

As one can see, in addition to the value of the double integral this also gives us a result concerning one-place integrals, to wit,

$$\mathscr{J}_{\eta\,0}^{\;1} \frac{\eta - 1}{\ln \eta} = \ln 2.$$

This result is unattainable by the elementary integration of one-place functions alone. (Note that both integrals, $\mathscr{J}_{\eta\,0}^{\;\xi} \frac{\eta}{\ln \eta}$ and $\mathscr{J}_{\eta\,0}^{\;\xi} \frac{1}{\ln \eta}$, improperly converge to ∞ as $\xi \rightsquigarrow 1$.)

According to the definitions given thus far, jump-continuous functions can only be integrated over intervals as domains of integration. For $\mathscr{J}_{\Xi}^{H}\phi$ we also write $\underset{[\Xi|H]}{\mathscr{J}}\phi$ (omitting now the superscript N). The question now arises, to what domains D can the definition of an integral ${}_D\mathscr{J}\phi$ be meaningfully extended? For bounded domains $D \subseteq [\Xi\,|H]$ the question can be attacked as follows. We construct the characteristic function χ_D of D:

$$\chi_D Z = \begin{cases} 1 & \text{if} \quad Z \in D \\ 0 & \text{if} \quad Z \notin D. \end{cases}$$

187 Then we examine the integral $\underset{[\Xi|H]}{\mathscr{J}}\chi_D$. If χ_D is integrable (by virtue, e.g., of being jump-continuous), then the integral exists and evidently depends upon our choice of the interval $[\Xi\,|H]$ containing D.

In case the product $\phi \cdot \chi_D$ is integrable, we can adopt the corresponding definition

$${}_D\mathscr{J}\phi = \underset{[\Xi|H]}{\mathscr{J}} \phi \cdot \chi_D.$$

If ϕ and χ_D are both jump-continuous, then of course so is their product; in any case, the integral ${}_D\mathscr{J}\phi$ exists under these assumptions.

The next thing to do would be to prove for this generalized integration the *transformation formula*,

$${}_{\Psi D}\mathscr{J}\phi = {}_D\mathscr{J}.\ \phi(\Psi)\,|\Psi'|.,$$

and *Stokes' formula* (cf. §22); particularly the validity conditions should be given for these formulas. Only in Chapter V will we carry out that task by way of a (seeming) digression into geometric considerations. However, that will spare us many of the (somewhat laborious) investigations that would otherwise be necessary.

At this point we shall only use the demonstrated interchangeability of the two integration operations to establish the interchangeability of differentiation and integration.

Again, it suffices to confine ourselves to the case $N = 2$. Let ϕ be a 2-place function differentiable with respect to the second place, and let $\mathcal{D}_2\phi$ be (mixed-)differentiable, e.g., jump-continuous. Then we have

$$\Delta_\eta \phi(\xi, \eta) = \mathcal{J}_\eta \mathcal{D}_2 \phi(\xi, \eta),$$

and thus

$$\phi(\xi, \eta_1) = \mathcal{J}_{\eta_0}^{\eta_1}{}_{\eta} \mathcal{D}_2 \phi(\xi, \eta) + \phi(\xi, \eta_0).$$

Now if $\phi(\xi, \eta)$ is integrable with respect to ξ, then we get

$$\mathcal{J}_{\xi_0}^{\xi_1}{}_{\xi} \phi(\xi, \eta_1) = \mathcal{J}_{\xi_0}^{\xi_1}{}_{\xi} \mathcal{J}_{\eta_0}^{\eta_1}{}_{\eta} \mathcal{D}_2 \phi(\xi, \eta) + \mathcal{J}_{\xi_0}^{\xi_1}{}_{\xi} \phi(\xi, \eta_0)$$

$$= \mathcal{J}_{\eta_0}^{\eta_1}{}_{\eta} \mathcal{J}_{\xi_0}^{\xi_1}{}_{\xi} \mathcal{D}_2 \phi(\xi, \eta) + \mathcal{J}_{\xi_0}^{\xi_1} \phi(\xi, \eta_0).$$

The right side shows that $\mathcal{J}_{\xi_0}^{\xi_1}{}_{\xi} \phi(\xi, \eta_1)$ is differentiable ω-everywhere with respect to η_1.

We get

$${}_{\eta}\mathcal{D}_{\eta_1} \mathcal{J}_{\xi_0}^{\xi_1}{}_{\xi} \phi(\xi, \eta_1) = \mathcal{J}_{\xi_0}^{\xi_1}{}_{\xi} \mathcal{D}_2 \phi(\xi, \eta) \quad \omega\text{-everywhere}$$

or, more symmetrically, 188

$${}_{\eta_0}\mathcal{D}_{\eta} \mathcal{J}_{\xi_0}^{\xi_1}{}_{\xi} \phi(\xi, \eta) = \mathcal{J}_{\xi_0}^{\xi_1}{}_{\xi}\, {}_{\eta_0}\mathcal{D}_{\eta} \phi(\xi, \eta) \quad \omega\text{-everywhere}, \qquad (17.6)$$

i.e., greatly abbreviated,

$$\mathcal{D}_2 \mathcal{J}_1 = \mathcal{J}_1 \mathcal{D}_2 \quad \omega\text{-everywhere}.$$

This *Leibniz formula* for the differentiation of integrals with respect to one *parameter*, as it is often called, thus presupposes —beyond the existence of $\mathcal{I}_1\phi$ and $\mathcal{D}_2\phi$—only the (mixed) differentiability of $\mathcal{D}_2\phi$. If we further assume the continuity of $\mathcal{D}_2\phi$ with respect to the second place, then the Leibniz formula $\mathcal{D}_2\mathcal{I}_1 = \mathcal{I}_1\mathcal{D}_2$ holds everywhere.

If we wish to use this formula to differentiate an integral term $\mathcal{I}_{\xi\,\alpha(\eta)}^{\;\beta(\eta)}\phi(\xi, \eta)$, where η also occurs in the limits, with respect to the parameter η (or, as we ought to say, with respect to the variable η), then

$$\mathcal{I}_{\xi\,\alpha(\eta)}^{\;\beta(\eta)}\phi(\xi, \eta) = \mathcal{I}_{\xi\,\xi_0}^{\;\beta(\eta)}\phi(\xi, \eta) - \mathcal{I}_{\xi\,\xi_0}^{\;\alpha(\eta)}\phi(\xi, \eta)$$

reduces it to the case where η occurs only in the upper limit.

$\mathcal{I}_{\zeta\,\xi_0}^{\;\xi}\phi(\zeta, \eta)$ is differentiable ω-everywhere with respect to ξ and everywhere with respect to η (under the above assumptions). The two-place function ψ such that

$$\psi(\xi, \eta) = \mathcal{I}_{\zeta\,\xi_0}^{\;\xi}\phi(\zeta, \eta)$$

is thus partially differentiable. But, where β is a differentiable function, the chain rule will establish the differentiability of $\psi(\beta(\eta), \eta)$ only if ψ is totally differentiable.

We get total differentiability by Theorem 15.2, provided the partial derivatives are continuous. If we assume, then, that ϕ is continuous, we thereby make $\mathcal{D}_1\psi = \phi$ continuous. For $\mathcal{D}_2\psi$ the Leibniz formula (17.6) gives us

$$\mathcal{D}_2\psi(\xi, \eta) = \mathcal{I}_{\zeta\,\xi_0}^{\;\xi}\mathcal{D}_2\phi(\zeta, \eta).$$

The simplest approach is to assume the continuity of $\mathcal{D}_2\phi$ too. Then $\mathcal{D}_2\psi$ is likewise continuous in view of

THEOREM 17.4 If a 2-place function ϕ is continuous in an open domain which includes the 1-dimensional interval $[\xi_0, \eta\,|\,\xi, \eta]$, then where

189

$$\psi(\xi, \eta) = \mathcal{I}_{\zeta\,\xi_0}^{\;\xi}\phi(\zeta, \eta),$$

the 2-place function ψ is also continuous at (ξ, η).

Proof. (By the covering theorem, Theorem 14.5) there is a δ such that, where

$$|(\xi_1, \eta_1) - (\xi, \eta)| < \delta,$$

the function ϕ is also continuous in the 2-dimensional interval $J = [\xi_0, \eta \mid \xi_1, \eta_1]$. Furthermore,

$$\psi(\xi_1, \eta_1) - \psi(\xi, \eta) = \underset{\xi_0}{\overset{\xi_1}{\mathcal{J}_\zeta}}\phi(\zeta, \eta_1) - \underset{\xi_0}{\overset{\xi}{\mathcal{J}_\zeta}}\phi(\zeta, \eta)$$

$$= \underset{\xi_0}{\overset{\xi_1}{\mathcal{J}_\zeta}}[\phi(\zeta, \eta_1) - \phi(\zeta, \eta)] + \underset{\xi_0}{\overset{\xi_1}{\mathcal{J}_\zeta}}\phi(\zeta, \eta) - \underset{\xi_{10}}{\overset{\xi}{\mathcal{J}_\zeta}}\phi(\zeta, \eta)$$

$$= \underset{\xi_0}{\overset{\xi_1}{\mathcal{J}_\zeta}}[\phi(\zeta, \eta_1) - \phi(\zeta, \eta)] + \underset{\xi}{\overset{\xi_1}{\mathcal{J}_\zeta}}\phi(\zeta, \eta).$$

Since ϕ is uniformly continuous in J and since $|\xi_1 - \xi_0| \leqslant |\xi - \xi_0| + \delta$, the first integral (by the Bound-Theorem 13.3), taken absolutely, approaches zero as $|\eta_1 - \eta|$ does. The second integral, taken absolutely, approaches zero as $|\xi_1 - \xi|$ does, for ϕ is bounded in J. Taking everything together, we get

$$\wedge_\varepsilon \vee_\delta \wedge_{\xi_1, \eta_1} \cdot |(\xi_1, \eta_1) - (\xi, \eta)| < \delta \rightarrow |\psi(\xi_1, \eta_1) - \psi(\xi, \eta)| < \varepsilon.,$$

i.e., ψ is continuous at (ξ, η).

Where ϕ is a continuous function continuously differentiable with respect to the second place, and α and β are differentiable functions, Theorem 17.4 together with the Leibniz formula (17.6) yields by the chain rule the formula

$$_{\eta_0}\mathcal{D}_\eta \underset{\alpha(\eta)}{\overset{\beta(\eta)}{\mathcal{J}_\xi}}\phi(\xi, \eta) = \underset{\alpha(\eta_0)}{\overset{\beta(\eta_0)}{\mathcal{J}_\xi}}\ {}_{\eta_0}\mathcal{D}_\eta\, \phi(\xi, \eta) + \phi(\beta(\eta_0), \eta_0) \cdot \beta'(\eta_0)$$

$$- \phi(\alpha(\eta_0), \eta_0) \cdot \alpha'(\eta_0). \qquad (17.7)$$

As for the importance of the differentiation of integrals (with respect to one parameter at a time), the generalization of the formula

$$\mathcal{J}\mathcal{D} = \Delta$$

to many-place functions mentioned at the beginning of this section is a case in point. For a one-place function ϕ, $\Delta\phi$ is uniquely determined by $\phi' = \mathcal{D}\phi$; for an N-place function, $\Delta^N\phi$ is similarly determined by $\mathcal{D}^N\phi$. But it also holds for N-place functions that $\Delta\phi$ is uniquely determined by ϕ', i.e., by the system of partial 190 derivatives $\phi'_{(1)}, \ldots, \phi'_{(N)}$. If the domain of ϕ is an *interval* of $\mathfrak{Z}^N$, then from

$$\phi'_{(i)} = 0 \qquad (i = 1, \ldots, N)$$

it immediately follows that $\Delta\phi = 0$, as (15.6) shows.

Again, the equation

$$\phi'_{(i)}(\xi_1, \ldots, \xi_i, \ldots, \xi_N) = 0$$

need only hold ω-everywhere: it may fail for denumerably many ξ_is, e.g., by virtue of ϕ's not being differentiable with respect to the ith place for these ξ_is. However, if ϕ is allowed to be less than everywhere differentiable, it must of course at least be continuous everywhere.

The unique determination of $\Delta\phi$ by ϕ' holds not only for intervals but also for regions as argument domains. For in a region (i.e., a connected domain), we can always get from a given point to any other point in finitely many steps all of which remain within some particular interval or other.

In the case of an interval, then, (15.6) immediately yields a formula for determining $\Delta\phi$, viz.,

$$\overset{\Xi}{\underset{\Xi^0}{\Delta}}\phi = \sum_{1}^{N}{}_i \,\mathcal{J}_{\xi_i^0}^{\xi_i} \phi'_{(i)}(\xi_1^0, \ldots, \xi_{i-1}^0, I, \xi_{i+1}, \ldots, \xi_N). \tag{17.8}$$

But in contrast to the cases treated thus far, it is by no means any longer the case that we have in general for a given system of functions $\psi_1, \ldots, \psi_N$—even if they are continuous—a primitive function, i.e., a continuous function ϕ differentiable ω-everywhere such that

$$\phi'_{(i)} = \psi_i \qquad (i = 1, \ldots, N).$$

This follows from the fact that, for differentiable ψ_i, $\phi'_{(i)} = \psi_i$ immediately implies

$$\psi'_{i(j)} = \psi'_{j(i)} \qquad (i, j = 1, \ldots, N) \tag{17.9}$$

in view of $\phi''_{(ji)} = \phi''_{(ij)}$. Only when (17.9) is satisfied, at least ω-everywhere, can the system $\psi_1, \ldots, \psi_N$ have a primitive function.

We now wish to show that under the condition (17.9) a primitive function indeed exists.

THEOREM 17.5 If the equations

$$\psi'_{i(j)} = \psi'_{j(i)} \qquad (i, j = 1, \ldots, N)$$

hold in some interval for a system of differentiable functions ψ_1, 191 $\ldots, \psi_N$, then there exists a primitive function ϕ of $\psi_1, \ldots, \psi_N$ such that

$$\overset{\Xi}{\underset{\Xi^0}{\Delta}}\phi = \sum_1^N {}_i \overset{\xi_i}{\underset{\xi_i^0}{\mathscr{J}}}\psi_i(\xi_1^0, \ldots, \xi^0_{i-1}, I, \xi_{i+1}, \ldots, \xi_N).$$

The proof consists simply of verifying that the function defined by

$$\phi(\xi_1, \ldots, \xi_N) = \sum_1^N {}_i \overset{\xi_i}{\underset{\xi_i^0}{\mathscr{J}}}\psi_i(\xi_1^0, \ldots, \xi^0_{i-1}, I, \xi_{i+1}, \ldots, \xi_N) \quad (17.10)$$

is really a primitive function. We differentiate partially with respect to the jth place. For $i > j$ we of course have

$$\mathscr{D}_{\xi_j} \overset{\xi_i}{\underset{\xi_i^0}{\mathscr{J}}}\psi_i(\xi_1^0, \ldots, \xi^0_{i-1}, I, \xi_{i+1}, \ldots, \xi_N) = 0.$$

For $i = j$ it will be the case that

$$\mathscr{D}_{\xi_j} \overset{\xi_i}{\underset{\xi_i^0}{\mathscr{J}}}\psi_i(\xi_1^0, \ldots, \xi^0_{i-1}, I, \xi_{i+1}, \ldots ; \xi_N)$$
$$= \psi_i(\xi_1^0, \ldots, \xi^0_{j-1}, I, \xi_{j+1}, \ldots, \xi_N).$$

Since the remaining summands ($i < j$) are partially differentiable —and since partial differentiations can be performed by means of the Leibniz formula—it turns out in view of (17.9) that

$$\phi'_{(j)}(\xi_1, \ldots, \xi_N) = \sum_1^{j-1} {}_i \overset{\xi_i}{\underset{\xi_i^0}{\mathscr{J}}}\psi'_{j(i)}(\xi_1^0, \ldots, \xi^0_{i-1}, I, \xi_{i+1}, \ldots, \xi_N)$$
$$+ \psi_j(\xi_1^0, \ldots, \xi^0_{j-1}, \xi_j, \ldots, \xi_N)$$
$$= \sum_1^{j-1} {}_i \overset{\xi_i}{\underset{\xi_i^0}{\Delta}}\psi_j(\xi_1^0, \ldots, \xi^0_{i-1}, I, \xi_{i+1}, \ldots, \xi_N)$$
$$+ \psi_j(\xi_1^0, \ldots, \xi^0_{j-1}, \xi_j, \ldots, \xi_N)$$
$$= \psi_j(\xi_1, \ldots, \xi_N).$$

Thus ϕ is indeed a primitive function of $\psi_1, \ldots, \psi_N$.

This Theorem 17.5 does not admit of generalization to arbitrary regions, but only to a special class of regions.

For an arbitrary region, it is true that the proof of Theorem 17.5 permits us—starting from a point Ξ^0—to specify a number $\phi_C\Xi$ for any point Ξ, provided that we have connected Ξ with Ξ^0 by a chain C of intervals. But it can happen that there is another chain $\overline{C}$ connecting Ξ with Ξ^0 such that the number $\phi_{\overline{C}}\Xi$ does not coincide with $\phi_C\Xi$. One example is provided by the inverse functions $\theta = \phi(\xi, \eta)$ of 192

$$\xi = \rho \cos \theta$$
$$\eta = \rho \sin \theta,$$

where $\xi,\eta \neq 0,0$. (Cf. polar coordinates in §16.) ϕ is continuously differentiable except at points such that $\xi = 0$, $\eta < 0$; at those points, ϕ is discontinuous $\left(\text{e.g., with a jump from } -\frac{\pi}{2} \text{ to } +\frac{3\pi}{2}\right)$. We get

$$\phi'_{(1)}(\xi, \eta) = -\frac{\eta}{\xi^2 + \eta^2}$$

$$\phi'_{(2)}(\xi, \eta) = \frac{\xi}{\xi^2 + \eta^2}.$$

Outside of the point 0,0 these functions ψ_1, ψ_2 are continuously differentiable throughout, and they satisfy $\psi'_{1(2)} = \psi'_{2(1)}$. If we connect the point 0,1 with 0,-1 by the interval chains C_+ and C_- (which must not contain the point 0,0) in the regions $\xi > 0$ and $\xi < 0$ respectively, then in the one case we get $\overset{0,-1}{\underset{0,1}{\Delta}} \phi_{C_+} = -\pi$; and in the other, $\overset{0,-1}{\underset{0,1}{\Delta}} \phi_{C_-} = +\pi$.

When, now, do two interval chains yield the same value? They will in any event do so if they both lie inside a common interval chain belonging to the region in question. In that case we say that the two chains are "*neighboring*" in the region. Thus, if we can get from one interval chain to a second by finitely many steps between neighboring chains, then the two former chains must again yield the same value $\phi\, \Xi$. In such a situation we shall say that the one chain is "*deformable*" into the other one in the region in question. A region is called *simple* if any two chains connecting points of the region are deformable into each other in that region. (The region just considered of all points $\neq 0,0$ is not simple.) Simple regions are also called *simply connected* domains. Theorem 17.5 holds for these, too, and not just for intervals.

V. DIFFERENTIAL GEOMETRY 193

§18. Analysis and Geometry

We have treated differentiation and integration in Chapters III and IV without making any use of *differentials*, the brilliant discovery of Leibniz. Often differentials are introduced merely as a handy but in reality inexact notation. We are told something to the effect that traditionally one writes ϕ' as $\frac{d\phi(\xi)}{d\xi}$, and that this permits a formulation of the chain rule

$$\phi(\psi)' = \phi'(\psi) \cdot \psi'$$

as

$$\frac{d\phi(\psi(\xi))}{d\xi} = \frac{d\phi(\eta)}{d\eta}(\psi) \cdot \frac{d\psi(\xi)}{d\xi}.$$

With $\eta = \psi(\xi)$ and $\zeta = \phi(\eta)$, this in turn becomes the imprecise—but catchy—

$$\frac{d\zeta}{d\xi} = \frac{d\zeta}{d\eta} \cdot \frac{d\eta}{d\xi}.$$

This notation clearly confounds all sorts of things. We can no longer tell whether η and ζ are variables, functions, or terms, nor do we know which arguments the functions $\frac{d\zeta}{d\eta}$ and $\frac{d\eta}{d\xi}$ are supposed to take.

It is therefore understandable that differentials are often gladly avoided altogether for purposes of differentiation. For integration, on the other hand, differentials are usually retained notationally. Rather than $\mathcal{J}\phi$ or $\mathcal{J}_{\xi}\,\phi(\xi)$, one usually writes $\int \phi(\xi)\,d\xi$. The substitution rule then appears as follows, with only a minor imprecision remaining:

$$\int_{\psi}^{\psi} \phi(\eta)d\eta = \int \phi[\psi(\xi)]\,\frac{d\psi(\xi)}{d\xi}\,d\xi.$$

With $\eta = \psi(\xi)$ this gives us 194

$$\int_{\psi}^{\psi} \phi(\eta) d\eta = \int \phi(\eta) \frac{d\eta}{d\xi} d\xi.$$

Where the left side has $d\eta$, the right side has $\frac{d\eta}{d\xi} d\xi$. Thus we are led to the equation

$$d\eta = \frac{d\eta}{d\xi} d\xi.$$

Formal division applied to both sides again yields the chain rule:

$$\frac{d\eta}{d\zeta} = \frac{d\eta}{d\xi} \cdot \frac{d\xi}{d\zeta}.$$

In the present chapter we shall show how one can arrive at a way of operating with differentials, a *differential calculus*, that is not just a notational convention. At the same time we shall take up the possibility of extending Leibniz's differential calculus discovered by E. Cartan.

We shall obviate remaining imprecisions in the notation quite naturally by shifting our consideration from functions of numbers (*number functions*) to functions with geometric points as arguments (*point functions*).

But what is a *geometric point*? For an answer to this question, let us recall analytic geometry. An analytic treatment makes use of only one feature of the points on a Euclidean plane, the fact that there is a particular way of mapping them onto pairs of numbers. The *coordinates* x and y are functions correlating numbers (the *coordinate values*) $\xi = x(P)$ and $\eta = y(P)$ with each point P of the plane.

A fundamental remark is in order here. Even in the one-dimensional case, that of Euclidean straight lines, one can ask how we know that the points on a line uniquely correspond to the real numbers. The question is ill-put, however. The real numbers (as members of an indefinite set) result from our constructions of rational numbers and sequences of rational numbers. The straight lines of geometry, on the other hand, are "ideas" arising out of our idealization of perceptible properties of actual bodies. This "idealization" is not a mental process; it consists rather in the erection of an abstract theory, to wit, Euclidean geometry. This is not the place to go into the grounding of the axioms required for the development of such a theory (cf. Dingler 1964).

195 This idealization includes the principle that a line segment shall admit of all the divisions we can construct (e.g., with the help of the rational or the real numbers). A line segment does not consist of points, but is rather divisible by points, ideally *ad infini-*

tum; i.e., every segment produced by division is itself divisible. (Aristotle was the first to formulate this idealization.) If, furthermore, the concept of geometric congruence of line segments is available, then the so-called Archimedean axiom (which is already found in Euclid and derives from Eudoxos) postulates as identical any two dividing points between which there is no rational dividing point. Thus a straight line can only be divided by real points, i.e., points corresponding to real numbers. The only answer to the question whether a straight line "contains" a point corresponding to *every* real number is that we *demand* of straight lines that they permit a corresponding division or partition for every real number. The class of points partitioning a straight line is (like the class of real numbers) an indefinite set.

The analytic theory of geometry deals exclusively with sets of partitioning points; continuous line segments as such do not appear. This makes it understandable that modern mathematics usually draws no distinction between the set of points partitioning a segment and the segment itself. We shall adopt this way of speaking. Thus we can also say of a line segment that it contains points: the points are the members of the segment (in this sense).

Similar remarks apply to the points of planes and solids.

If we are dealing with the Cartesian coordinate system x, y of a plane, we can "translate" geometric statements about the points into statements about the coordinate values. For example, the statement that the distance between the points P and Q is δ,

$$||P, Q|| = \delta,$$

is translated into the statement

$$\sqrt{[(x(P) - x(Q))^2 + (y(P) - y(Q))^2]} = \delta.$$

However, there is not just one Cartesian coordinate system, but rather infinitely many, all on an equal footing. All Cartesian coordinate systems result from one another upon the application of certain *transformations*:

$$\bar{x} = ax + by + e$$

$$\bar{y} = cx + dy + f.$$

The coefficients here must form a so-called *orthogonal* matrix $\begin{pmatrix} a & b \\ c & d \end{pmatrix}$:

$$\begin{pmatrix} a & b \\ c & d \end{pmatrix}\begin{pmatrix} a & c \\ b & d \end{pmatrix} = \begin{pmatrix} 1 & 0 \\ 0 & 1 \end{pmatrix}.$$

196 Now, a statement A about coordinate values which translates a geometric statement is *invariant* under the *coordinate transformations*; i.e.,

$$A(x(P), y(P);\; x(Q), y(Q); \ldots) \leftrightarrow A(\bar{x}(P), \bar{y}(P);\; \bar{x}(Q), \bar{y}(Q); \ldots),$$

or

$$A(\xi_1, \eta_1;\; \xi_2, \eta_2; \ldots) \leftrightarrow A(\bar{\xi}_1, \bar{\eta}_1;\; \bar{\xi}_2, \bar{\eta}_2; \ldots),$$

if we set

$$\xi_1 = x(P), \quad \eta_1 = y(P), \quad \bar{\xi}_1 = \bar{x}(P), \quad \bar{\eta}_1 = \bar{y}(P),$$
$$\xi_2 = x(Q), \quad \eta_2 = y(Q), \quad \bar{\xi}_2 = \bar{x}(Q), \quad \bar{\eta}_2 = \bar{y}(Q).$$

Conversely, every statement about number pairs

$$A(\xi_1, \eta_1;\; \xi_2, \eta_2; \ldots)$$

which is invariant under the mappings

$$\begin{aligned}\bar{\xi} &= a\xi + b\eta + e\\ \bar{\eta} &= c\xi + d\eta + f,\end{aligned} \quad \text{where} \quad \begin{pmatrix} a & b \\ c & d \end{pmatrix} \quad \text{is orthogonal,}$$

yields a geometric statement about points on the Euclidean plane.

In the *Erlangen Program* (of F. Klein, 1871), geometric statements are classified according to the "*groups*" of mappings with respect to which the analytic statements

$$A(\xi_1, \eta_1;\; \xi_2, \eta_2; \ldots)$$

are invariant.

The two most important such groups for our present purposes are the *affine mappings*

$$\bar{\xi} = a\xi + b\eta + e$$
$$\bar{\eta} = c\xi + d\eta + f$$

with an invertible matrix, $\begin{pmatrix} a & b \\ c & d \end{pmatrix}$, and the *locally affine mappings*

$$\bar{\xi} = \phi(\xi, \eta)$$
$$\bar{\eta} = \psi(\xi, \eta)$$

with continuously differentiable functions ϕ, ψ and an invertible derivative-matrix

$$\begin{pmatrix} \phi'_{(1)} & \phi'_{(2)} \\ \psi'_{(1)} & \psi'_{(2)} \end{pmatrix}.$$

If we also relinquish the differentiability of ϕ, ψ, requiring only continuity (including continuous invertibility), then we get *topological mappings.* 197

Any analytic statement that is invariant with respect to, say, all topological mappings can be translated into a geometric statement. The theory of such geometric statements is then called "*topology*."

Similarly, by (affine) *differential geometry* we mean the theory of the geometric statements translating analytic statements that are invariant under all locally affine mappings.

These analytic theories, such as topology and differential geometry, are logically independent of Euclidean geometry. It is nevertheless convenient to retain the idiom of geometry. For dimensions 2 and 3 this has the added advantage of suggesting an "intuitive" interpretation of the statements of the theory, i.e., an interpretation in Euclidean geometry.

Only in the language of differential geometry will Leibniz's differentials now emerge as a fully appropriate conceptual construct, and not just as a suggestive notation.

§19. Curves, Surfaces, and Figures

DEFINITION 19.1 By a *simple curve segment* we mean a set K (the members of which are called *points* $P, Q, \ldots$) which is mapped onto sets of real numbers by invertible mappings (called *coordinates* of K) under the following conditions:

(1) There are coordinates x such that the x-image of K is a closed interval.

(2) If x and $\bar{x}$ are coordinates of K, then the real function ϕ defined by

$$\bar{x}(P) = \phi x(P) \text{ for all } P$$

is continuously differentiable on the x-image of K, and $\phi' \neq 0$ there.

We denote the *x-image of K* for short by $x[K]$. By the continuous differentiability of ϕ on $x[K]$ let us then mean that ϕ is applicable in an open interval containing $x[K]$ and that ϕ is continuously differentiable throughout that interval.

If x is a coordinate, we shall always assume that, for every continuously differentiable function ϕ such that $\phi' \neq 0$, ϕx is a coordinate too.

Since $\phi' \neq 0$ in the interval, all and only properly monotonic continuously differentiable functions ϕ occur here as "*transformations*."

198 Notice that we do not presuppose the points of the curve segment to be points in a Euclidean space. Thus we are dealing with abstract curve segments, as the phrase goes—without regard to the possibility of imbedding them in a Euclidean space. This abstract treatment has the advantage of making it clear from the outset that the possibility of imbedding is irrelevant for differentiation and integration on the curve.

It follows from (1) and (2) that $x[K]$ is a closed interval for every x. If $x[K] = [\alpha \mid \beta]$, then the points P, Q such that $x(P) = \alpha$ and $x(Q) = \beta$ are called the *boundary points* of the simple curve segment K.

With respect to a coordinate x, the two boundary points P and Q of the curve segment can be distinguished as lower and upper boundary points. If $x(P) < x(Q)$, then P is the *lower boundary point* with respect to x, and Q is the *upper boundary point* with respect to x.

Independently of the choice of coordinate, we can define *sameness of sense* for point pairs. We say of two ordered pairs of distinct points P_1, Q_1 and P_2, Q_2 that they have the *same sense* if for some coordinate x (and hence for all!) the differences $x(P_1) - x(Q_1)$ and $x(P_2) - x(Q_2)$ have the same sign; i.e., if

$$\frac{x(P_1) - x(Q_1)}{x(P_2) - x(Q_2)} > 0.$$

Sameness of sense of point pairs is an equivalence relation with just two equivalence classes. Each of the two equivalence classes is called an *orientation* of the curve segment. Thus an orientation is represented by a point pair.

If one of the two orientations of a curve segment K is singled out (by designating one orientation as "*positive*" and the other as "*negative*"), then the curve segment will be called an *oriented curve segment*, denoted by $\mathfrak{K}$. The point pairs making up the positive orientation are then said to have *positive sense*, and the others, *negative sense*.

For oriented curve segments the boundary points P and Q can be distinguished as *initial point* and *end point*. If the sense of the pair P, Q, in that order, is positive, then P is the *initial point* and Q the *end point*.

Thus an orientation also singles out those coordinates x for which the initial point is a lower boundary point with respect to x. Such coordinates will also be said to have *positive sense*, and the rest, *negative sense*. Two coordinates $x, \overline{x}$ have the *same sense* if and only if for $\overline{x} = \phi x$, $\phi' > 0$.

The coordinates are mappings of the points onto real numbers. We shall call such mappings "*point functions*" as opposed to the functions we have worked with so far (mapping real numbers onto real numbers), which we shall call "*number functions.*" 199

If ϕ is any number function applicable to $x[K]$, then $f = \phi x$ is a point function:

$$fP = \phi \urcorner x(P).$$

On the other hand, a point function f and a coordinate x uniquely determine a number function ϕ such that

$$f = \phi x.$$

If we denote the *inverse* of the mapping x by $\check{x}$ ($\check{x}$ maps the real numbers in $x[K]$ onto the points of K), then

$$\phi = f\check{x},$$

i.e.,

$$\phi\xi = f(\check{x}\xi),$$

or

$$\phi\xi = \eta \leftrightarrow \bigvee_P . x(P) = \xi \wedge fP = \eta.$$

$$\phi\xi = f \urcorner \iota_P (x(P) = \xi).$$

We call a function f *continuous* if for some coordinate x—and thus for all coordinates!—the number function $f\check{x}$ is continuous on $x[K]$. Since the transformations of the coordinates are continuously differentiable number functions, a corresponding definition of differentiability is possible. A point function f is (continuously) *differentiable* if for some (and hence again for every) coordinate x the number function $f\check{x}$ is (continuously) differentiable on $x[K]$. In §20 differentiation of the point functions will lead us to differentials. For the time being, however, we shall extend our considerations to compound curve segments and then generalize them to N-dimensional figures.

By a *curve segment* we mean a set that can be represented as the union of *finitely* many *simple* curve segments. If $K_1, \ldots, K_n$ are the simple curve segments of which a curve segment K is composed, and if these simple segments are *oriented* "coherently" (i.e., the end point of $\mathfrak{K}_{i+1}$ is in each case the initial point of $\mathfrak{K}_i$), then the segment K is called an *oriented curve segment* $\mathfrak{K}$.

For the N-dimensional case we again define "abstract" figures in analogy to Definition 19.1.

Definition 19.2 A *simple N-dimensional figure R* is a set (of points) mapped into the N-dimensional number space by invertible mappings (*coordinate systems*) in such a way that

200 (1) There are coordinate systems $X = x_1, \ldots, x_N$ such that $X[R]$, the X-image of R, is a closed interval.

(2) If X and $\overline{X} = \overline{x}_1, \ldots, \overline{x}_N$ are coordinate systems, then the invertible mapping Φ (from the N-dimensional number space into itself) uniquely determined by

$$\overline{X} = \Phi(X) \quad \text{in} \quad X[R]$$

is continuously differentiable, and

$$|\Phi'| \neq 0$$

in $X[R]$.

Here too the continuous differentiability of the mapping is to be understood to mean that Φ is continuously differentiable in an open domain containing $X[R]$. If X is a coordinate system, then for every invertible mapping Φ which is continuously differentiable in $X[R]$ such that $|\Phi'| \neq 0$, we count the system $\Phi(X)$ of point functions as one of the coordinate systems of R.

These coordinate systems are in every case N-member systems of point functions. Each such point function is called a coordinate. By the results of §16, the inverse of the mapping Φ is also continuous; thus Φ is a topological mapping.

The orientation of curve segments can be generalized to figures. This can be done simply be ascribing to two coordinate systems $x_1, \ldots, x_N$ and $\overline{x}_1, \ldots, \overline{x}_N$ the *same sense* if their *transformation determinant* $|\Phi'| > 0$. Coordinate systems will then fall into exactly two classes, the members of each of which are alike in sense. An "*orientation*" involves singling out one of those classes by assigning "*positive sense*" to its member coordinate systems and "*negative sense*" to the others.

To appreciate the geometric significance of such an orientation, we must first take up the notion of the *tangent space* of a figure R.

We consider curve segments K in R, i.e., curve segments whose points are also points of R. We must then distinguish the coordinate systems of R from the coordinates of K, which for the sake of clarity we shall now call "*parameters*." If t is a parameter of K, then every coordinate system $x_1, \ldots, x_N$ of R has number functions

$$\phi_1, \ldots, \phi_N \quad \text{such that} \quad x_i = \phi_i t,$$

i.e.,

$$x_i(P) = \phi_i t(P) \quad \text{for all} \quad P \in K.$$

We call K a *smooth* curve segment in R if these functions ϕ_i are continuously differentiable and if there is no point at which $\phi_i' = 0$ for all $i = 1, \ldots, N$. If $t[K]$ is the interval $[\alpha|\beta]$, then with every number $\tau \in [\alpha|\beta]$ there will be associated the coordinate values 201

$$\xi_1 = \phi_1 \tau, \ldots, \xi_N = \phi_N \tau$$

of a point P of K.

Now let there be two *oriented smooth* curve segments $\mathfrak{K}_1$, $\mathfrak{K}_2$ in R with a common initial point P_0. Let t_1 and t_2 be parameters of $\mathfrak{K}_1$ and $\mathfrak{K}_2$ respectively. Then the curve segments (with these parameters) are said to be *"vector-identical"* if the functions ϕ_{1i} and ϕ_{2i} uniquely determined by

$$\begin{aligned} x_i &= \phi_{1i}(t_1) \\ x_i &= \phi_{2i}(t_2) \end{aligned} \qquad (i = 1, \ldots, N)$$

are such that

$$\phi_{1i}'(t_1(P_0)) = \phi_{2i}'(t_2(P_0)) \qquad (i = 1, \ldots, N). \tag{19.1}$$

Intuitively speaking, the two curve segments then have the same *direction* at P_0.

The vector-identity of curve segments with parameters is an equivalence relation and is evidently independent of our choice of the coordinate system $x_1, \ldots, x_N$. We may thus say that by abstraction every curve segment (with a parameter) which has P_0 as its initial point represents a *vector* (called the *tangent vector* at P_0). The numbers defined by (19.1) (which are always distinct from $0, \ldots, 0$) are called the *coordinate values* of the vector in the coordinate system $x_1, \ldots, x_N$. If we switch from one coordinate system $x_1, \ldots, x_N$ to another $\bar{x}_1, \ldots, \bar{x}_N$ by way of

$$\bar{X} = \Phi(X),$$

then we can obtain the new coordinate values of the vector by a linear transformation with the derivative-matrix Φ'. This follows immediately from the chain rule.

We can thus perform calculations with the tangent vectors at P_0 just as we can with Euclidean vectors: all we ever have to use are the coordinate values.

In particular, the *addition* of vectors at P_0 and the *multiplication* of a vector by a real number can be defined in terms of coordinate values. These operations are invariant with respect to coordinate transformations.

A system of N linearly independent vectors at P_0 is referred to for short as an N-leg at P_0. The existence of such N-legs follows directly from the existence of coordinate systems, for every 202 coordinate system $x_1, \ldots, x_N$ defines N curve segments with P_0 as their initial point. Take, e.g., the set of points P in an appropriate neighborhood of P_0 such that

$$x_1(P) \geqslant x_1(P_0), x_2(P) = x_2(P_0), \ldots, x_N(P) = x_N(P_0).$$

This gives us a curve segment with a parameter x_1. That curve segment with its parameter represents a vector at P_0, the x_1-*vector* $\mathfrak{x}_1$ of the N-leg at P_0 associated with the coordinate system $x_1, \ldots, x_N$.

In similar fashion we define the x_2-vector $\mathfrak{x}_2, \ldots$, and the x_N-vector $\mathfrak{x}_N$. These N *basis vectors* indeed form an N-leg, for they are linearly independent. The coordinate values (with respect to $x_1, \ldots, x_N$), after all—written as columns—are the *unit columns*

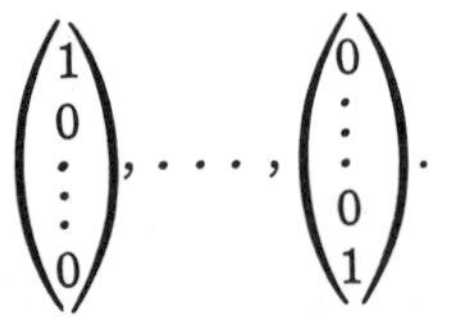

The determinant of the coordinate values of an N-leg at P_0 is never 0. Two N-legs at P_0 can therefore be said to have *like sense* if their determinants of the coordinate values have the same sign. Identity of sense is independent of the particular coordinate system employed.

Having introduced vectors and N-legs (always relative to a point P_0), we now see that in defining orientation as we did above by designating one class of coordinate systems we have simultaneously singled out one class of N-legs of like sense, viz., those with the same sense as an N-leg associated with a coordinate system of positive sense at P_0.

Now, the distinction of positive sense thus conferred on certain N-legs by an orientation can in fact be traced back inductively to the orientation of curve segments by point pairs. In this way the attribution of sense to N-legs, at least for $N = 2$ and $N = 3$, becomes geometrically visualizable.

Before going on, let us note—as an immediate consequence of the properties of determinants—that an N-leg switches to opposite sense if one of its vectors is multiplied by -1 or if two of its vectors are interchanged. Any even permutation of the vectors of an N-leg yields an N-leg of like sense.

A simple 2-dimensional figure is mapped by a coordinate system x_1, x_2 onto an interval which we represent—on a Euclidean plane—by a rectangle:

Fig. 16

A point P is called a *boundary point* if the system $x_1(P)$, $x_2(P)$ of coordinate values belongs to the boundary of the interval. By §16, the *boundary* of the figure is thereby defined invariantly with respect to coordinates.

The *boundary* consists of four simple curve segments with x_1 and x_2 as respective parameters. These can be oriented coherently by the assignment of initial and end points. Pictorially we do this by drawing in arrows with the head always pointing toward the end point. 203

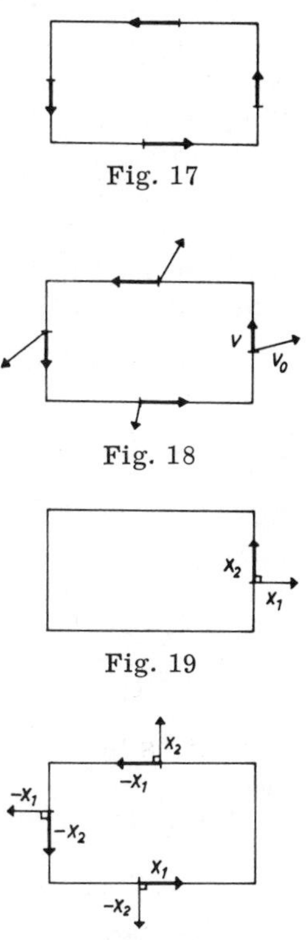

Fig. 17

Fig. 18

Fig. 19

Fig. 20

These arrows also represent vectors (at their initial points). To each of the vectors $\mathfrak{v}$ already drawn, let us now join a linearly independent vector $\mathfrak{v}_0$ pointing *outward* in each case (i.e., a curve segment representing the vector $-\mathfrak{v}_0$ shall contain only interior points of the figure in addition to the boundary point).

The four 2-legs $\mathfrak{v}_0$, $\mathfrak{v}$ thus formed (in that order, i.e., with the *outward*-pointing vector always *first*) all have the same sense. To see that, let the one 2-leg have the same sense as the 2-leg associated with x_1, x_2. The vectors $\mathfrak{v}_0$, $\mathfrak{v}$ need not have the coordinates $\begin{pmatrix}1\\0\end{pmatrix}$, $\begin{pmatrix}0\\1\end{pmatrix}$ with respect to x_1, x_2, but they do have coordinates

$$\begin{pmatrix}a\\b\end{pmatrix}, \begin{pmatrix}0\\c\end{pmatrix}$$

with positive a and c, so they have a positive determinant of the coordinate values relative to x_1, x_2. Now the remaining 2-legs clearly have the same sense as the 2-legs associated with the coordinate systems $x_2, -x_1$ and $-x_1, -x_2$ and $-x_2, x_1$ respectively: hence all 2-legs have the same sense as x_1, x_2. Thus coherent orientation of the boundary distinguishes this coordinate system (and any of like sense) as having positive sense. The procedure just employed (supplementation of a boundary vector of positive sense by a *prefixed* vector pointing outward) correlates a unique orientation of the plane figure with any orientation of the boundary. We call this correlation the *canonical orientation convention.*

If we imagine the boundary of a simple plane figure distorted into a circle, we can represent the orientation pictorially for $N = 2$ by a circle with an arrowhead.

204

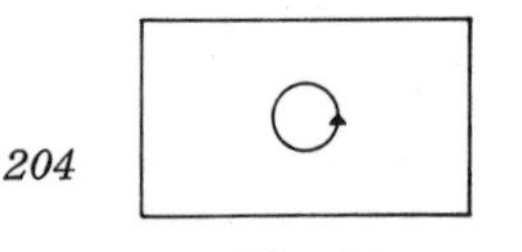

Fig. 21

Conversely, the canonical orientation convention yields an orientation of the boundary for every orientation of a plane figure as follows. At some boundary point, choose a 2-leg $\mathfrak{v}_0, \mathfrak{v}_1$ with positive sense such that $\mathfrak{v}_0$ points *outward* and $\mathfrak{v}_1$ is a vector of the boundary curve. The vector $\mathfrak{v}_1$ then orients the boundary curve.

Finitely many simple 2-dimensional figures can be joined together to form another (compound) 2-dimensional *figure* (a "manifold"). All the simple figures are to have in common are (whole) boundary segments whose parameters are transformable into one another. A *coherent orientation* of the simple figures means in this case that coincident boundary segments must have opposite sense of orientation. A simple example is furnished by a disk (composed of 5 simple figures):

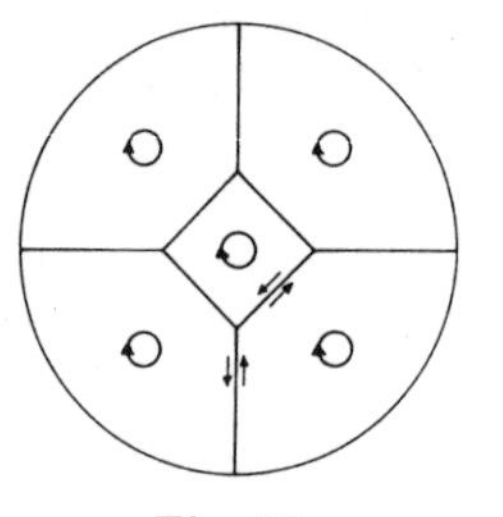

Fig. 22

A further example of a (compound) 2-dimensional figure is the boundary of a 3-dimensional figure. We represent the 3-dimensional interval corresponding to the coordinate values of a simple figure by means of a Euclidean rectangular parallelepiped. The faces of the parallelepiped then correspond to the segments of the boundary.

If we orient one face of the parallelepiped arbitrarily, we thereby also orient the edges of that face—as well as, by the coherence requirement, the four bordering faces.

This also orients the four edges of the remaining face coherently, as is easily seen. Thus the entire boundary is coherently oriented. [The familiar Möbius loop shows that a (compound) 2-dimensional figure need not always be coherently orientable.]

If the boundary of the 3-dimensional figure is oriented, then if we choose a 2-leg $\mathfrak{v}_1, \mathfrak{v}_2$ in any of the boundary segments and join it to a linearly independent vector $\mathfrak{v}_0$—in the leftmost position!—which is directed into the *exterior* of the figure (i.e., $-\mathfrak{v}_0$ is represented by curve segments made up of interior points), the resulting 3-legs will again have the same sense in every case. By the *canonical orientation convention* these 3-legs are assigned positive sense. We refrain from going into further details here; the same goes for the extension of these considerations to $N > 3$.

We have yet to generalize the case considered above of a smooth curve segment lying within a figure. Let R be an N-dimensional figure. For $1 \leqslant p \leqslant N$, let S be a p-dimensional figure whose points fall within R. Let $x_1, \ldots, x_N$ be a coordinate system of R. The coordinate systems $u_1, \ldots, u_p$ of S we shall call *parameter systems.*

S is called a *smooth surface segment* in R if the p-place functions ϕ_i $(i = 1, \ldots, N)$ uniquely determined by a coordinate system X and a parameter system U such that

$$x_i = \phi_i(u_1, \ldots, u_p)$$

205

are continuously differentiable and the p columns

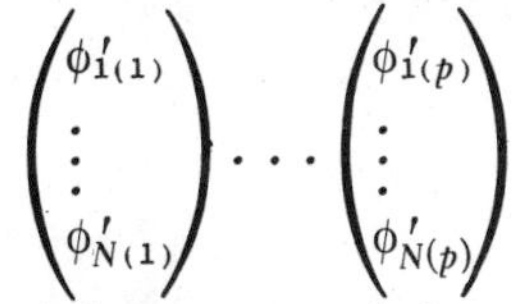

are linearly independent for every point.

Thus the p-leg associated with the parameter system $u_1, \ldots, u_p$ must always be linearly independent as a p-leg made up of the vectors of the figure R.

This concept of a smooth surface segment within a figure is independent of the particular coordinate and parameter systems used in the definition.

Finitely many smooth surface segments in a figure can be joined together into "*piecewise smooth surfaces*." We construe the notion of a *surface* within a figure quite generally so as to include the special cases $p = 1$ (a curve within a figure) and $p = N$.

If in the definition of a smooth surface segment S in a simple figure R the functions ϕ_i such that

$$x_i = \phi_i(u_1, \ldots, u_p)$$

are required to be continuously differentiable not just once but up to the rth order (or even analytic), the result is a surface segment that is *smooth up to the rth order* (or *analytic*, as the case may be). In that case, however, in order for the concept in question to be independent of particular coordinate and parameter systems, the coordinate transformations in R and the parameter transformations in S must also be restricted to functions that are continuously differentiable up to the rth order (or analytic). In connection with differentials there will be places where it is important to confine ourselves at least to doubly smooth surface segments (those that are smooth at least up to the 2nd order).

§20. Tensors and Differential Forms

Let K be a simple curve segment, x a coordinate of K, and f a point function on K. f and x uniquely determine a number function ϕ such that $f = \phi x$; i.e.,

$$fP = \phi x(P) \quad \text{for all } P.$$

As already noted, f is called (continuously) differentiable whenever ϕ is (continuously) differentiable.

206 If f is differentiable, then the *derivative of f with respect to x at P_0* can be defined by the following limit:

$$\lim_{P \to P_0} \frac{fP - fP_0}{xP - xP_0}.$$

The approach to the limit $P \to P_0$ is of course defined as $xP \to xP_0$ (x being an arbitrary coordinate). Thus for $\xi_0 = xP_0$ we get

$$\lim_{P \to P_0} \frac{fP - fP_0}{xP - xP_0} = \lim_{\xi \to \xi_0} \frac{\phi\xi - \phi\xi_0}{\xi - \xi_0}. \tag{20.1}$$

This limit is dependent upon f, x, and P_0. We denote it as $f_x'(P_0)$; thus

$$f_x'(P_0) = \lim_{P \to P} \frac{fP - fP_0}{xP - xP_0}.$$

By (20.1) we also have

$$f_x'(P_0) = \phi' x(P_0),$$

so that

$$f_x' = \phi' x. \tag{20.2}$$

The point function f_x' is known as the *derivative of f with respect to x*. The notation f' as opposed to f_x' is imprecise and should be used only when there can be no mistake about the coordinate x with respect to which f is being differentiated. (In contrast, ϕ' is appropriate for number functions ϕ, since in that case there is no variable to which to refer.)

In place of ϕ' we also wrote $\mathscr{D}\phi$ and $\mathscr{D}_\xi\phi(\xi)$. Similarly, we could now write $\mathscr{D}_x f$ for f_x'. But to avoid confusion, we will reserve $\mathscr{D}$ for numerical functions.

The possibility remains of representing the derivative f_x' as a *"differential quotient"*

$$\frac{df}{dx}.$$

So as not simply to introduce this as traditional notation, we proceed as follows. We consider what $f_{\bar{x}}'$ might be for a second coordinate $\bar{x}$. If $x = \psi\bar{x}$, then it follows directly from $f = \phi x$ that $f = \phi\psi\bar{x}$, whence by (20.2)

$$f_{\bar{x}}' = (\phi \urcorner \psi)'\bar{x},$$

and by the chain rule,

$$\begin{aligned} f_{\bar{x}}' &= \phi'(\psi\bar{x}) \cdot \psi'\bar{x} \\ &= \phi'x \cdot \psi'\bar{x} \\ &= f_x' \cdot x_{\bar{x}}'. \end{aligned}$$

The resulting equation,

207

$$f_{\bar{x}}' = f_x' \cdot x_{\bar{x}}', \tag{20.3}$$

which is an equation between point functions, means that for any point $P \in K$,

$$f_{\bar{x}}'(P) = f_x'(P) \cdot x_{\bar{x}}'(P).$$

But let us stay with equation (20.3) and consider $f_{\bar{x}}'$ and $x_{\bar{x}}'$ now as terms for point functions in their dependence on the variable $\bar{x}$ (which ranges over coordinates of K). The term $f_{\bar{x}}'$ represents a *"functional,"* i.e., a function the arguments of which are themselves functions, in this case, coordinates. The values here are once again point functions. We call the functional thus represented the *differential* df; i.e., we stipulate that

$$df \urcorner \bar{x} = f_{\bar{x}}'. \tag{20.4}$$

Then we can write in place of (20.3) the equation

$$df = f_x' \cdot dx. \tag{20.5}$$

On the right we now have the differential dx, which is likewise a functional. Multiplication by the point function f_x' of course again yields a functional—its value for the argument $\bar{x}$ is $f_x' \cdot dx \urcorner \bar{x}$—and as (20.5) asserts, this new functional is none other than the differential df. Since df arises from the multiplication of dx by f_x', f_x' is uniquely determined as the quotient of df and dx:

$$f_x' = \frac{df}{dx}, \tag{20.6}$$

with the sole proviso that $x'_{\bar{x}}(P) \neq 0$ for any P.

Taking everything together, we now have

$$\frac{df}{dx}(P_0) = \lim_{P \to P_0} \frac{fP - fP_0}{xP - xP_0}.$$

If we further define

$$\overset{P}{\underset{P_0}{\Delta}} f = fP - fP_0,$$

and similarly

$$\overset{P}{\underset{P_0}{\Delta}} x = xP - xP_0,$$

then we also have—in abbreviation—

$$\frac{df}{dx}(P_0) = \lim_{P_0} \frac{\underset{P_0}{\Delta f}}{\underset{P_0}{\Delta x}}$$

208 or, abbreviated still further,

$$\frac{df}{dx} = \lim \frac{\Delta f}{\Delta x}. \tag{20.7}$$

This notation originated in the use of d in place of Δ when the approach to the limit was intended. But since, of course,

$$\lim_{P \to P_0} \overset{P}{\underset{P_0}{\Delta}} f = 0,$$

the purported explanation of (20.7) to the effect that "df and dx stand for the limits of Δf and Δx" explains nothing at all. The above definition (20.4) of differentials as functionals, on the other hand, yields an unobjectionable derivation of (20.7).

In connection with differential quotients, we might take note in passing of the way Leibniz's symbolism has been extended to higher derivatives. For $f = \phi x$ we laid down $f'_x = \phi' x$; similarly (in the case of multiple differentiability) we can define

$$f''_x = \phi'' x$$

$$\vdots$$

$$f^{(r)}_x = \phi^{(r)} x.$$

Writing $\frac{df}{dx}$ alternatively as $\frac{d}{dx} f$, we get

$$f''_x = \frac{d}{dx}\frac{df}{dx}.$$

The right-hand side can then be abbreviated as $\frac{d^2f}{dx^2}$. In similar fashion we get

$$f_x^{(r)} = \frac{d^rf}{dx^r}.$$

We shall make no use of this symbolism for higher differential quotients, however, for it is not compatible with operations on differentials we have yet to introduce (viz., alternating multiplication and differentiation). The notation $f_x^{(r)}$ for the higher derivatives of point functions is already adequate, anyway.

We now turn again to differentials. These are, as we said, functionals. The chain rule yields the *transformation law*, i.e., the law for computing $df \rceil \bar{x}$ from $df \rceil x$, viz., by multiplication with $\frac{dx}{d\bar{x}}$.

The addition of differentials again gives rise to functionals subject to the same transformation law, as does the multiplication of a differential by a point function (provided the latter is applicable on the curve segment). The linear combinations ω of differentials thus produced are functionals obeying the same transformation law,

209

$$\omega \rceil \bar{x} = \omega \rceil x \cdot \frac{dx}{d\bar{x}}. \tag{20.8}$$

Such linear combinations of differentials are called "*differential forms*." Their arguments—like those of differentials—are coordinates of the curve, and their values are point functions on the curve.

If ω is a differential form, $\omega \rceil x$ is a point function. Only $\omega \rceil x(P)$ is a real number, where P is a point on the curve.

For any coordinate x, any differential form ω can be represented by

$$\omega = gdx, \tag{20.9}$$

where g is an appropriate point function. For, since this holds of differentials

$$df = \frac{df}{dx} \cdot dx,$$

it also holds of all linear combinations.

$\omega = gdx$ of course entails $\omega \rceil x = g$, so that the representation (20.9) is unique provided x is specified.

If g is "integrable," i.e., representable in the form $g = \frac{df}{dx}$, then $\omega = df$; i.e., the differential form is a differential. In the one-dimensional case, therefore, the distinction between differentials and differential forms is not very important. The situation changes in the multidimensional case. Only then does it become advisable to define differentials as "tensors" rather than as functionals.

So that we may investigate the multidimensional case, let R be an N-dimensional figure and $X = x_1, \ldots, x_N$ a coordinate system of R.

If f is a point function on R, then there is a uniquely determined number function ϕ such that

$$fP = \phi(x_1 P, \ldots, x_N P) \quad \text{for all } P \in R.$$

We write this for short as

$$f = \phi(x_1, \ldots, x_N).$$

If ϕ is continuously differentiable, then we can form the partial derivatives of ϕ, and

$$\phi'_{(i)}(x_1, \ldots, x_N) \quad (i = 1, \ldots, N)$$

defines new point functions on R, the partial derivatives of f with respect to the coordinate system X.

210 The usual notation for these partial derivatives is

$$\frac{\partial f}{\partial x_i}.$$

This notation is formed by analogy to the differential quotients $\frac{df}{dx}$ used in the one-dimensional case, but it no longer pretends to be a quotient. Notwithstanding its unpretentiousness, however, the notation is still incomplete. In order to designate

$$\phi'_{(i)}(x_1, \ldots, x_N)$$

fully, the symbol must at least contain occurrences of f, X, and i (or x_i). There must be a reference to the entire coordinate system X.

So as to preserve the connection with the customary $\frac{\partial f}{\partial x_i}$ as closely as possible, let us write

$$\frac{\partial_X f}{\partial x_i} = \phi'_{(i)}(x_1, \ldots, x_N). \tag{20.10}$$

When the coordinate system X with respect to which we are differentiating is understood in a particular context, we can of course omit the subscript X from the upper ∂ by way of abbreviation.

In cases of multiple partial differentiability, we can represent the higher partial derivatives by iterating $\frac{\partial_X}{\partial x_i}$—again supplementing the usual notation by X; e.g., for $X = (x, y, z)$ we can write

$$\frac{\partial_X}{\partial z}\,\frac{\partial_X}{\partial y}\,\frac{\partial_X f}{\partial x}$$

$$\frac{\partial_X}{\partial y}\,\frac{\partial_X}{\partial x}\,\frac{\partial_X f}{\partial x}.$$

We can then abbreviate these symbols as

$$\frac{\partial_X^3 f}{\partial z\,\partial y\,\partial x}$$

$$\frac{\partial_X^3 f}{\partial y\,\partial x^2}.$$

If f is continuously differentiable up to a sufficiently high order, then, as we know, the order of the partial differentiations, and hence the order of ∂x, ∂y, ∂z, does not matter.

In the one-dimensional case we were brought from derivatives to differentials by consideration of the chain rule. Similarly, we now have to investigate how to get from the partial derivatives 211 $\frac{\partial_X f}{\partial x_i}$ with respect to one coordinate system to the partial derivatives $\frac{\partial_{\overline{X}} f}{\partial \overline{x}_i}$ with respect to another coordinate system.

If $X = \Psi(\overline{X})$, i.e., $x_i = \psi_i(\overline{x}_1, \ldots, \overline{x}_N)$, $(i = 1, \ldots, N)$, then from $f = \phi(x_1, \ldots, x_N)$ it follows here that

$$f = \phi(\psi_1(\overline{x}_1, \ldots, \overline{x}_N), \ldots, \psi_N(\overline{x}_1, \ldots, \overline{x}_N))$$

and thence by the chain rule that

$$\begin{aligned}\frac{\partial_{\overline{X}} f}{\partial \overline{x}_i} &= \sum_j \phi'_{(j)}(X) \cdot \psi'_{j(i)}(\overline{X}) \\ &= \sum_j \frac{\partial_X f}{\partial x_j} \cdot \frac{\partial_{\overline{X}} x_j}{\partial \overline{x}_i}.\end{aligned}$$

If we form N-element columns from the partial derivatives, we get

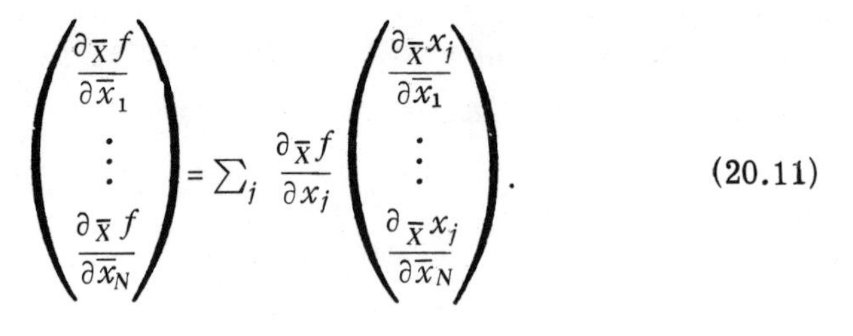

$$\begin{pmatrix} \dfrac{\partial_{\overline{X}} f}{\partial \overline{x}_1} \\ \vdots \\ \dfrac{\partial_{\overline{X}} f}{\partial \overline{x}_N} \end{pmatrix} = \sum_j \frac{\partial_{\overline{X}} f}{\partial x_j} \begin{pmatrix} \dfrac{\partial_{\overline{X}} x_j}{\partial \overline{x}_1} \\ \vdots \\ \dfrac{\partial_{\overline{X}} x_j}{\partial \overline{x}_N} \end{pmatrix}. \tag{20.11}$$

The system

$$\frac{\partial_{\overline{X}} f}{\partial \overline{x}_1}, \ldots, \frac{\partial_{\overline{X}} f}{\partial \overline{x}_N}$$

is uniquely determined by f and the coordinate system $\overline{X}$. Thus once again we can form a functional, with the coordinate systems X as arguments and (written as columns) the systems

$$\begin{pmatrix} \dfrac{\partial_X f}{\partial x_1} \\ \vdots \\ \dfrac{\partial_X f}{\partial x_N} \end{pmatrix}$$

as the respective values. This functional we may call the *differential* df; i.e., we define

$$df \urcorner X = \begin{pmatrix} \dfrac{\partial_X f}{\partial x_1} \\ \vdots \\ \dfrac{\partial_X f}{\partial x_N} \end{pmatrix}. \tag{20.12}$$

212 The chain rule then gives us

$$df = \sum_j \frac{\partial_X f}{\partial x_j} dx_j. \tag{20.13}$$

Here we have on the right a linear combination of differentials, and hence in any case a functional once again.

The chain rule also yields the transformation law for differentials:

$$df \urcorner \overline{X} = \begin{pmatrix} \dfrac{\partial_{\overline{X}} x_1}{\partial \overline{x}_1} & \cdots & \dfrac{\partial_{\overline{X}} x_N}{\partial \overline{x}_1} \\ \vdots & & \vdots \\ \dfrac{\partial_{\overline{X}} x_1}{\partial \overline{x}_N} & \cdots & \dfrac{\partial_{\overline{X}} x_N}{\partial \overline{x}_N} \end{pmatrix} df \urcorner X. \tag{20.14}$$

The matrix on the right here is for $X = \Psi(\overline{X})$ none other than the matrix $\Psi'(\overline{X})$—except that it is *transposed*, reflected around the main diagonal.

Linear combinations of differentials—by addition and by multiplication with point functions—are again functionals ω with (20.14) as their transformation law:

$$\omega \rceil \overline{x} = \begin{pmatrix} \frac{\partial_{\overline{X}} x_1}{\partial \overline{x}_1} & \cdots & \frac{\partial_{\overline{X}} x_N}{\partial \overline{x}_1} \\ \vdots & & \vdots \\ \frac{\partial_{\overline{X}} x_1}{\partial \overline{x}_N} & \cdots & \frac{\partial_{\overline{X}} x_N}{\partial \overline{x}_N} \end{pmatrix} \omega \rceil x.$$

Linear combinations of differentials may again be called *differential forms.*

For any coordinate system X, a differential form can always be represented as

$$\omega = \textstyle\sum_i g_i \, dx_i .$$

By (20.13) this holds for differentials and thus extends to all linear combinations. From $\omega = \sum_i g_i \, dx_i$ it follows that

$$\omega \rceil x = \begin{pmatrix} g_1 \\ \vdots \\ g_N \end{pmatrix}.$$

The g_is are thus uniquely determined by ω and X.

In contrast to the one-dimensional case, a differential form ω in the multidimensional situation is not in general a differential. For if it were the case that

$$df = \textstyle\sum_i g_i \, dx_i ,$$

then it would follow that 213

$$\frac{\partial_X f}{\partial x_i} = g_i \qquad (i = 1, \ldots, N),$$

and hence for $f = \phi(x_1, \ldots, x_N)$ that

$$\phi'_{(i)} (x_1, \ldots, x_N) = g_i .$$

However, these conditions on ϕ, in the case of double continuous differentiability at any rate, are only satisfiable when

$$\frac{\partial_X g_i}{\partial x_j} = \frac{\partial_X g_j}{\partial x_i}.$$

Those differential forms ω which are differentials, i.e., for which there is a function f such that $\omega = df$, are called *total differential forms.*

This introduction of differentials may seem to be nothing more than a tricky exploitation of the chain rule. The notion of identity on which it is based,

$$df = dg \Leftrightarrow \frac{\partial_X f}{\partial x_i} = \frac{\partial_X g}{\partial x_i} \quad (i = 1, \ldots, N),$$

reveals its geometric significance only when the vectors of the tangent space are brought into the picture.

To that end—for reasons to be explained presently—we shall from now on always write the indices of coordinates as superscripts, thus: $x^1, \ldots, x^N$, rather than $x_1, \ldots, x_N$. If t is a parameter of a curve segment, then similarly the indices of the coordinate values of the tangent vector $\mathfrak{v}$ at P_0 represented by the curve segment are to be written as superscripts:

$$v^i(P_0) = \frac{dx^i}{dt}(P_0).$$

As this applies to every point P_0 of the curve segment, we write

$$v^i = \frac{dx^i}{dt}.$$

If, as is customary, we call v^i the coordinate values of the vector $\mathfrak{v}$, then a point P_0 as argument (reference point) is always to be tacitly understood. The same goes for the basis vectors $\mathfrak{x}_1, \ldots, \mathfrak{x}_N$ of the coordinate system $x^1, \ldots, x^N$: they are always understood to be the vectors

$$\mathfrak{x}_1(P_0), \ldots, \mathfrak{x}_N(P_0)$$

at some point P_0.

Every vector $\mathfrak{v}$ is a linear combination of the basis vectors:

$$\mathfrak{v} = \sum_i v^i \mathfrak{x}_i.$$

214 Here we sum over an index i which appears once as a superscript and once as a subscript. This situation comes up again and again in tensor algebra, viz., that summation is to be carried out over an index that occurs twice, once as a subscript and once as a superscript. We can therefore save ourselves the trouble of writing the summation sign in such cases; we simply write

$$\mathfrak{v} = v^i \mathfrak{x}_i .$$

The index appearing in $\frac{\partial f}{\partial x^i}$ is treated as a *subscript* for the purposes of this *summation convention*.

Where

$$a_i = \frac{\partial f}{\partial x^i} \quad \text{and} \quad v^i = \frac{dx^i}{dt},$$

the chain rule

$$\sum_i \frac{\partial f}{\partial x^i} \frac{dx^i}{dt} = \frac{df}{dt}$$

can now be rewritten as

$$a_i v^i = \frac{df}{dt} .$$

Similarly, we have for any other coordinate system $\overline{X}$

$$\overline{a}_i \overline{v}^i = \frac{df}{dt},$$

where

$$\overline{a}_i = \frac{\partial f}{\partial \overline{x}^i} \quad \text{and} \quad \overline{v}^i = \frac{d\overline{x}^i}{dt} .$$

Since v^i and $\overline{v}^i$ are coordinate values of the vector $\mathfrak{v}$, and a_i and $\overline{a}_i$ are the respective values of the functional df, what the above equations show is that $\frac{df}{dt}$ is determined by the differential df and the vector $\mathfrak{v}$ independently of the coordinate system.

Considering now the function which can take any vector as an argument, and which takes the value $\frac{df}{dt}$ for the argument $\mathfrak{v}$, we see that this function is uniquely determined by the differential df.

If we write

$$df\,[\mathfrak{v}] = \frac{df}{dt},$$

then we even have

$$df = dg \longleftrightarrow \Lambda_{\mathfrak{v}}\, df\,[\mathfrak{v}] = dg[\mathfrak{v}].$$

(For we of course defined

$$df = dg \rightleftharpoons df\,[\mathfrak{x}_i] = dg[\mathfrak{x}_i], \quad \text{where} \quad i = 1, \ldots, N.)$$

215 We may thus say that differentials are these *vector functions*, rather than saying that they are functionals.

Differentials are clearly *linear vector functions*, and conversely every linear vector function is a linear combination of differentials. For if ω is a linear vector function and

$$\omega[\mathfrak{x}_j] = c_j,$$

then it follows that

$$\omega = c_i\, dx^i,$$

since

$$\omega[\mathfrak{x}_j] = c_i\, dx^i[\mathfrak{x}_j] \quad \text{in view of} \quad c_j = c_i \frac{\partial x^i}{\partial x^j}.$$

Now, linear vector functions in general are known as "*tensors*"; more precisely, multilinear p-place vector functions are called "*tensors of degree p*." The differential forms treated so far are thus none other than the first-degree tensors of the tangent space.

At the same time, we have in the differentials an alternative symbolism for these tensors.

If $\omega = a_i dx^i$, then $a_1, \ldots, a_N$ are called the (*covariant*) *coordinate values* of the tensor ω with respect to the coordinate system X. It is customary—and extremely useful!—to speak of the tensor a_i for short. This designation is in fact correct if we use the "i" not as a free variable but rather as a place marker (like *, † for sequences). Of course, a_i only indicates the coordinate values, and one must know what coordinate system it is to which one has reference. The coordinate values are called "covariant" because the coordinate values of a system X are converted to those of another system $\overline{X}$ in accordance with the following transformation equations:

$$\bar{a}_j = a_i \frac{\partial_{\overline{X}} x^i}{\partial \bar{x}^j}.$$

The same matrix appears here as in the transformation equations for the basis vectors:

$$\bar{\mathfrak{x}}_j = \mathfrak{x}_i \frac{\partial_{\overline{X}} x^i}{\partial \bar{x}^j}.$$

For the coordinates v^i of a vector $\mathfrak{v} = v^i \mathfrak{x}_i$, on the other hand, we have

$$\bar{v}^j = v^i \frac{\partial_X \bar{x}^j}{\partial x^i}.$$

It is the inverse transposed matrix that appears here, so the coordinates v^i of the vector $\mathfrak{v}$ are called *contravariant* coordinate values. The indices for covariant coordinate values are written as subscripts; those for contravariant coordinate values, as superscripts. That is of course a convention (we could just as well do it the other way around); but once established, it entails writing the indices on coordinate functions as superscripts. One point in favor of the convention is that the coordinate values of the differential df, i.e., $\frac{\partial f}{\partial x^i}$, then come out with subscripts. For example, we can use $f'_{(i)}$ —the customary notation $f_{|i}$ is still simpler— and this accords well with the rest of our notation, e.g., $\phi'_{(i)}$. *216*

In what follows we shall employ not only the *analytic notation* —which simply gives the coordinate values—but also the so-called *symbolic notation*, with special letters, such as ω. For general tensor analysis the analytic notation is indispensable; but for the special case of differential forms, the symbolic notation is occasionally handier. Not until §23 will we use a combination of both notations.

Our earlier considerations surrounding the introduction of differential forms can be generalized from functions to mappings, i.e., to systems of functions.

To begin with, let $F = f_1, \ldots, f_N$ be a system of N point functions on an N-dimensional figure R with a coordinate system $X = (x^1, \ldots, x^N)$.

Then

$$f_i = \phi_i(x^1, \ldots, x^N)$$

uniquely determines a system of number functions $\Phi = \phi_1, \ldots, \phi_N$. The derivative determinant $\psi = |\Phi'|$ is a number function, so that

$$\psi(x^1, \ldots, x^N)$$

is a point function on R which is uniquely determined by F and X. This point function, which we could designate as F_X'—

$$F_X' = |\Phi'(X)| \tag{20.15}$$

—is again representable as a quotient of differentials or a *differential quotient*:

$$F_X' = \frac{d(f_1, \ldots, f_N)}{d(x^1, \ldots, x^N)}. \tag{20.16}$$

In order for this to be more than just traditional notation, the differentials $d(f_1, \ldots, f_N)$—*differentials of degree N*, as they are called—must in turn be defined as functionals, or better still as tensors.

To that end let $\mathfrak{v}_1, \ldots, \mathfrak{v}_N$ be a system of vectors

$$\mathfrak{v}_n = v_n{}^i \mathfrak{x}_i \qquad (n = 1, \ldots, N)$$

217 such that

$$v_n{}^i = \frac{dx^i}{dt_n} \quad (t_n \text{ being a curve parameter}).$$

By the chain rule, the determinant

$$\begin{vmatrix} \frac{df_1}{dt_1} & \cdots & \frac{df_1}{dt_N} \\ \vdots & & \vdots \\ \frac{df_N}{dt_1} & \cdots & \frac{df_N}{dt_N} \end{vmatrix}$$

admits of representation as a product:

$$\begin{vmatrix} \frac{\partial f_1}{\partial x^1} & \cdot & \frac{\partial f_1}{\partial x^N} \\ & \cdot & \\ \frac{\partial f_N}{\partial x^1} & \cdot & \frac{\partial f_N}{\partial x^N} \end{vmatrix} \cdot \begin{vmatrix} \frac{dx^1}{dt_1} & \cdot & \frac{dx^1}{dt_N} \\ & \cdot & \\ \frac{dx^N}{dt_1} & \cdot & \frac{dx^N}{dt_N} \end{vmatrix}.$$

The mapping determined by $f_1, \ldots, f_N$ of the vector system $\mathfrak{v}_1, \ldots, \mathfrak{v}_N$ onto the determinant

$$\begin{vmatrix} \frac{df_1}{dt_1} & \cdot & \frac{df_1}{dt_N} \\ & \cdot & \\ \frac{df_N}{dt_1} & \cdot & \frac{df_N}{dt_N} \end{vmatrix}$$

is accordingly a multilinear vector function, i.e., a tensor. This tensor is denoted by $d(f_1, \ldots, f_N)$. Thus we write

$$d(f_1, \ldots, f_N)[\mathfrak{v}_1, \ldots, \mathfrak{v}_N] = \begin{vmatrix} \frac{df_1}{dt_1} & \cdot & \frac{df_1}{dt_N} \\ & \cdot & \\ \frac{df_N}{dt_1} & \cdot & \frac{df_N}{dt_N} \end{vmatrix}. \qquad (20.17)$$

$d(f_1, \ldots, f_N)$ is a tensor of degree N; furthermore, in view of familiar properties of the determinant, it is alternating; i.e., it changes sign when two arguments $\mathfrak{v}_i, \mathfrak{v}_j$ $(i \neq j)$ are interchanged.

In particular, the basis vectors $\mathfrak{x}_1, \ldots, \mathfrak{x}_N$ of a coordinate system $X = x^1, \ldots, x^N$ are such that

$$d(f_1, \ldots, f_N)[\mathfrak{x}_1, \ldots, \mathfrak{x}_N] = \begin{vmatrix} \dfrac{\partial f_1}{\partial x^1} & \cdot & \dfrac{\partial f_1}{\partial x^N} \\ & \cdot & \\ \dfrac{\partial f_N}{\partial x^1} & \cdot & \dfrac{\partial f_N}{\partial x^N} \end{vmatrix}.$$

The numbers

218

$$d(f_1, \ldots, f_N)[\mathfrak{x}_{i_1}, \ldots, \mathfrak{x}_{i_N}]$$

are known as the *coordinate values* of the differential, i.e., of the tensor $d(f_1, \ldots, f_N)$. Except for their sign, all the coordinate values are equal. If we let

$$A_{i_1 \ldots i_N} = d(f_1, \ldots, f_N)[\mathfrak{x}_{i_1}, \ldots, \mathfrak{x}_{i_N}], \qquad (20.18)$$

then in view of (20.17) we have

$$d(f_1, \ldots, f_N)[\mathfrak{v}_1, \ldots, \mathfrak{v}_N] = A_{i_1 \ldots i_N} \frac{dx^{i_1}}{dt_1} \cdots \frac{dx^{i_N}}{dt_N}.$$

The advantage of this representation is that it no longer expressly involves determinants.

On the other hand, the chain rule also enables us to write (20.17) in the form

$$d(f_1, \ldots, f_N)[\mathfrak{v}_1, \ldots, \mathfrak{v}_N] = A_{1 \ldots N}\, d(x^1, \ldots, x^N)[\mathfrak{v}_1, \ldots, \mathfrak{v}_N]$$

or, more briefly,

$$d(f_1, \ldots, f_N) = A_{1 \ldots N}\, d(x^1, \ldots, x^N). \qquad (20.19)$$

This equation justifies representing the derivative determinant $A_{1 \ldots N} = F_X'$ as a quotient of differentials:

$$A_{1 \ldots N} = \frac{d(f_1, \ldots, f_N)}{d(x^1, \ldots, x^N)}.$$

For a second coordinate system $\overline{X} = \overline{x}^1, \ldots, \overline{x}^N$, the chain rule gives us the transformation law for computing the coordinate values

$$\overline{A}_{i_1 \ldots i_N} = \frac{d(f_1, \ldots, f_N)}{d(\overline{x}^{i_1}, \ldots, \overline{x}^{i_N})}.$$

The law is

$$\overline{A}_{i_1 \ldots i_N} = A_{j_1 \ldots j_N} \frac{\partial x^{j_1}}{\partial \overline{x}^{i_1}} \cdots \frac{\partial x^{j_N}}{\partial \overline{x}^{i_N}}. \qquad (20.20)$$

For, to take an example, we have

$$\bar{A}_{1\ldots N} = \begin{vmatrix} \frac{\partial f_1}{\partial \bar{x}^1} & \cdot & \frac{\partial f_1}{\partial \bar{x}^N} \\ & \cdot & \\ \frac{\partial f_N}{\partial \bar{x}^1} & \cdot & \frac{\partial f_N}{\partial \bar{x}^N} \end{vmatrix} = \begin{vmatrix} \frac{\partial f_1}{\partial x^1} & \cdot & \frac{\partial f_1}{\partial x^N} \\ & \cdot & \\ \frac{\partial f_N}{\partial x^1} & \cdot & \frac{\partial f_N}{\partial x^N} \end{vmatrix} \cdot \begin{vmatrix} \frac{\partial x^1}{\partial \bar{x}^1} & \cdot & \frac{\partial x^1}{\partial \bar{x}^N} \\ & \cdot & \\ \frac{\partial x^N}{\partial \bar{x}^1} & \cdot & \frac{\partial x^N}{\partial \bar{x}^N} \end{vmatrix}$$

$$= A_{1\ldots N} \operatorname{sgn}(j_1, \ldots, j_N) \frac{\partial x^{j_1}}{\partial \bar{x}^1} \cdots \frac{\partial x^{j_N}}{\partial \bar{x}^N}$$

$$= A_{j_1 \ldots j_N} \frac{\partial x^{j_1}}{\partial \bar{x}^1} \cdots \frac{\partial x^{j_N}}{\partial \bar{x}^N}.$$

219 The expression $\operatorname{sgn}(j_1, \ldots, j_N)$ denotes here the number +1 or -1 according as $j_1, \ldots, j_N$ is an even or an odd permutation of $1, \ldots, N$; otherwise $\operatorname{sgn}(j_1, \ldots, j_N) = 0$.

Every linear combination of differentials is itself an alternating tensor of degree N with the same transformation law (20.20) for its coordinate values. By (20.19) every linear combination ω of differentials of degree N can be represented as a differential form:

$$\omega = f d(x^1, \ldots, x^N).$$

In fact, every alternating tensor ω of degree N is a differential form of this type. For from

$$\omega[\mathfrak{v}_1, \ldots, \mathfrak{v}_N] = A_{i_1 \ldots i_N} \frac{dx^{i_1}}{dt_1} \cdots \frac{dx^{i_N}}{dt_N},$$

with alternating coordinate values $A_{i_1 \ldots i_N}$ (i.e., where $A_{i_1 \ldots i_N}$ changes its sign whenever two of its subscripts are interchanged), it follows that

$$\omega[\mathfrak{v}_1, \ldots, \mathfrak{v}_N] = A_{1\ldots N} d(x^1, \ldots, x^N)[\mathfrak{v}_1, \ldots, \mathfrak{v}_N];$$

so that for $f = A_{1\ldots N}$,

$$\omega = f d(x^1, \ldots, x^N).$$

If f is partially integrable, representable, let us say, as $f = \frac{\partial_X g}{\partial x^1}$, then the differential form ω is a differential, viz.,

$$\omega = d(g, x^2, \ldots, x^N).$$

In fact, we have

$$\frac{d(g, x^2, \ldots, x^N)}{d(x^1, x^2, \ldots, x^N)} = \begin{vmatrix} \frac{\partial g}{\partial x^1} & \frac{\partial x^2}{\partial x^1} & \cdots & \frac{\partial x^N}{\partial x^1} \\ \vdots & & & \vdots \\ \frac{\partial g}{\partial x^N} & \frac{\partial x^2}{\partial x^N} & \cdots & \frac{\partial x^N}{\partial x^N} \end{vmatrix} = \begin{vmatrix} f & 0 & \cdots & & 0 \\ & 1 & 0 & \cdots & 0 \\ & & 1 & \ddots & \\ & & \cdots & 0 & 1 \end{vmatrix}$$

$$= f.$$

The distinction between differentials and differential forms does not assume importance again until we take up differentials of degree m in figures of dimensionality $N > m$. For that purpose let $F = f_1, \ldots, f_m$ be a system of point functions on an N-dimensional figure R. If $X = x^1, \ldots, x^N$ is a coordinate system of R, then we can form *partial derivatives* of F with respect to X. These will be determinants formed from the partial derivatives

$$\frac{\partial_X f_j}{\partial x^i} \quad \begin{pmatrix} j = 1, \ldots, m \\ i = 1, \ldots, N \end{pmatrix}.$$

An m-element system $i_1, \ldots, i_m$ of numbers from $\{1, \ldots, N\}$ *220* gives rise to the determinant

$$\begin{vmatrix} \frac{\partial f_1}{\partial x^{i_1}} & \cdots & \frac{\partial f_m}{\partial x^{i_1}} \\ \vdots & & \vdots \\ \frac{\partial f_1}{\partial x^{i_m}} & \cdots & \frac{\partial f_m}{\partial x^{i_m}} \end{vmatrix}.$$

Only if the numbers $i_1, \ldots, i_m$ are pairwise distinct can this determinant be other than 0. Furthermore, since the interchange of two numbers within the system $i_1, \ldots, i_m$ only changes the sign of the determinant, we confine ourselves to systems $i_1, \ldots, i_m$ *in natural order*, i.e., with $i_1 < i_2 < \ldots < i_m$. For these we set

$$\frac{\partial_X(f_1, \ldots, f_m)}{\partial(x^{i_1}, \ldots, x^{i_m})} = \begin{vmatrix} \frac{\partial f_1}{\partial x^{i_1}} & \cdots & \frac{\partial f_m}{\partial x^{i_1}} \\ \vdots & & \vdots \\ \frac{\partial f_1}{\partial x^{i_m}} & \cdots & \frac{\partial f_m}{\partial x^{i_m}} \end{vmatrix}.$$

The *"partial differential quotients"* of systems thus introduced are not quotients either. Unlike the operators $\frac{\partial}{\partial x^i}$, the operators

$$\frac{\partial}{\partial(x^{i_1}, \ldots, x^{i_m})}$$

cannot be iterated; for while their arguments are systems f_1, $\ldots, f_m$, their values are not m-element systems but rather point functions.

Partial differential quotients are a special case of determinants formed from a system $f_1, \ldots, f_m$ and a system of curve parameters $t_1, \ldots, t_m$:

$$\begin{vmatrix} \frac{df_1}{dt_1} & \cdot & \frac{df_m}{dt_1} \\ & \cdot & \\ \frac{df_1}{dt_m} & \cdot & \frac{df_m}{dt_m} \end{vmatrix}.$$

By the chain rule, this determinant can be represented as a linear combination:

$$\sum\nolimits_{i_1 \ldots i_m}^{<} \begin{vmatrix} \frac{\partial f_1}{\partial x^{i_1}} & \cdot & \frac{\partial f_m}{\partial x^{i_1}} \\ & \cdot & \\ \frac{\partial f_1}{\partial x^{i_m}} & \cdot & \frac{\partial f_m}{\partial x^{i_m}} \end{vmatrix} \cdot \begin{vmatrix} \frac{dx^{i_1}}{dt_1} & \cdot & \frac{dx^{i_m}}{dt_1} \\ & \cdot & \\ \frac{dx^{i_1}}{dt_m} & \cdot & \frac{dx^{i_m}}{dt_m} \end{vmatrix}. \qquad (20.21)$$

221 We only sum here over the combinations $i_1 < \ldots < i_m$ of indices from $\{1, \ldots, N\}$. That is why we use $\sum^{<}$ rather than $\sum$. If we wish to sum over indices which appear both as superscripts and as subscripts, then we retain the $\sum^{<}$ but refrain from giving the indices.

The representation (20.21) shows that the determinant

$$\begin{vmatrix} \frac{df_1}{dt_1} & \cdot & \frac{df_m}{dt_1} \\ & \cdot & \\ \frac{df_1}{dt_m} & \cdot & \frac{df_m}{dt_m} \end{vmatrix}$$

depends only on the functions $f_1, \ldots, f_m$ and the vectors $\mathfrak{v}_1, \ldots, \mathfrak{v}_m$ $\left(\text{with coordinate values } \frac{dx^i}{dt_1}, \ldots, \frac{dx^i}{dt_m}\right)$. Thus

$$d(f_1, \ldots f_m)[\mathfrak{v}_1, \ldots, \mathfrak{v}_m] = \begin{vmatrix} \frac{df_1}{dt_1} & \cdot & \frac{df_m}{dt_1} \\ & \cdot & \\ \frac{df_1}{dt_m} & \cdot & \frac{df_m}{dt_m} \end{vmatrix} \qquad (20.22)$$

defines an—obviously alternating—tensor of degree m.

By (20.21), the coordinate values $A_{i_1 \ldots i_m}$ in the representation

$$d(f_1, \ldots, f_m)[\mathfrak{v}_1, \ldots, \mathfrak{v}_m] = A_{i_1 \ldots i_m} \frac{dx^{i_1}}{dt_1} \cdots \frac{dx^{i_m}}{dt_m} \tag{20.23}$$

are determined by

$$A_{i_1 \ldots i_m} = \frac{\partial \mathrm{x}(f_1, \ldots, f_m)}{\partial(x^{i_1}, \ldots, x^{i_m})}.$$

Furthermore, we again get

$$d(f_1, \ldots, f_m)[\mathfrak{v}_1, \ldots, \mathfrak{v}_m]$$

$$= \sum^{<} A_{i_1 \ldots i_m} d(x^{i_1}, \ldots, x^{i_m})[\mathfrak{v}_1, \ldots, \mathfrak{v}_m],$$

or

$$d(f_1, \ldots, f_m) = \sum^{<} A_{i_1 \ldots i_m} d(x^{i_1}, \ldots, x^{i_m}). \tag{20.24}$$

An analogous representation as a linear combination of differentials, i.e., as a *differential form* of degree m, results for every alternating tensor of degree m. For it must hold of any such tensor ω that

$$\omega[\mathfrak{v}_1, \ldots, \mathfrak{v}_m] = A_{i_1 \ldots i_m} \frac{dx^{i_1}}{dt_1} \cdots \frac{dx^{i_m}}{dt_m}$$

with an alternating system of coefficients (the *coordinate values* of the tensor with respect to X). 222

As for the transformation of coordinate values from a coordinate system X to another coordinate system $\overline{X}$, we immediately infer from

$$\omega[\mathfrak{v}_1, \ldots, \mathfrak{v}_m] = A_{i_1 \ldots i_m} \frac{dx^{i_1}}{dt_1} \cdots \frac{dx^{i_m}}{dt_m}$$

$$= A_{i_1 \ldots i_m} \frac{\partial x^{i_1}}{\partial \overline{x}^{j_1}} \cdots \frac{\partial x^{i_m}}{\partial \overline{x}^{j_m}} \frac{d\overline{x}^{j_1}}{dt_1} \cdots \frac{d\overline{x}^{j_m}}{dt_m}$$

that

$$\overline{A}_{j_1 \ldots j_m} = A_{i_1 \ldots i_m} \frac{\partial x^{i_1}}{\partial \overline{x}^{j_1}} \cdots \frac{\partial x^{i_m}}{\partial \overline{x}^{j_m}}.$$

This transformation law for tensors of degree m can be replaced for alternating tensors by an equivalent procedure of substitution:

$$\omega = \sum^{<}_{i_1 \ldots i_m} A_{i_1 \ldots i_m} d(x^{i_1}, \ldots, x^{i_m}) \tag{20.25}$$

$$= \sum\nolimits^{<}_{i_1 \ldots i_m} A_{i_1 \ldots i_m} \sum\nolimits^{<}_{j_1 \ldots j_m} \frac{\partial(x^{i_1}, \ldots, x^{i_m})}{\partial(\bar{x}^{j_1}, \ldots, \bar{x}^{j_m})} d(\bar{x}^{j_1}, \ldots, \bar{x}^{j_m})$$

$$= \sum\nolimits^{<}_{j_1 \ldots j_m} \left(\sum\nolimits^{<}_{i_1 \ldots i_m} A_{i_1 \ldots i_m} \frac{\partial(x^{i_1}, \ldots, x^{i_m})}{\partial(\bar{x}^{j_1}, \ldots, \bar{x}^{j_m})} \right) d(\bar{x}^{j_1}, \ldots, \bar{x}^{j_m}).$$

This explains why it usually suffices to employ the representations (20.25), and in particular (20.24), without explicit recourse to the definition of differential forms as tensors.

Many textbooks therefore refrain from giving a geometric definition of differential forms. They simply define calculating procedures for the "formal" expressions, as they call those in (20.25). One can also use definitions that single out a particular coordinate system. In place of point functions one can then use number functions—and the transformations of coordinates appear as "*substitutions for variables*."

The experienced worker easily recognizes these different routes as equivalent. For the beginner, however, the path we have taken has the advantage of not requiring him to learn equations of differential forms as a merely traditional notation for something that ought strictly speaking to be denoted quite differently.

In particular, by using point functions we guarantee that differential forms are defined invariantly with respect to coordinates, and not in terms of any special coordinate system.

Interestingly enough, the invariance of differential forms is even preserved when we shift our attention from an N-dimensional figure R to a p-dimensional surface segment S contained in R.

223 The coordinate systems of surface segments S in R we called "parameter systems."

If $U = u^1, \ldots, u^p$ is a parameter system, then it will hold on S that

$$x^i = \psi_i(u^1, \ldots, u^p).$$

If now ω is a differential form of degree m on R, i.e.,

$$\omega = \sum\nolimits^{<}_{i_1 \ldots i_m} d(x^{i_1}, \ldots, x^{i_m}),$$

then the $A_{i_1 \ldots i_m}$s and $x^1, \ldots, x^N$ are at the same time point functions on the surface segment S. Of course, one can distinguish a function f on R from the subfunction f_S which is only applicable to the subset $S \subseteq R$ but which for every $P \in S$ takes precisely the value fP. Instead, we shall simply say that every function on R is *simultaneously* a function on every surface segment S contained in R.

When f_S is distinguished from f, f_S is called the "function f *restricted* to S."

Now, the differential forms on R can likewise be restricted to any surface segment contained in R. The situation is somewhat more complicated than with point functions, however, for the differential forms are tensors with vectors as arguments. Only the values of a tensor (at a certain point) are numbers.

Suppose first that $m \leqslant p$. Since $x^{i_1}, \ldots, x^{i_m}$ are simultaneously functions on S, $d(x^{i_1}, \ldots, x^{i_m})$ is simultaneously a differential form on S:

$$d(x^{i_1}, \ldots, x^{i_m}) = \sum_{1}^{p} {}^{<}_{k_1 \ldots k_m} \frac{\partial_U (x^{i_1}, \ldots, x^{i_m})}{\partial (u^{k_1}, \ldots, u^{k_m})} d(u^{k_1}, \ldots, u^{k_m}).$$

Hence also every differential form ω on R (being a linear combination of differentials) simultaneously represents a differential form ω_S on S. We do, of course, have to prove here that any other representation of ω, e.g., as a linear combination of differentials $d(\bar{x}^{j_1}, \ldots, \bar{x}^{j_m})$, will lead to the same differential form ω_S on S. Since

$$d(x^{i_1}, \ldots, x^{i_m}) = \sum{}^{<}_{j_1 \ldots j_m} \frac{\partial_{\bar{X}} (x^{i_1}, \ldots, x^{i_m})}{\partial (\bar{x}^{j_1}, \ldots, \bar{x}^{j_m})} d(\bar{x}^{j_1}, \ldots, \bar{x}^{j_m}),$$

all we have to prove is that

$$\frac{\partial_U (x^{i_1}, \ldots, x^{i_m})}{\partial (u^{k_1}, \ldots, u^{k_m})} = \sum{}^{<}_{j_1 \ldots j_m} \frac{\partial_{\bar{X}} (x^{i_1}, \ldots, x^{i_m})}{\partial (\bar{x}^{j_1}, \ldots, \bar{x}^{j_m})} \frac{\partial_U (\bar{x}^{j_1}, \ldots, \bar{x}^{j_m})}{\partial (u^{k_1}, \ldots, u^{k_m})}.$$

But this is simply a consequence of the chain rule.

For $m > p$, no system $k_1 < k_2 < \ldots < k_m$ can be formed out of $\{1, \ldots, p\}$, so that $\omega_S = 0$. 224

The simplest example of the restricting of differential forms is the case of a first-degree differential form which is restricted to a curve segment K.

If u is a parameter of the curve, then for

$$\omega = f_i \, dx^i$$

the restricted differential form will be

$$\omega_K = f_i \frac{dx^i}{du} du.$$

Ordinarily, we shall abbreviate ω_S as ω.

§21. Alternating Multiplication and Differentiation

So far we have introduced the higher-degree differentials $d(f_1, \ldots, f_m)$ only as a generalization of first-degree differentials df. There is, however, a close algebraic connection between the higher-degree differentials and those of the first degree. All differentials are tensors the coordinate values of which are determinants formed from the partial differential quotients of the first degree:

$$\frac{\partial x(f_1, \ldots, f_m)}{\partial(x^{i_1}, \ldots, x^{i_m})}$$

is the determinant of the matrix with the columns

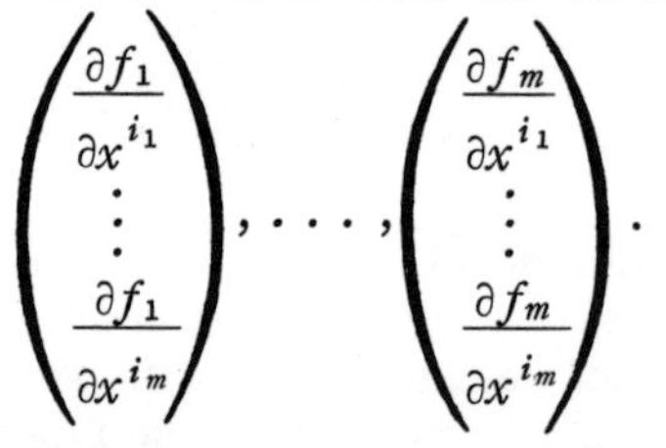

Now, every determinant, viewed as a function of its columns, is multilinear and alternating (i.e., with the interchange of two columns, the determinant changes sign).

Thus in order to obtain the representation

$$d(f_1, \ldots, f_m) = \sum\nolimits^{<}_{i_1 \ldots i_m} \frac{\partial(f_1, \ldots, f_m)}{\partial(x^{i_1}, \ldots, x^{i_m})} \, d(x^{i_1}, \ldots, x^{i_m})$$

from the representations

$$df_j = \sum\nolimits_i \frac{\partial f_j}{\partial x^i} \, dx^i,$$

225 we have only to form the product

$$df_1 \ldots df_m = \sum\nolimits_{i_1 \ldots i_m} \frac{\partial f_1}{\partial x^{i_1}} \cdots \frac{\partial f_m}{\partial x^{i_m}} \, dx^{i_1} \ldots dx^{i_m} \quad (21.1)$$

(by the usual algebraic rules) and to stipulate the following for the index systems $i_1, \ldots, i_m$ occurring here:

(1) If $i_1, \ldots, i_m$ are not pairwise distinct, then let

$$dx^{i_1} \ldots dx^{i_m} = 0.$$

(2) If $i_1, \ldots, i_m$ are pairwise distinct and $i_{j_1} < i_{j_2} < \ldots < i_{j_m}$, where $j_1, \ldots, j_m$ is a permutation of $1, \ldots, m$, then let

$$dx^{i_1} \ldots dx^{i_m} = \operatorname{sgn}(j_1, \ldots, j_m) d(x^{i_{j_1}}, \ldots, x^{i_{j_m}}).$$

By the definition of the determinant, these conventions yield the simple result that

$$d(f_1, \ldots, f_m) = df_1 \ldots df_m. \tag{21.2}$$

Thus at this point we can even do away with the notation $d(f_1, \ldots, f_m)$. The products of the differentials of the coordinates, $dx^{i_1} \ldots dx^{i_m}$, then yield $\pm dx^{i_{j_1}} \ldots dx^{i_{j_m}}$, with $i_{j_1} < \ldots < i_{j_m}$. The sign is determined by the fact that the product $dx^{i_1} \ldots dx^{i_m}$ is to be transformed by the interchange of adjacent differentials: we need only set

$$dx^k dx^l = -dx^l dx^k \tag{21.3}$$

(always shifting the sign to the left of the entire product).

If we apply (21.3) to $k = l$, we get

$$dx^k dx^k = 0$$

and in general

$$dx^{i_1} \ldots dx^{i_m} = 0$$

for $i_1, \ldots, i_m$ which are not pairwise distinct.

In essence, then, the multiplication of differentials introduced here is uniquely established by (21.3); hence it is called "alternating multiplication." The advantage of alternating multiplication is that the trivially derived equation (21.1) can take the place of the earlier definition of $d(f_1, \ldots, f_m)$. In particular, no special mention of determinants is any longer necessary.

However, if we wish to introduce alternating multiplication for differential forms of arbitrary degree (multiplication follows the usual rules, with order taken into account), then the proof of invariance with respect to coordinates (and, with the differential forms restricted to surface segments, the proof of invariance with respect to parameters) is equivalent to the proof of familiar determinant theorems, such as Laplace's expansion theorem. *226*

Alternating multiplication is then uniquely determined by the fact that it is

(1) associative

(2) distributive

and the fact that for functions f and first-degree differential forms ω, ω_1, ω_2

(3) $f\omega = \omega f$

(4) $\omega_1\omega_2 = -\omega_2\omega_1.$

We refrain here from verifying the invariance of this operation with respect to coordinate systems, which comes down algebraically to invariance with respect to the basis representations

$$\omega = a_i dx^i$$

of differential forms of the first degree.

The purely algebraic transition from first-degree differential forms to those of higher degree (which need not be interpreted as alternating tensors) makes use only of the fact that the first-degree differential forms constitute a vector space. Thus—and this will become important in §23—the vectors of the tangent space of a figure (at a fixed point P_0) can also be alternately multiplied by one another. The resulting products (as well as their linear combinations) are called "*multivectors.*"

Using now V, V_1, V_2, . . . as variables for these multivectors, we may denote the alternating product of V_1 and V_2 by $[V_1 V_2]$. *Alternating multiplication* of multivectors is associative and distributive; furthermore, for functions f and vectors (i.e., first-degree multivectors) $\mathfrak{v}$, $\mathfrak{v}_1$, $\mathfrak{v}_2$ we have

$$[f\mathfrak{v}] = [\mathfrak{v}f]$$

$$[\mathfrak{v}_1\mathfrak{v}_2] = -[\mathfrak{v}_2\mathfrak{v}_1].$$

Just as every vector $\mathfrak{v}$ can be interpreted as a linear function of first-degree tensors ω—by setting

$$\mathfrak{v}[\omega] = \omega[\mathfrak{v}]$$

—so also every multivector of degree p can be interpreted as a linear function of alternating tensors of degree p.

227 For vectors

$$\mathfrak{v}_1 = v_1^i \mathfrak{x}_i, \ldots, \mathfrak{v}_p = v_p^{\,i}\mathfrak{x}_i$$

(with basis vectors $\mathfrak{x}_1, \ldots, \mathfrak{x}_N$) we get as the alternating product (with inner brackets omitted in view of associativity)

$$[\mathfrak{v}_1 \ldots \mathfrak{v}_p] = v_1^{i_1} \ldots v_p^{\,i_p}\,[\mathfrak{x}_{i_1} \ldots \mathfrak{x}_{i_p}]$$

$$= \sum\nolimits^{<}_{i_1 \ldots i_p}\Big(\sum\nolimits_1^p{}_{j_1 \ldots j_p}\operatorname{sgn}(j_1, \ldots, j_p)\, v^{i_{j_1}} \ldots v^{i_{j_p}}\Big)[\mathfrak{x}_{i_1} \ldots \mathfrak{x}_{i_p}].$$

We use the coefficients

$$\sum_{j_1 \ldots j_p} \operatorname{sgn}(j_1, \ldots, j_p) v_1^{i_{j_1}} \ldots v_p^{i_{j_p}}$$

which appear here as (*contravariant*) *coordinate values*

$$v^{i_1 \ldots i_p}$$

of the multivector $[\mathfrak{v}_1 \ldots \mathfrak{v}_p]$.

If ω is an alternating tensor of degree p with coordinate values $A_{i_1 \ldots i_p}$, then

$$\omega[\mathfrak{v}_1, \ldots, \mathfrak{v}_p] = A_{i_1 \ldots i_p} v_1^{i_1} \ldots v^{i_p},$$

and it follows that

$$\omega[\mathfrak{v}_1, \ldots, \mathfrak{v}_p] = \sum{}^{<} A_{i_1 \ldots i_p} v^{i_1 \ldots i_p}.$$

Accordingly, $\omega[\mathfrak{v}_1, \ldots, \mathfrak{v}_p]$ is indeed dependent only on ω and on the product $[\mathfrak{v}_1 \ldots \mathfrak{v}_p]$.

Hence we may set

$$[\mathfrak{v}_1 \ldots \mathfrak{v}_p][\omega] = \omega[\mathfrak{v}_1, \ldots, \mathfrak{v}_p].$$

That makes $[\mathfrak{v}_1 \ldots \mathfrak{v}_p]$ a linear function of the alternating tensors of degree p, and by forming linear combinations we obtain all linear functions of the alternating tensors of degree p.

From the multivectors, which as we said will not be needed until §23, we turn our attention back again to the alternating tensors (differential forms) alone. Having already characterized alternating multiplication, we proceed now to introduce an operation of *alternating differentiation* for these tensors.

So far we only have a differentiation operation d that takes us from functions (invariants) to differentials (first-degree tensors), i.e., from f to df, or in analytic notation from f to $\frac{\partial f}{\partial x^i}$.

It seems natural to try to generalize this operation by differentiating the coordinate values so as to take an arbitrary tensor of degree p into a tensor of degree $p + 1$.

Such an operation, however, is not invariant with respect to coordinates. Even for a first-degree tensor A_j, the numbers $\frac{\partial A_j}{\partial x^i}$ are not transformed covariantly. For it is after all the case that 228

$$\frac{\partial \bar{A}_j}{\partial \bar{x}^i} = \frac{\partial}{\partial \bar{x}^i}\left(A_k \frac{\partial x^k}{\partial \bar{x}^j}\right).$$

Only if $\frac{\partial x^k}{\partial \bar{x}^j}$ is constant (i.e., in cases of affine, and not just locally affine, coordinate transformations) do we get

$$\frac{\partial \bar{A}_j}{\partial \bar{x}^i} = \frac{\partial A_k}{\partial x^l} \frac{\partial x^l}{\partial \bar{x}^i} \frac{\partial x^k}{\partial \bar{x}^j}.$$

For nonaffine coordinate transformations, on the other hand, we get

$$\frac{\partial \bar{A}_j}{\partial \bar{x}^i} = \frac{\partial A_k}{\partial x^l} \frac{\partial x^k}{\partial \bar{x}^j} \frac{\partial x^l}{\partial \bar{x}^i} + A_k \frac{\partial^2 x^k}{\partial \bar{x}^j \partial \bar{x}^i}.$$

From this it follows in the case of twice continuously differentiable coordinate transformations that, since

$$\frac{\partial^2 x^k}{\partial \bar{x}^j \partial \bar{x}^i} = \frac{\partial^2 x^k}{\partial \bar{x}^i \partial \bar{x}^j},$$

$$\frac{\partial \bar{A}_j}{\partial \bar{x}^i} - \frac{\partial \bar{A}_i}{\partial \bar{x}^j} = \left(\frac{\partial A_k}{\partial x^l} - \frac{\partial A_l}{\partial x^k}\right) \frac{\partial x^k}{\partial \bar{x}^j} \frac{\partial x^l}{\partial \bar{x}^i}.$$

Thus $\frac{\partial A_k}{\partial x^l} - \frac{\partial A_l}{\partial x^k}$ is a tensor of the second degree.

Accordingly, by differentiating the coordinate values of a first-degree tensor, we can form an *alternating* second-degree tensor. With the help of differential forms, we may formulate this result as follows.

If $\omega = \sum_i a_i dx^i$ is a differential form of the first degree, it can be correlated with the second-degree differential form

$$\sum^{<}_{j,i} \left(\frac{\partial a_i}{\partial x^j} - \frac{\partial a_j}{\partial x^i}\right) dx^j dx^i.$$

Since

$$da_i = \sum_j \frac{\partial a_i}{\partial x^j} dx^j,$$

that is just another way of writing

$$\sum_i da_i dx^i = \sum_{j,i} \frac{\partial a_i}{\partial x^j} dx^j dx^i.$$

229 Thus every first-degree differential form

$$\omega = \sum_i a_i dx^i$$

can be correlated invariantly with respect to coordinates with the second-degree differential form

$$\sum_i da_i dx^i.$$

We denote this differential form by $d\omega$; i.e.,

$$d\omega = \sum_i da_i dx^i.$$

With that we have extended the *differentiation* d of functions f to first-degree differential forms ω.

Before proceeding to higher-degree differential forms, we note the following:

(1) For any function f which is twice continuously differentiable we have

$$ddf = 0.$$

Proof: $ddf = \sum_i d \frac{\partial f}{\partial x^i} dx^i = \sum_{j,i}^{<} \frac{\partial^2 f}{\partial x^j \partial x^i} - \frac{\partial^2 f}{\partial x^i \partial x^j} dx^j dx^i = 0.$

(2) The operation d is linear.

Proof: trivial.

(3) Denoting alternating multiplication for the sake of clarity with a dot, we get the following rules for products:

$$d(f \cdot \omega) = df \cdot \omega + f \cdot d\omega$$

$$d(\omega \cdot f) = d\omega \cdot f - \omega \cdot df.$$

Proof. Where $\omega = \sum a_i dx^i$, we have

$$\begin{aligned} d\sum fa_i dx^i &= \sum d(fa_i)dx^i \\ &= \sum df \cdot a_i dx^i + \sum f \cdot da_i dx^i \\ &= df \cdot \omega + f \cdot d\omega \end{aligned}$$

and

$$\omega \cdot f = f \cdot \omega, \quad d\omega \cdot f = f \cdot d\omega, \quad \omega \cdot df = -df \cdot \omega.$$

In order to extend the operation d to differential forms of arbitrary degree, it suffices to lay down the following general multiplication rules (for functions f, first-degree differential forms ω, and arbitrary differential forms Ω) in addition to (1) and (2):

$$(3') \qquad \begin{aligned} d(f \cdot \Omega) &= df \cdot \Omega + f \cdot d\Omega \\ d(\omega \cdot \Omega) &= d\omega \cdot \Omega - \omega \cdot d\Omega. \end{aligned}$$

These postulates uniquely determine the operation d. Every differential form is the sum of products $fdg_1 \ldots dg_m$, and by (3′) we get *230*

$$d(fdg_1 \ldots dg_m) = dfdg_1 \ldots dg_m + fd(dg_1 \ldots dg_m)$$

and

$$d(dg_1 \ldots dg_m) = ddg_1 \cdot dg_2 \ldots dg_m - dg_1 d(dg_2 \ldots dg_m).$$

Since $ddg_1 = 0$, the first addend on the right vanishes. Induction on m therefore yields the general result that

$$d(dg_1 \ldots dg_m) = 0.$$

It thus follows necessarily from our postulates for

$$\Omega = \sum^{<} f_{i_1 \ldots i_m} dx^{i_1} \ldots dx^{i_m}$$

that

$$d\Omega = \sum^{<} df_{i_1 \ldots i_m} dx^{i_1} \ldots dx^{i_m}. \tag{21.4}$$

If we now take (21.4) as our definition of *alternating differentiation* d for arbitrary differential forms Ω, we shall thereby have generalized the previous cases $m = 0$ and $m = 1$. The invariance of definition (21.4) with respect to coordinates follows from the fact that the rules (1), (2), and (3′) are satisfied. For (1) and (2) this is trivial.

For the multiplication rules (3′), we get first of all

$$\begin{aligned} d(f\Omega) &= d\sum^{<} ff_{i_1 \ldots i_m} dx^{i_1} \ldots dx^{i_m} \\ &= \sum^{<} df \cdot f_{i_1 \ldots i_m} dx^{i_1} \ldots dx^{i_m} \\ &\qquad + \sum^{<} f \cdot df_{i_1 \ldots i_m} dx^{i_1} \ldots dx^{i_m} \\ &= df \cdot \Omega + fd\Omega. \end{aligned}$$

For the second multiplication rule—since ω is the sum of products fdg, and $ddg = 0$—it is obvious that the special case

$$d(dg \cdot \Omega) = -dg \cdot d\Omega$$

suffices. And indeed, by (21.4),

$$\begin{aligned} d(dg \cdot \Omega) &= d\sum^{<} f_{i_1 \ldots i_m} dgdx^{i_1} \ldots dx^{i_m} \\ &= \sum^{<} df_{i_1 \ldots i_m} dgdx^{i_1} \ldots dx^{i_m} \\ &= -\sum^{<} dgdf_{i_1 \ldots i_m} dx^{i_1} \ldots dx^{i_m} \\ &= -dg \cdot d\Omega. \end{aligned}$$

With that we have shown that alternating differentiation as defined by (21.4) is invariant with respect to coordinates—under the assumption of double continuous differentiability, that is. From now on we shall always make this assumption when we use alternating differentiation: 231

(1) The coordinate systems shall be transformable into each other by twice continuously differentiable mappings.

(2) The coefficients of differential forms shall be twice continuously differentiable functions of some (and hence of every) coordinate system.

(3) If the differential forms are restricted to surface segments S, the latter shall be doubly smooth surface segments.

Alternating differentiation d, like alternating multiplication (for which we did not explicitly mention the fact), is interchangeable with its subfunction as restricted to S. That is, if

$$\omega = \sum^{<} f_{i_1 \ldots i_m} dx^{i_1} \ldots dx^{i_m}$$

and

$$\omega_S = \sum^{<}_{i_1 \ldots i_m} f_{i_1 \ldots i_m} \sum^{<}_{j_1 \ldots j_m} \frac{\partial(x^{i_1}, \ldots, x^{i_m})}{\partial(u^{j_1}, \ldots, u^{j_m})} du^{j_1} \ldots du^{j_m},$$

then we have

$$(d\omega)_S = d\omega_S.$$

Proof. From

$$\omega_S = \sum^{<} f_{i_1 \ldots i_m} \left(\sum \frac{\partial x^{i_1}}{\partial u^{j_1}} du^{j_1}\right) \ldots \left(\sum \frac{\partial x^{i_m}}{\partial u^{j_m}} du^{j_m}\right)$$

it follows by the multiplication rules, since $ddx^i = 0$, that

$$d\omega_S = \sum^{<} df_{i_1 \ldots i_m} \left(\sum \frac{\partial x^{i_1}}{\partial u^{j_1}} du^{j_1}\right) \ldots \left(\sum \frac{\partial x^{i_m}}{\partial u^{j_m}} du^{j_m}\right) = (d\omega)_S.$$

The properties by which we characterized alternating differentiation are easily generalized:

THEOREM 20.1 For alternating differentiation d we have

(1) $d(\omega_1 + \omega_2) = d\omega_1 + d\omega_2$

(2) $dd\omega = 0$

(3) $d(\omega_1 \cdot \omega_2) = d\omega_1 \cdot \omega_2 + (-1)^m \omega_1 \cdot d\omega_2$,

where m is the degree of ω_1.

232 While (1) and (2) follow trivially from what we have already done, to get (3) we must use induction on m. The cases $m = 0$ and $m = 1$ are already taken care of. For a representation $\omega_1 = \sum_i df_i \cdot \overline{\omega}_i$, where $\overline{\omega}_i$ are differential forms of degree $m - 1$, the inductive hypothesis yields

$$\begin{aligned} d(\omega_1 \cdot \omega_2) &= d\sum df_i\, \overline{\omega}_i \cdot \omega_2 \\ &= -\sum df_i\, d(\overline{\omega}_i \cdot \omega_2) \\ &= -\sum df_i\, d\overline{\omega}_i \cdot \omega_2 - (-1)^{m-1} \sum df_i\, \overline{\omega}_i\, d\omega_2 \\ &= d\left(\sum df_i \cdot \overline{\omega}_i\right) \cdot \omega_2 + (-1)^m \left(\sum df_i\, \overline{\omega}_i\right) \cdot d\omega_2 \\ &= d\omega_1 \cdot \omega_2 + (-1)^m \omega_1 \cdot d\omega_2 . \end{aligned}$$

Generalizing the notion as it applied to first-degree differential forms, we call any differential form ω a *total differential form* if there is a differential form Ω such that $\omega = d\Omega$. We saw that a first-degree differential form

$$\omega = \sum f_i\, dx^i$$

is total in a simply connected domain if and only if

$$\frac{\partial f_i}{\partial x^j} = \frac{\partial f_j}{\partial x^i} \qquad (i, j = 1, \ldots, N).$$

Hence, since

$$\begin{aligned} d\omega &= \sum df_i\, dx^i \\ &= \sum_{j,i}^{<} \left(\frac{\partial f_i}{\partial x^j} - \frac{\partial f_j}{\partial x^i}\right) dx^j dx^i , \end{aligned}$$

a first-degree differential form is total just in case $d\omega = 0$.

In simple figures (which can be mapped by coordinate systems onto intervals of the number space), this theorem holds for arbitrary differential forms.

POINCARÉ'S THEOREM. In a simple figure, a differential form ω of degree m is total (i.e., there is a differential form Ω of degree $m - 1$ such that $\omega = d\Omega$) if and only if $d\omega = 0$.

Proof. Let

$$\omega = \sum_{i_1 \ldots i_m}^{<} f_{i_1 \ldots i_m}\, dx^{i_1} \ldots dx^{i_m},$$

where the coordinate system $x^1, \ldots, x^N$ is a mapping onto an N-dimensional interval. In cases where $N < m$ there is nothing to prove, since then $\omega = 0$.

We proceed with the proof by induction on N. The hypothesis of induction will thus be Poincaré's theorem for differential forms of degree m on simple $(N - 1)$-dimensional figures. 233

We break ω down in the following fashion:

$$\omega = \sum_{2}^{N}{}^{<}_{i_2 \ldots i_m} f_{1 i_2 \ldots i_m}\, dx^1 dx^{i_2} \ldots dx^{i_m}$$

$$+ \sum_{2}^{N}{}^{<}_{i_1 \ldots i_m} f_{i_1 \ldots i_m}\, dx^{i_1} \ldots dx^{i_m} \qquad (21.5)$$

and we choose functions $F_{i_2 \ldots i_m}$ such that

$$\frac{\partial_X F_{i_2 \ldots i_m}}{\partial x^1} = f_{1 i_2 \ldots i_m}.$$

If

$$f_{i_1 \ldots i_m} = \phi_{i_1 \ldots i_m}(x^1, \ldots, x^N),$$

then (letting the lower end point of the interval of coordinate values be $0, \ldots, 0$) we can simply set

$$F_{i_2 \ldots i_m} = \mathcal{J}_{\xi\,0}^{\;x^i} \phi_{1 i_2 \ldots i_m}(\xi, x^2, \ldots, x^N). \qquad (21.6)$$

We want to show for

$$\Omega = \sum_{2}^{N}{}^{<}_{i_2 \ldots i_m} F_{i_2 \ldots i_m}\, dx^{i_2} \ldots dx^{i_m}$$

that the differential form $\omega - d\Omega$ is a differential form on the simple $(N - 1)$-dimensional figure of points P such that $x^1(P) = 0$.

To begin with, we have

$$d\Omega = \sum_{2}^{N}{}^{<}_{i_2 \ldots i_m} \sum_{1}^{N}{}_{i} \frac{\partial F_{i_2 \ldots i_m}}{\partial x^i} dx^i dx^{i_2} \ldots dx^{i_m}$$

$$= \sum_{2}^{N}{}^{<}_{i_2 \ldots i_m} f_{1 i_2 \ldots i_m}\, dx^1 dx^{i_2} \ldots dx^{i_m} \qquad (21.7)$$

$$+ \sum_{2}^{N}{}_{i_2 \ldots i_m} \sum_{2}^{N}{}_{i} \frac{\partial F_{i_2 \ldots i_m}}{\partial x^i} dx^i dx^{i_2} \ldots dx^{i_m}.$$

We now make use of the assumption that $d\omega = 0$; i.e.,

$$\sum{}^{<}_{i_1 \ldots i_m} \sum{}_{i} \frac{\partial f_{i_1 \ldots i_m}}{\partial x^i} dx^i dx^{i_1} \ldots dx^{i_m} = 0.$$

234 For a system $i_0 < i_1 < \ldots < i_m$, the coefficient of

$$dx^{i_0} \ldots dx^{i_m}$$

is in this case given by

$$\sum_{0}^{m}{}_j (-1)^j \frac{\partial f_{i_0 \ldots \hat{i}_j \ldots i_m}}{\partial x^{i_j}},$$

where we use $i_0 \ldots \hat{i}_j \ldots i_m$ to denote the system that results from deleting i_j from $i_0, \ldots, i_m$.

These coefficients are thus all equal to 0; i.e., for $i_0 = 1$ in particular,

$$\frac{\partial f_{i_1 \ldots i_m}}{\partial x^1} + \sum_{1}^{m}{}_j (-1)^j \frac{\partial f_{1 i_1 \ldots \hat{i}_j \ldots i_m}}{\partial x^{i_j}} = 0. \qquad (21.8)$$

Similarly, in the second summand of (21.7) the coefficient of $dx^{i_1} \ldots dx^{i_m}$ for a system

$$i_1 < \ldots < i_m$$

is

$$\sum_{1}^{m}{}_j (-1)^{j-1} \frac{\partial F_{i_1 \ldots \hat{i}_j \ldots i_m}}{\partial x^{i_j}}.$$

By (21.8) and the definition of $F_{i_2 \ldots i_m}$, this coefficient is given by

$$\mathcal{J}_{\xi\,0}^{x^1} \sum_{1}^{m}{}_j (-1)^{j-1} \phi'_{1 i_1 \ldots \hat{i}_j \ldots i_m (i_j)} (\xi, x^2, \ldots, x^N)$$

$$= \mathcal{J}_{\xi\,0}^{x^1} \phi'_{i_1 \ldots i_m (1)}(\xi, x^2, \ldots, x^N)$$

$$= \phi_{i_1 \ldots i_m} (x^1, x^2, \ldots, x^N) - \phi_{i_1 \ldots i_m} (0, x^2, \ldots, x^N)$$

$$= f_{i_1 \ldots i_m} - \phi_{i_1 \ldots i_m} (0, x^2, \ldots, x^N).$$

A comparison of (21.5) with (21.7) now yields

$$\omega - d\Omega = \sum_{2}^{N}{}^{<}_{i_1 \ldots i_m} \phi_{i_1 \ldots i_m} (0, x^2, \ldots, x^N) dx^{i_1} \ldots dx^{i_m}.$$

This is a differential form on an $(N - 1)$-dimensional figure. From $d\omega = 0$ it follows that $d(\omega - d\Omega) = 0$; hence by the hypothesis of induction $\omega - d\Omega$ is a total differential form, i.e., ω is total.

§22. Multidimensional Integrals

235

The condition that $d\Omega = \omega$ does not determine Ω uniquely, but only up to the addition of a differential form Ω_0 such that $d\Omega_0 = 0$. This corresponds exactly to the simple case of number functions: $\Phi' = \phi$ does not determine Φ uniquely, but only up to the addition of a function Φ_0 such that $\Phi_0' = 0$ (i.e., in this case, up to a constant).

Since it will always hold of a constant function Φ_0 that $\Delta\Phi_0 = 0$, the integral $\mathcal{J}\phi = \Delta\Phi$ is uniquely determined by $\Phi' = \phi$. Poincaré's theorem enables us to extend this definition of the integral from number functions to differential forms. To begin with, let ω be a first-degree differential form on an oriented curve segment $\mathfrak{K}$. If x is a coordinate of $\mathfrak{K}$, then

$$\omega = fdx.$$

For example, if f is continuous, then ω is a total differential form

$$\omega = dF.$$

The function F is determined by ω up to an additive constant F_0 (since $dF_0 = 0$).

Now let P be the initial point and Q the end point of $\mathfrak{K}$. We define

$$\Delta_{\mathfrak{K}} F = F(Q) - F(P). \tag{22.1}$$

$\Delta_{\mathfrak{K}} F$ is uniquely determined by ω and the oriented curve segment. Finally, we define

$$\int_{\mathfrak{K}} \omega = \Delta_{\mathfrak{K}} F \quad \text{where} \quad \omega = dF. \tag{22.2}$$

$\int_{\mathfrak{K}} \omega$ is called the *integral* of ω over $\mathfrak{K}$. Since $\omega = dF$, we can also write

$$\int_{\mathfrak{K}} dF = \Delta_{\mathfrak{K}} F. \tag{22.3}$$

We can formulate this more explicitly as

$$\frac{dF}{dx} = f \rightarrow \int_{\mathfrak{K}} fdx = \Delta_{\mathfrak{K}} F,$$

or quite briefly as

$$\int d = \Delta .$$

The relation between the *geometric integral* $\int$ and the integral $\mathcal{J}$ for numerical functions is simple. If we let $F = \Phi(x)$ and $f = \phi(x)$, then we have

$$\frac{dF}{dx} = f \leftrightarrow \Phi' = \phi$$

236 and furthermore

$$\int_{\mathfrak{R}} f dx = \mathop{\mathcal{J}}_{x(P)}^{x(Q)} \phi, \quad \Delta_{\mathfrak{R}} F = \mathop{\Delta}_{x(P)}^{x(Q)} \Phi .$$

At this point, however, we have no need to resort to earlier results concerning the *numerical integral* $\mathcal{J}$. For example, the *transformation formula*

$$\int_{\mathfrak{R}} f dx = \int_{\mathfrak{R}} f \frac{dx}{d\bar{x}} d\bar{x} \qquad (22.4)$$

now follows from the fact that the integrands are equal differential forms:

$$f dx = f \frac{dx}{d\bar{x}} d\bar{x} .$$

In the Leibnizian notation for the integral $\int f dx$, the differential dx is no mere traditional flourish: the integrand is not f but rather the differential form $f dx$.

Note that in (22.4) integration is to be performed over the same oriented curve segment in each case. Only when the geometric integral—or *curve integral,* as it is called in the one-dimensional case—is transformed into a numerical integral do we get as lub and glb $x(P)$, $x(Q)$ on the left and $\bar{x}(P)$, $\bar{x}(Q)$ on the right respectively.

The transformation formula (22.4) holds for arbitrary coordinates x and $\bar{x}$. The coordinates need not be of like sense; in particular, they need not have positive sense (with respect to the orientation of $\mathfrak{R}$). The restriction to coordinates of positive sense is only necessary if it is always presupposed in the definition of the numerical integral $\mathop{\mathcal{J}}_{\alpha}^{\beta} \phi$, which of course is usually written as $\int_{\alpha}^{\beta} \phi(\xi) d\xi$, that $\alpha < \beta$.

Linearity is proved for curve integrals in just the same way as for numerical integrals. For constants c_1, c_2 we have

$$\int (c_1 \omega_1 + c_2 \omega_2) = c_1 \int \omega_1 + c_2 \int \omega_2 .$$

(For the sake of brevity we omit here the reference to the curve segment under the integral sign.) For if $dF_1 = \omega_1$ and $dF_2 = \omega_2$, then it follows that

$$d(c_1F_1 + c_2F_2) = c_1\omega_1 + c_2\omega_2$$

and

$$\Delta(c_1F_1 + c_2F_2) = c_1\Delta F_1 + c_2\Delta F_2.$$

For a coordinate x of positive sense, the property of *monotony* holds: 237

$$f_1 \leqslant f_2 \rightarrow \int_{\mathfrak{K}} f_1\,dx \leqslant \int_{\mathfrak{K}} f_2\,dx;$$

in particular,

$$f \geqslant 0 \rightarrow \int_{\mathfrak{K}} f dx \geqslant 0. \tag{22.5}$$

For the proof we must revert to the bound-theorem: if $f = \phi(x)$ and $\Phi' = \phi$, then where $x(P) \leqslant x(Q)$, $\underset{x(P)}{\overset{x(Q)}{\Delta}} \Phi \geqslant 0$ follows from $\phi \geqslant 0$. We therefore have

$$\int_{\mathfrak{K}} f dx = \underset{\mathfrak{K}}{\Delta}\Phi(x) \geqslant 0.$$

One advantage of curve integrals over numerical integrals is that now even first-degree differential forms defined in an N-dimensional figure are integrable over curve segments contained in the figure as smooth curve segments.

Let

$$\omega = \sum_1^N {}_i f_i dx^i$$

be a differential form on a figure R and $\mathfrak{K}$ a smooth oriented curve segment in R. If the differential form ω is restricted to $\mathfrak{K}$, then ω will be a differential form on $\mathfrak{K}$, which means that the integral of ω over $\mathfrak{K}$ will also be defined. If u is a parameter of $\mathfrak{K}$, then of course

$$\int_{\mathfrak{K}} \omega = \int_{\mathfrak{K}} \left(\sum_i f_i \frac{dx^i}{du}\right) du.$$

From simple curve segments, integration can in fact be extended to *oriented piecewise smooth curves*. By that we mean systems of finitely many oriented smooth curve segments $\mathfrak{K}_1, \ldots, \mathfrak{K}_n$ satisfying the condition that the end point of $\mathfrak{K}_i$ is the initial point of $\mathfrak{K}_{i+1}$. Otherwise the curve segments are to have no points in common.

Let us again denote this curve, i.e., system, with the letter $\mathfrak{K}$. Then the initial point P of $\mathfrak{K}_1$ shall also be the initial point of $\mathfrak{K}$, and the end point Q of $\mathfrak{K}_n$ shall be the end point of $\mathfrak{K}$. The *integral over* $\mathfrak{K}$ is defined as

$$\int_{\mathfrak{K}} \omega = \sum_1^n {}_i \int_{\mathfrak{K}_i} \omega . \tag{22.6}$$

238 This definition would be of little interest if it were to hold only for curves introduced as systems of simple curve segments. However, if we think of a piecewise smooth curve now as the union of the sets of points in its component simple curve segments, then conversely we can call the system of simple curve segments a *decomposition* of the curve. The question then arises whether definition (22.6) always yields the same integral for two different decompositions of the same curve $\mathfrak{K}$ (with the same orientation around each point).

To decompose a curve segment, of course, we need give only finitely many partition points $P_0 = P, P_1, \ldots, P_m = Q$ on the curve segment. Denoting the integral over a *partial segment* by

$$\int_{P_{j-1}}^{P_j} \omega,$$

we have

$$\int_{\mathfrak{K}} \omega = \sum_1^m {}_j \int_{P_{j-1}}^{P_j} \omega$$

for *simple* curve segments $\mathfrak{K}$. This property of *additivity* is not a definition, but follows rather—as in the case of numerical integrals—from the equation

$$\Delta_{\mathfrak{K}} F = \sum_1^m {}_j \overset{P_j}{\underset{P_{j-1}}{\Delta}} F.$$

Now, because of the additivity of simple curve segments, the definition (22.6) is indeed independent of the decomposition $\mathfrak{K}_1, \ldots, \mathfrak{K}_n$. For if we have another decomposition $\mathfrak{K}_1', \ldots, \mathfrak{K}_m'$ of the same curve $\mathfrak{K}$, then all the boundary points of the $\mathfrak{K}_i$s and $\mathfrak{K}_j'$s together yield a decomposition of $\mathfrak{K}$: a *common refinement* of the original decompositions, as it is called. From the existence of a common refinement and the additivity of all simple curve segments, it follows trivially that

$$\sum_1^n {}_i \int_{\mathfrak{K}_i} \omega = \sum_1^m {}_j \int_{\mathfrak{K}_j'} \omega;$$

for both sums can be broken down into double sums with the same summands.

We can now use definition (22.6) to define the integral for the circle, for example. The *circle* $\mathfrak{K}$ (being part of a plane with Cartesian coordinates x, y) has as its points the points P such that

$$x^2(P) + y^2(P) = r^2.$$

Any arbitrary pair of points will divide the circle into two simple curve segments $\mathfrak{K}_1$, $\mathfrak{K}_2$. If the circle is oriented according to the canonical orientation convention (and the coordinate system x, y has positive sense in the plane, let us say), then the curve segments are also oriented. We have here the special case in which the initial point (i.e., the initial point of $\mathfrak{K}_1$) coincides with the end point (i.e., that of $\mathfrak{K}_2$). 239

Curves whose initial and end points coincide are called *closed* curves. In view of additivity (22.6), it is immaterial for the integral $\int_{\mathfrak{K}} \omega$ in the case of closed curves which of the points P of $\mathfrak{K}$ is chosen as the single initial and end point.

For total differential forms $\omega = dF$ and for closed curves $\mathfrak{K}$ we have

$$\int_{\mathfrak{K}} \omega = 0$$

since $\Delta_{\mathfrak{K}} F = F(P) - F(P) = 0$.

The disappearance of the integral $\int \omega$ over closed curves is equivalent to the so-called property of *path-independence*: if $\mathfrak{K}_1$, $\mathfrak{K}_2$ are two curves with the same initial point and the same end point, then

$$\int_{\mathfrak{K}_1} \omega = \int_{\mathfrak{K}_2} \omega. \tag{22.7}$$

For the curve $\mathfrak{K}_1 + (-\mathfrak{K}_2)$, composed of the curve $\mathfrak{K}_1$ and the reoriented curve $\mathfrak{K}_2$, is closed, so that

$$\int_{\mathfrak{K}_1 - \mathfrak{K}_2} \omega = \int_{\mathfrak{K}_1} \omega - \int_{\mathfrak{K}_2} \omega = 0.$$

Thus the integral over a total differential form is path-independent. The converse also holds: if $\int \omega$ is path-independent, then ω is a total differential form. We shall content ourselves with the case where ω is defined in a simply connected domain. Then by Theorem 17.5 the totality of ω requires only a proof of the condition $d\omega = 0$ in the domain. The satisfaction of this condition will result as a corollary of (22.18) below if we turn our attention now to the integration of differential forms of higher degree.

Let ω be a second-degree differential form on a simple 2-dimensional figure R. If x, y is a coordinate system of R, then ω can be represented as

$$\omega = f dx dy.$$

240 Even without recourse to Poincaré's theorem ω is easily seen to be total—at least for continuously differentiable fs—since every function F such that

$$\frac{\partial F}{\partial x} = f$$

satisfies

$$d(Fdy) = dFdy = \left(\frac{\partial F}{\partial x}\, dx + \frac{\partial F}{\partial y}\, dy\right) dy = fdxdy.$$

Where Ω is a first-degree differential form such that

$$d\Omega = \omega,$$

Ω is uniquely determined up to a differential form Ω_0 such that $d\Omega_0 = 0$. Consequently, for an arbitrary closed curve $\mathfrak{K}$ in R,

$$\int_{\mathfrak{K}} \Omega$$

is uniquely determined by ω and $\mathfrak{K}$ alone.

Now let the 2-dimensional figure $\mathfrak{R}$ be oriented. As $\mathfrak{K}$ we take the *boundary curve* of $\mathfrak{R}$ (a closed, piecewise smooth curve to be oriented according to the canonical orientation convention). So as to dispense with a special notation for the boundary curve of $\mathfrak{R}$, we write

$$\oint_{\mathfrak{R}} \Omega$$

for the curve integral $\int_{\mathfrak{K}} \Omega$. $\oint_{\mathfrak{R}} \Omega$ is called the *boundary integral* over $\mathfrak{R}$.

We are now in a position to define the integral of ω over $\mathfrak{R}$.

DEFINITION 22.1 Let $\mathfrak{R}$ be a simple 2-dimensional oriented figure, and ω a second-degree differential form. For all differential forms Ω such that $d\Omega = \omega$, the boundary integral $\oint_{\mathfrak{R}} \Omega$ then takes the same value; it is called the *integral* of ω over $\mathfrak{R}$:

$$\int_{\mathfrak{R}} \omega = \oint_{\mathfrak{R}} \Omega.$$

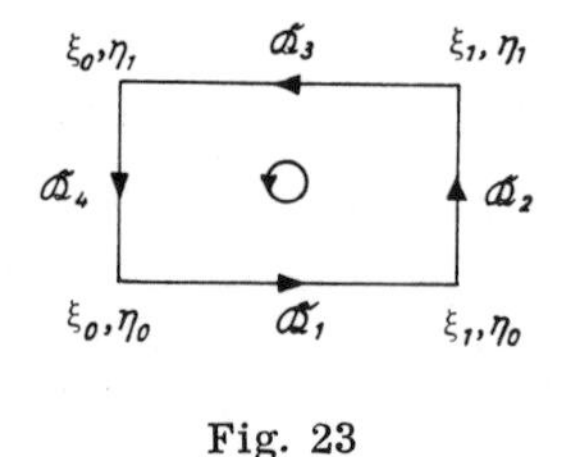

Fig. 23

The connection between the 2-dimensional *geometric integral* thus defined and the *numerical integral* $\mathcal{J}^2$ is quite simply established, just as in the one-dimensional case.

Let $f = \phi(x, y)$ be twice continuously differentiable, and let the coordinate system x, y be a mapping of $\mathfrak{R}$ onto the numerical interval $[\xi_0, \eta_0 | \xi_1, \eta_1]$.

If x, y has positive sense, then the boundary segments $\mathfrak{K}_1$, . . . , $\mathfrak{K}_4$ are to be oriented as shown in Figure 23. We have 241

$$\frac{\partial^2 f}{\partial x \partial y} = \mathscr{D}^2 \phi(x, y),$$

and we claim that

$$\int_{\mathfrak{R}} \frac{\partial^2 f}{\partial x \partial y}\, dxdy = \mathscr{T}^2_{\xi_0, \eta_0}{}^{\xi_1, \eta_1} \mathscr{D}^2 \phi \tag{22.8}$$

also holds. The right-hand numerical integral is (by §17)

$$\Delta^2{}_{\xi_0, \eta_0}^{\xi_1, \eta_1} \phi = \phi(\xi_1, \eta_1) - \phi(\xi_1, \eta_0) - \phi(\xi_0, \eta_1) + \phi(\xi_0, \eta_0).$$

In order to compute the left-hand geometric integral, we take

$$\Omega = \frac{\partial f}{\partial y}\, dy,$$

of which

$$d\Omega = \frac{\partial^2 f}{\partial x \partial y}\, dxdy$$

is true. Hence by Definition 22.1

$$\int_{\mathfrak{R}} \omega = \sum_1^4{}_i \int_{\mathfrak{K}_i} \frac{\partial f}{\partial y}\, dy.$$

On the curve segments $\mathfrak{K}_1$ and $\mathfrak{K}_3$, y is constant, so that $dy = 0$; i.e.,

$$\int_{\mathfrak{K}_1} \Omega = \int_{\mathfrak{K}_3} \Omega = 0.$$

On the curve segments $\mathfrak{K}_2$ and $\mathfrak{K}_4$, y is a parameter, so that

$$\int_{\mathfrak{K}_2} \frac{\partial f}{\partial y}\, dy = \int_{\mathfrak{K}_2} \phi'_{(2)}(\xi_1, y) dy = \Delta_{\mathfrak{K}_2} \phi(\xi_1, y) = \phi(\xi_1, \eta_1) - \phi(\xi_1, \eta_0).$$

Similarly for $\mathfrak{K}_4$ (y now having negative sense),

$$\int_{\mathfrak{K}_4} \frac{\partial f}{\partial y}\, dy = \int_{\mathfrak{K}_4} \phi'_{(2)}(\xi_0, y) dy = \Delta_{\mathfrak{K}_4} \phi(\xi_0, y) = \phi(\xi_0, \eta_0) - \phi(\xi_0, \eta_1).$$

Addition immediately yields (22.8).

However, just as in one dimension, we have no need here of resorting to the connection with numerical integrals. The theory of geometric integrals can be developed on the basis of Definition 22.1 independently of the numerical integral.

242 First of all, Definition 22.1 yields

$$\int d\Omega = \oint \Omega,$$

i.e., *Gauss's formula* (for $\Omega = Fdx + Gdy$):

$$\int \left(-\frac{\partial F}{\partial y} + \frac{\partial G}{\partial x}\right) dxdy = \oint (Fdx + Gdy), \tag{22.9}$$

with which we here begin, whereas it usually appears only at the end of the theory of integration.

Furthermore, *linearity*,

$$\int (c_1\omega_1 + c_2\omega_2) = c_1 \int \omega_1 + c_2 \int \omega_2, \tag{22.10}$$

follows from the linearity of the operations d and $\oint$ in similar fashion as the linearity of the curve integral.

In contrast to the situation with the numerical integral, we immediately have here a *transformation rule* at our disposal:

$$\int_{\mathfrak{R}} fdxdy = \int_{\mathfrak{R}} f \frac{d(x, y)}{d(\bar{x}, \bar{y})} d\bar{x}d\bar{y}, \tag{22.11}$$

for it is the same differential form that is being integrated on both sides.

Since the integrands are differential forms, integration can also be performed over 2-dimensional oriented surface segments $\mathfrak{S}$ which are contained in a simple N-dimensional figure. If, let us say,

$$\omega = \sum_{i,j}^{<} f_{ij} dx^i dx^j$$

for some coordinate system $x^1, \ldots, x^N$ of the figure, and if u, v is a parameter system of the surface segment $\mathfrak{S}$, then

$$\omega = \sum_{i,j}^{<} f_{ij} \frac{d(x^i, x^j)}{d(u, v)} dudv$$

and

$$\int_{\mathfrak{S}} \omega = \int_{\mathfrak{S}} \sum_{i,j}^{<} f_{ij} \frac{d(x^i, x^j)}{d(u, v)} dudv \tag{22.12}$$

on $\mathfrak{S}$. Gauss's theorem ($N = 2$) gives way to Stokes's theorem ($N > 2$):

$$\int_{\mathfrak{S}} d\Omega = \oint_{\mathfrak{S}} \Omega.$$

On the right we now have a curve integral over a closed curve in
243 the N-dimensional figure. For $\Omega = \sum F_i dx^i$ we explicitly get

$$\int_{\mathfrak{S}} \sum dF_i\, dx^i = \oint_{\mathfrak{S}} \sum F_i\, dx^i$$

$$\int_{\mathfrak{S}} \sum_{i,j}^{<} \left(\frac{\partial F_j}{\partial x^i} - \frac{\partial F_i}{\partial x^j}\right) dx^i dx^j = \oint_{\mathfrak{S}} \sum F_i \, dx^i .$$

The task remains of defining the 2-dimensional integral $\int \omega$ not just for simple 2-dimensional figures, but also for *2-dimensional figures* (*surfaces*) composed of finitely many simple segments. Let us consider such assemblies as, e.g.,

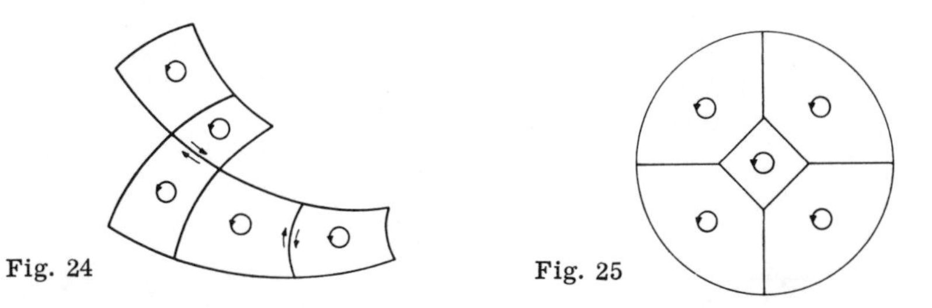

Fig. 24 Fig. 25

Let the simple segments, as in these examples, be so assembled that two segments with common boundary points (except for a vertex) always have an entire boundary segment in common. On the other hand, let there be no common interior points. The segments are to be so oriented that every boundary segment that appears twice has positive sense in the one case and negative sense in the other. In algebraic terms, the boundary of the total surface is the *sum* of the boundaries of the simple segments.

Accordingly, if $\mathfrak{S}_1, \ldots, \mathfrak{S}_m$ are the simple segments which are assembled into a surface $\mathfrak{S}$, then when we take the sum over all boundary integrals $\oint_{\mathfrak{S}_j} \Omega$, each "interior" boundary segment drops out. The only boundary segments of the $\mathfrak{S}_j$s that are left are those which form the boundary of $\mathfrak{S}$. Hence

$$\sum_j \oint_{\mathfrak{S}_j} \Omega = \oint_{\mathfrak{S}} \Omega . \tag{22.13}$$

Accordingly, we also write $\mathfrak{S} = \sum_j \mathfrak{S}_j$, and we call $\mathfrak{S}$ a *combinatory sum* of the $\mathfrak{S}_j$s.

If the assembly of the simple segments $\mathfrak{S}_j$ yields another simple segment $\mathfrak{S}$, then we have for $d\Omega = \omega$ by definition

$$\int_{\mathfrak{S}} \omega = \oint_{\mathfrak{S}} \Omega$$

and

$$\int_{\mathfrak{S}_j} \omega = \oint_{\mathfrak{S}_j} \Omega .$$

That gives us the *additivity* of the 2-dimensional integral *244*

$$\int_{\mathfrak{S}} \omega = \sum_j \int_{\mathfrak{S}_j} \omega. \tag{22.14}$$

The assumption here is that not only every component segment $\mathfrak{S}_j$ but also the composite surface $\mathfrak{S}$ is a simple surface segment.

For composite surfaces that are not simple surface segments, e.g., the surface of a sphere, one would like to use (22.14) as a definition. In contrast to the one-dimensional case, however, the difficulty arises here that for two *decompositions* of a simple segment into simple segments there need be no common *refinement*. An example is the decomposition of a rectangle by means of two curves that intersect infinitely many times.

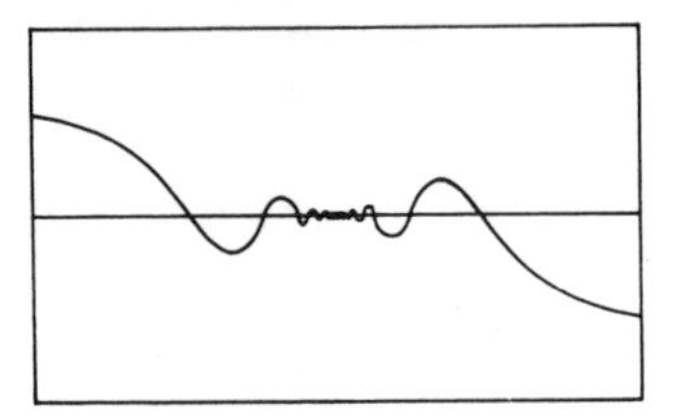

Fig. 26

If the one curve used in the decomposition is the straight line $y = 0$ (in a coordinate system x, y), then, e.g.,

$$y = x^4 \sin \frac{1}{x}$$

generates a twice continuously differentiable curve intersecting the line at infinitely many points. The curve decomposes each half of the rectangle into infinitely many pieces.

We can avoid this difficulty by restricting ourselves to analytic curves, which (by §10) cannot intersect infinitely many times in a finite interval without coinciding. However, this restriction is not in fact required by the nature of the situation, for we can show that the integral is also additive for certain decompositions into infinitely many parts. Additivity in this sense is called "*complete additivity.*"

If we confine ourselves to analytic segments, we can skip over the entire discussion of complete additivity that follows. We can proceed immediately to definition (22.25) on page 267,

$$\int_{\mathfrak{R}} \omega = \sum_i \int_{\mathfrak{R}_i} \omega,$$

for a decomposition of a composite figure $\mathfrak{R}$ into simple analytic segments $\mathfrak{R}_i$.

For suppose

$$\int_{\mathfrak{R}} \omega = \sum_j \int_{\mathfrak{R}_j'} \omega$$

is another decomposition of $\mathfrak{R}$. Then (by decomposition of the intersections $\mathfrak{R}_j' \cap \mathfrak{R}_i$ into *finitely* many simple analytic segments) there will be a common refinement of both decompositions, 245 whence it follows that the integral $\int_{\mathfrak{R}} \omega$ is independent of the decomposition of $\mathfrak{R}$. However, in order to take care of nonanalytic segments as well, an excursus through (22.25) will be necessary on the complete additivity of the integral for certain decompositions.

The simplest case arises if we confine ourselves to open domains in an oriented figure $\mathfrak{R}$. An *open domain* in number space is defined as a union of open intervals. We shall always assume here that it is a definite sequence of open intervals with rational end points that we are uniting. The definition of an *open domain in the figure* $\mathfrak{R}$ (we are dealing here specifically with 2-dimensional figures, but the following discussion applies generally) makes reference to a coordinate system X: a subset D of the figure $\mathfrak{R}$ is an *open domain* if the X-image of D is an open domain of the number space. Since, however, for any other coordinate system $\overline{X}$ the $\overline{X}$-image will always be a topological map of the X-image, this definition of open domains is invariant with respect to coordinates (assuming—tacitly—the rational definiteness of the transformation of X onto $\overline{X}$; cf. §16).

Let D be an open domain and U_* a definite sequence of x, y-neighborhoods (the x, y-images of which, remember, are rational intervals) such that

$$D = \bigcup U_*.$$

Moving now from U_* to the sequence

$$V_1 = U_1, \; V_2 = U_1 \cup U_2, \; V_3 = U_1 \cup U_2 \cup U_3, \; \ldots\ldots$$

we obtain

$$D = \bigcup V_*$$

with an increasing sequence

$$V_1 \subseteq V_2 \subseteq V_3 \subseteq \ldots\ldots$$

of *open interval sums*. (For present purposes *interval sum* shall always mean that the x, y-image is a finite interval sum in number space.)

For the sums V_n, let the integral be defined as in (22.14). We will now define the integral $\int \omega$ over the open domain D as

$$\int_D \omega = \lim \int_{V_*} \omega = \lim_{n \to \infty} \int_{V_n} \omega.$$

We must now show for these definitions that the sequence $\int_{V_*} \omega$ converges, and that for two increasing sequences V_* and W_* of open interval sums it always follows from $\bigcup V_* = \bigcup W_*$ that

$$\lim \int_{V_*} \omega = \lim \int_{W_*} \omega.$$

246 For the required proofs we first of all confine our attention to

$$\omega = dx^1 \ldots dx^N$$

for a positively-sensed coordinate system $X = x^1, \ldots, x^N$ of $\mathfrak{R}$. We abbreviate this as $\omega = dX$, and we write

$$X(D) = \int_D dX.$$

We call $X(D)$ the *X-measure* of D.

For any X-interval the X-measure is a positive number, viz., the product of the lengths of the edges: if the end points of the interval U have as X-coordinate values $\xi_1, \ldots, \xi_N$ and $\eta_1, \ldots, \eta_N$ (with $\xi_i < \eta_i$), then we have

$$X(U) = \prod_{1}^{N}{}_i (\eta_i - \xi_i).$$

Proof. By the definition of the integral (for $N = 2$, though we give the formulation for arbitrary N),

$$\int_U dx^1 \ldots dx^N = \oint_U x^1 dx^2 \ldots dx^N.$$

Only on the boundary segments $x^1 = \xi_1$ and $x^1 = \eta_1$ is $dx^2 \ldots dx^N \neq 0$. Hence

$$\oint_U x^1 dx^2 \ldots dx^N = \xi_1 \int_{x^1 = \xi_1} dx^2 \ldots dx^N + \eta_1 \int_{x^1 = \eta_1} dx^2 \ldots dx^N.$$

By canonical orientation (the seeming arbitrariness of which can be seen here to be justified for the first time), in view of $\xi_1 < \eta_1$, the system $x^2, \ldots, x^N$ has positive sense on the boundary segment $x^1 = \xi_1$. Thus it follows that

$$\int_U dx^1 \ldots dx^N = (\eta_1 - \xi_1) \int_{U_1} dx^2 \ldots dx^N,$$

where U_1 is an $(N - 1)$-dimensional interval for which $x^2, \ldots, x^N$ has positive sense. If we take the assertion

$$\int_{U_1} dx^2 \ldots dx^N = \prod_{2}^{N}{}_i (\eta_i - \xi_i)$$

as proved for U_1 (the case $N = 1$ of the assertion is trivial: $\int_U dx = \eta - \xi$), then it indeed follows that

$$\int_U dx^1 \ldots dx^N = \prod_{1}^{N}{}_i (\eta_i - \xi_i).$$

247 For X-interval sums $V = \sum_{1}^{k}{}_j U_j$ with pairwise disjoint X-intervals U_k we can define

$$X(V) = \sum_{1}^{k} {}_j X(U_j). \qquad (22.15)$$

Because of additivity (22.14), and because for two decompositions of V into X-intervals there are always common refinements (the intersection of two X-intervals is always another X-interval), $X(V)$ is independent of the particular decomposition selected of V into X-intervals. The X-measure is *additive* for interval sums too, as a consequence of (22.15).

It also follows from (22.15) that

$$X(V) > 0,$$

so that the X-measure is *monotonic* for X-interval sums:

$$V_1 \subset V_2 \rightarrow X(V_1) < X(V_2).$$

Proof. $V_1 \subset V_2$ entails $V_2 = V_1 + V$ for an X-interval sum V; hence

$$X(V_2) = X(V_1) + X(V) > X(V_1).$$

Every open domain D is the union of an increasing definite sequence V_* of X-interval sums. If D is bounded, i.e., contained in an X-interval U, then the increasing sequence

$$X(V_1) \leqslant X(V_2) \leqslant \ldots\ldots$$

is also bounded, viz., by $X(U)$. Hence $\lim X(V_*)$ exists.

For different representations of an open bounded domain as the union of increasing sequences of X-interval sums

$$D = \bigcup V_* = \bigcup W_*$$

we now get

$$\lim X(V_*) = \lim X(W_*)$$

by a theorem of Borel's that is fundamental for measure theory.

BOREL'S THEOREM. For definite increasing sequences V_*, W_* of X-interval sums,

$$\bigcup V_* \subseteq \bigcup W_* \rightarrow \lim X(V_*) \leqslant \lim X(W_*). \qquad (22.16)$$

Proof. It suffices to show that for any ε and any m

$$X(V_m) \leqslant \lim X(W_*) + \varepsilon.$$

An X-interval sum V can be chosen in V_m such that

248

$$\overline{V} \subseteq V_m$$

$$X(V_m) \leqslant X(V) + \varepsilon.$$

It is enough to appreciate this fact for an interval U rather than V_m —but this special case is trivial. By the covering theorem, $\overline{V} \subseteq \mathbf{U}W_*$ then implies $\overline{V} \subseteq W_n$ for an appropriate n, for W_* is a definite open covering of the closed interval sum $\overline{V}$ (cf. Theorem 14.5).

From

$$X(V) \leqslant X(W_n) \leqslant \lim X(W_*)$$

the desired inequality now follows.

$\mathbf{U}V_* = \mathbf{U}W_*$ accordingly implies that $\lim X(V_*) = \lim X(W_*)$.

Corollary. It follows for Boolean sums $V_m \sqcup W_n$ (cf. §3; these sums are likewise interval sums) under the above assumption that they have arbitrarily small X-measure for sufficiently large m, n. For let m be large enough that

$$X(W_n) \leqslant X(V_m) + \varepsilon \quad \text{for all } n.$$

For V_m a V can again be chosen as above. The result is that for sufficiently large ns

$$V_m \sqcup W_n \subseteq V_m \llcorner \overline{V} \mathbin{\dot\cup} W_n \llcorner \overline{V},$$

where

$$X(V_m \llcorner V) = X(V_m) - X(V) \leqslant \varepsilon$$

and

$$\begin{aligned} X(W_n \llcorner \overline{V}) &= X(W_n) - X(V) \\ &= X(W_n) - X(V_m) + X(V_m) - X(V) \leqslant 2\varepsilon. \end{aligned}$$

In view of Borel's theorem, the X-measure of open domains D with a representation $D = \mathbf{U}V_*$ for a definite increasing sequence of X-interval sums can be defined as

$$X(D) = \lim X(V_*). \tag{22.17}$$

This X-measure is *completely additive*: if D_* is a definite increasing sequence of open domains with a bounded union

$$D = \mathbf{U}D_*,$$

then we have

$$X(D) = \lim X(D_*).$$

249 For the representations $D = \mathbf{U}V_{n\dagger}$ yield

$$D = \mathbf{U}_n \overset{n}{\underset{1}{\mathbf{U}_i}} V_{in},$$

so that

$$X(D) = \lim_{n \to \infty} X\left(\bigcup_1^n {}_i V_{in}\right) \leqslant \lim_{n \to \infty} X\left(\bigcup_1^n {}_i D_i\right) = \lim X(D_*),$$

while $X(D_n) \leqslant X(D)$ is trivial.

For complete additivity the only thing we have to watch is that D_* be a definite sequence; i.e., the double sequence $V_{*\dagger}$ of interval sums in $D_n = \bigcup V_{n\dagger}$ must be definite. (Thus, ultimately, the sequences of rational end points of intervals involved must be definite.)

We now extend these considerations to integrands

$$\omega = fdX = fdx^1 \ldots dx^N.$$

For every X-interval U, $\int_U \omega$ is defined. We select an orientation for U such that X will have positive sense. We then have

$$f \geqslant 0 \to \int_U fdX \geqslant 0. \tag{22.18}$$

Proof. Suppose the end points of the X-interval take as X-coordinate values $\xi_1, \ldots, \xi_N$ and $\eta_1, \ldots, \eta_N$. By Definition 22.1 of the integral,

$$\int_U fdX = \oint_U Fdx^2 \ldots dx^N \quad \text{where} \quad \frac{\partial F}{\partial x^1} = f.$$

Since $dx^2 \ldots dx^N$ vanishes everywhere on the boundary of U other than on the boundary segments $x^1 = \xi_1$ and $x^1 = \eta_1$, we get for $F = \Phi(x^1, \ldots, x^N)$

$$\oint_U Fdx^2 \ldots dx^N = \int_{x^1=\xi_1} \Phi(\xi_1, x^2, \ldots, x^N)dx^2 \ldots dx^N + \int_{x^1=\eta_1} \Phi(\eta_1, x^2, \ldots, x^N)dx^2 \ldots dx^N.$$

By canonical orientation of the boundary, where $\xi_1 < \eta_1$, the coordinate system $x^2, \ldots, x^N$ has positive sense for the boundary segment $x^1 = \eta_1$, and negative sense for the boundary segment $x^1 = \xi_1$. As a result, we obtain

$$\oint_U Fdx^2 \ldots dx^N = \int_{x^1=\eta_1} (\Phi(\eta_1, x^2, \ldots, x^N) - \Phi(\xi_1, x^2, \ldots, x^N))\, dx^2 \ldots dx^N.$$

By hypothesis $f \geqslant 0$; i.e., $\Phi'_{(1)}(x^1, \ldots, x^N) \geqslant 0$ in U.

It follows from the bound-theorem that

$$\Phi(\eta_1, x^2, \ldots, x^N) - \Phi(\xi_1, x^2, \ldots, x^N) \geqslant 0,$$

250 so that if we assume the assertion to be proved for $N - 1$, it is thereby also proved for N. In particular, it is proved for the case $N = 2$, which is the only one we need for the time being, since by (22.5) it already holds for $N = 1$.

COROLLARY. The same line of argument also yields

$$f > 0 \rightarrow \int_U f dX > 0.$$

From this result, the theorem for curve integrals mentioned in connection with (22.7) can be derived without further ado. If ω is a first-degree differential form and the integral $\int \omega$ is path-independent in a simply connected domain, then ω is total.

Proof. All we have to show is that $d\omega = 0$. Since

$$\int_U d\omega = \oint_U \omega = 0,$$

$\int_U d\omega = 0$ holds for all intervals U. If $d\omega = f dx dy$, then f must equal 0 in the entire domain. For if at some point $f \neq 0$, but say $f > 0$, then it would also be the case in a neighborhood of this point that $f > 0$, whence $\int_U f dx dy > 0$. The same goes for $f < 0$.

From (22.18) monotony follows as a general result:

$$f_1 \leq f_2 \text{ in } U \rightarrow \int_U f_1 dX \leq \int_U f_2 dX. \tag{22.19}$$

In particular, for any constant c we get in view of

$$\int_U c dX = cX(U)$$

the important estimate

$$|f| \leq c \rightarrow \left| \int_U f dX \right| \leq cX(U). \tag{22.20}$$

Beyond the integral $\int_U \omega$ for X-intervals U, we obtain the integral $\int_V \omega$ for X-interval sums $V = \sum_1^k {}_j U_j$ on the basis of additivity by the definition

$$\int_V \omega = \sum_1^k {}_j \int_{U_j} \omega. \tag{22.21}$$

251 Again, as in the case of the X-measure, $\int_V \omega$ is independent of the particular decomposition of V into X-intervals chosen (though the integral still depends on the coordinate system X).

Estimate (22.20) extends immediately to interval sums V:

$$|f| \leq c \text{ in } V \rightarrow \left| \int_V f dX \right| \leq cX(V).$$

Still retaining a designated coordinate system X, let us now consider representations

$$D = \bigcup V_*,$$

where D is an open bounded domain and V_* is an increasing definite sequence of X-interval sums.

The existence of $\lim \int\limits_{V_*} \omega$ (by which, of course, $\int\limits_{D} \omega$ is ultimately to be defined) results as follows. The sequence $X(V_*)$ is monotonic and bounded, hence convergent and in particular concentrated: for every ε there is an n such that for all $m_1 > m_2 \geqslant n$,

$$X(V_{m_1}) - X(V_{m_2}) < \varepsilon.$$

Hence also (where $\omega = f dX$ and $|f| \leqslant c$ in D)

$$\left| \int\limits_{V_{m_1}} \omega - \int\limits_{V_{m_2}} \omega \right| = \left| \int\limits_{V_{m_1} \llcorner V_{m_2}} \omega \right| \leqslant cX\left(V_{m_1} \llcorner V_{m_2}\right) < c\varepsilon;$$

i.e., the sequence $\int\limits_{V_*} \omega$ is concentrated and therefore convergent.

This proof makes use of the fact that the difference of two X-interval sums is always another X-interval sum.

For two representations

$$D = \bigcup V_* = \bigcup W_*$$

with increasing definite sequences V_* and W_* of X-interval sums (we still retain the coordinate system X), it now follows that

$$\lim \int\limits_{V_*} \omega = \lim \int\limits_{W_*} \omega. \tag{22.22}$$

Proof. By the corollary to Borel's theorem (22.16), we have for every ε and sufficiently large m, n

$$X(V_m \sqcup W_n) < \varepsilon.$$

It thus follows for $\omega = f dX$, where $|f| \leqslant c$ in D, that

$$\left| \int\limits_{V_m} \omega - \int\limits_{W_n} \omega \right| \leqslant cX(V_m \sqcup W_n) < c\varepsilon$$

because of

252

$$\int\limits_{V_m} \omega = \int\limits_{V_m \llcorner W_n} \omega + \int\limits_{V_m \cap W_n} \omega$$

and

$$\int\limits_{W_n} \omega = \int\limits_{W_n \llcorner V_m} \omega + \int\limits_{V_m \cap W_n} \omega.$$

The differences $V_m \llcorner W_n$, $W_n \llcorner V_m$, and the intersections $V_m \cap W_n$ here are in each case X-interval sums.

The definition

$$\int_D \omega = \lim \int_{V_*} \omega$$

for increasing sequences V_* of X-interval sums still depends on the coordinate system X despite (22.22). To establish independence from the coordinate system, we must first show that for any simple figure U (for which $\int_U \omega$ is defined as $\oint_U \Omega$, with $d\Omega = \omega$) and for any representation of the open core

$$\underline{U} = \bigcup V_*,$$

where V_* is an increasing definite sequence of X-interval sums, we have

$$\int_U \omega = \lim \int_{V_*} \omega. \tag{22.23}$$

The essential lemma for proving (22.23) is the

APPROXIMATION THEOREM. For every simple figure U and for every ε there are X-interval sums V_i, V_a such that

$$\overline{V_i} \subseteq \underline{U}, \quad \overline{U} \subseteq \underline{V_a} \quad \text{and} \quad X(V_a) - X(V_i) < \varepsilon.$$

Proof. It suffices to show that the boundary segments of U can be covered by finitely many open X-intervals, which together have an arbitrarily small X-measure. For then the covering of the boundary segments can be represented in the form $\underline{V_a \llcorner V_i}$.

If $\overline{X} = \overline{x}^1, \ldots, \overline{x}^N$ (all we need for now is the case $N = 2$, but we are giving the general formulation right off) is a coordinate system mapping U onto an interval of the N-dimensional number space, then the boundary segments of U are sets with points which on the one hand satisfy some such equation as, say,

$$\overline{x}^N = \phi(x^1, \ldots, x^N) = \gamma,$$

where ϕ is a continuously differentiable function (and there is no 253 point of the boundary segment where $\phi'_{(i)} = 0$ for all $i = 1, \ldots, N$), and which on the other hand satisfy inequalities such as

$$\alpha_i \leqslant \overline{x}^i \leqslant \beta_i \quad (i = 1, \ldots, N-1).$$

If we let $\Phi = \max_i |\phi'_{(i)}|$, then $\Phi(X) > 0$ everywhere on the boundary segment. In fact, since the boundary segment is compact, $\Phi(X) \geqslant \varepsilon_0$ for an appropriate $\varepsilon_0 > 0$. Each of the functions $\phi'_{(i)}(X)$ is continuous on the boundary segment and hence uniformly continuous: there is a δ such that for arbitrary points P_0, P_1 of the boundary segment,

$$|\overline{X}(P_1) - \overline{X}(P_0)| < \delta$$

always entails

$$|\phi'_{(i)}(X(P_1)) - \phi'_{(i)}(X(P_0))| < \varepsilon_0.$$

The $(N - 1)$-dimensional $\overline{X}$-interval $[\alpha_1, \ldots, \alpha_{N-1} | \beta_1, \ldots, \beta_{N-1}]$ can thus be decomposed into finitely many, say k, subintervals so that in each subinterval,

$$|\phi'_{(i)}(X(P_1)) - \phi'_{(i)}(X(P_0))| < \varepsilon_0.$$

Consequently there is at least one i such that

$$\phi'_{(i)}(X) \neq 0$$

in each of these $(N - 1)$-dimensional $\overline{X}$-subintervals. (Were $\phi'_{(i)}(X) = 0$ for every i at some point, then $|\phi'_{(i)}(X)| < \varepsilon_0$ would follow for all points, so that $\Phi(X) < \varepsilon_0$.) In a subinterval where, let us say,

$$\phi'_{(N)}(x^1, \ldots, x^N) \neq 0,$$

we get a representation of the subsegment of the boundary in an equation

$$x^N = \psi(x^1, \ldots, x^{N-1})$$

[where, by the inversion theorem (Theorem 15.4), ψ is a continuously differentiable function]. The coordinates $x^1, \ldots, x^{N-1}$ are thus a parameter system of the subsegment, which is also an interval U_0 with respect to $\overline{x}^1, \ldots, \overline{x}^{N-1}$.

We want to show that this subsegment can be covered by finitely many X-intervals in such a way that the sum of their X-measures is $< \frac{\varepsilon}{k}$. For $N = 2$, an $\overline{x}_1$-interval is always an x_1-interval too. Let the x_1-measure of U_0 be $< M$. The continuous function ψ can be approximated by step functions τ_0, τ_1 such that $\tau_0 < \psi < \tau_1$ and $\tau_1 - \tau_0 < \frac{\varepsilon}{kM}$ in the interval U_0. Thus the X-interval sum lying between the step functions covers the subsegment in question and has an X-measure $< \frac{\varepsilon}{k}$.

For $N > 2$, the approximation theorem for $N - 1$ can be used as the inductive hypothesis. Thus the $\overline{x}^1, \ldots, \overline{x}^{N-1}$-interval U_0 can be approximated arbitrarily closely by an $x^1, \ldots, x^{N-1}$-interval sum. Over such an interval sum approximating from within with an $x^1, \ldots, x^{N-1}$-measure $< M$, the continuous function ψ can again be approximated by step functions τ_0, τ_1 such that $\tau_0 <$ 254

$\psi < \tau_1$ and, say, $\tau_1 - \tau_0 < \frac{\varepsilon}{2kM}$. The result is open X-intervals with a total X-measure $< \frac{\varepsilon}{2k}$. The remainder that still has to be covered lies over the boundary of the $\bar{x}^1, \ldots, \bar{x}^{N-1}$-interval. Consequently, by the induction hypothesis—and because ψ is continuous—the approximation of U_0 can be so chosen that an open covering of the remainder is possible with an X-measure $< \frac{\varepsilon}{2k}$. So in this case too the subsegment under consideration is totally covered with an X-measure $< \frac{\varepsilon}{k}$.

The approximation used in this proof now yields a proof of (22.23). All we have to do is to give an approximation of the $\bar{X}$-interval U for arbitrary ε from within by means of an X-interval sum V such that

$$\left| \int_U \omega - \int_V \omega \right| < \varepsilon .$$

For $N = 2$ the integrals $\int_U \omega$ and $\int_V \omega$ are representable on the basis of (22.13) and (22.21) as boundary integrals $\oint_U \Omega$ and $\oint_V \Omega$. We now show (for arbitrary N) that these boundary integrals differ from one another arbitrarily little for appropriate V. For this purpose we can restrict our attention to a subsegment of the boundary with a representation

$$x^N = \psi(x^1, \ldots, x^{N-1})$$

which is also an $\bar{x}^1, \ldots, \bar{x}^{N-1}$-interval. This subsegment can be approximated with the help of a step function $\tau_0 < \psi$.

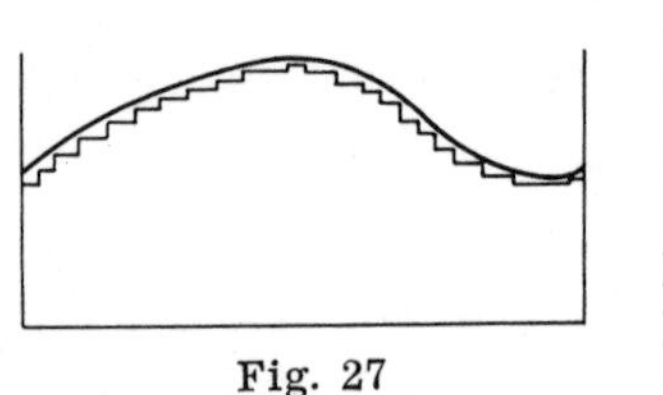

Fig. 27

Thus, for the boundary integrals such that $\Omega = F dx^1 \ldots dx^{N-1}$, the subsegment yields the components $\int F dx^1 \ldots dx^{N-1}$ once over the boundary of U and once over the boundary of the approximating interval sum V (where the pieces of the boundary of V on which $dx^1 \ldots dx^{N-1} = 0$ can be omitted). For

$$F = \Phi(x^1, \ldots, x^N)$$

we have in the integrand one occurrence of $\Phi(x^1, \ldots, x^{N-1}, \psi(x^1, \ldots, x^{N-1}))$ and one occurrence of $\Phi(x^1, \ldots, x^{N-1}, \tau_0(x^1, \ldots, x^{N-1}))$. Since Φ is continuous and $\psi - \tau_0$ is arbitrarily small, the integrals also differ by arbitrarily little. (For $N > 2$ this can so far be said only hypothetically, as we have not yet even defined the integrals.)

255

(22.23) also establishes that for simple figures U the representations $\underline{U} = \bigcup V_*$ with increasing sequences V_* of X-interval sums yield the same $\lim \int_{V_*} \omega$ for any coordinate system X, viz., $\int_U \omega$.

For a segment $V = \sum_1^k{}_i U_i$ composed of simple segments U_i that lies within a simple figure U, we can next define

$$\int_V \omega = \sum_1^k{}_i \int_{U_i} \omega .$$

This definition is independent of any particular decomposition, since it follows from

$$V = \sum_1^k{}_i U_i = \sum_1^l{}_j U_j{}'$$

with the help of representations of the $\underline{U}_i$s and $\underline{U}_j$'s as unions of X-intervals (with respect to an arbitrary coordinate system X of the simple segment $U \supseteq V$) that

$$\sum{}_i \int_{U_i} \omega = \sum{}_j \int_{U_j{}'} \omega .$$

Finally, for an open domain D representable as a union of simple segments, or more precisely as a union $D = \bigcup V_*$ of an increasing definite sequence V_* of finitely composite segments V_n within a simple figure U, we can set

$$\int_D \omega = \lim \int_{V_*} \omega . \tag{22.24}$$

The limit on the right is the same for any representing sequence V_*. This can be shown as above with the aid of representations of the $\underline{V_n}$s as unions of X-intervals for some coordinate system X of U.

Only now have we attained a definition of the integral over open domains D within a simple figure U which is independent of the choice of coordinate system.

The integral thus defined is additive, indeed *completely additive*: for $D = \bigcup D_*$, where D_* is an increasing definite sequence of open domains, we have

$$\int_D \omega = \lim \int_{D_*} \omega .$$

To prove this we have only to take the representations 256

$$D_n = \bigcup V_{n\dagger} ,$$

where $V_{n\dagger}$ are increasing sequences of interval sums, and form the representation $D = \bigcup W_*$ such that $W_n = \bigcup_1^n{}_i V_{in}$. W_* is increasing, and we have

$$\lim \int_{W_*} \omega = \lim \int_{D_*} \omega,$$

since (for an arbitrary coordinate system X)

$$\lim X(D_* \sqcup W_*) = 0$$

holds on the basis of

$$D_m \sqcup W_n = D_m \llcorner W_n \dot{\cup} W_n \llcorner D_m \subseteq D_m \llcorner V_{mn} \dot{\cup} D \llcorner D_m$$

$$\text{for} \quad m \leqslant n.$$

Thus appropriate choices (first of m and then of n) will make $X(D_m \sqcup W_n)$ arbitrarily small.

The only reason we introduced integrals over open domains here was so that we could extend the integral from simple figures to arbitrary composite figures (composed of finitely many simple figures), first of all for $N = 2$. We now accomplish this by moving from the *combinatory composition* (sum) of the simple figures (with no common interior points and with common boundaries oriented oppositely) to the *set-theoretic union*. Every composite figure, considered as a set of points, is a union of open domains and certain boundary segments. If we supplement the point set with an orientation, the figure will be uniquely determined.

By the approximation theorem, the relevant boundary segments, each of which belongs to a simple figure, have open coverings (consisting in fact of finitely many X-intervals) with a total X-measure $< \varepsilon$ for every ε and for every coordinate system X. Within a simple figure, those sets which for every ε and for some coordinate system X (and hence by the approximation theorem for any coordinate system) have an open covering of *finitely* many X-intervals with an X-measure $< \varepsilon$ are called *Jordan null sets*.

A *Lebesgue null set* (or simply *null set*), on the other hand, is one with open coverings (possibly by *infinitely* many intervals) of arbitrarily small X-measure (for some and hence for all X). For every null set K there is a decreasing sequence D_* of open domains approximating it from without such that

$$K = \bigcap D_* \quad \text{and} \quad \lim X(D_*) = 0.$$

257 Using Lebesgue null sets has the advantage that the union of a definite sequence K_* of such null sets is always another null set. (For K_* to be definite means that the double sequence $D_{*\dagger}$ such that $K_i = \bigcap D_{i\dagger}$ is definite.) For K_i has an open covering D_i for every ε such that $X(D_i) < \frac{\varepsilon}{2^i}$; hence $\bigcup K_*$ is covered by $\bigcup D_*$ in such a way that

$$X(\bigcup D_*) < \sum_i \frac{\varepsilon}{2^i} = \varepsilon .$$

Accordingly, any definitely denumerable set in particular will be a null set, e.g., the set of rational numbers in the interval $[0 | 1]$, which however is not a Jordan null set.

The intervals, interval sums, and open domains considered so far as domains of integration always have a positive X-measure and hence are not null sets. Thus we can define the X-measure $X(K)$ and the integrals $\int_K \omega$ anew for null sets K.

We set

$$\int_K \omega = 0$$

for null sets; in particular, $X(K) = 0$.

More generally, we introduce as *"almost open"* domains those sets $D \cup K$ which can be represented as the union of an open domain D and a null set K. Among them are the closed intervals (with reference to some coordinate system). We could also call differences $D \llcorner K$ "almost open," but that would be of no advantage in what follows.

For almost open domains $C = D \cup K$ we define

$$\int_C \omega = \int_{D \cup K} \omega = \int_D \omega .$$

This definition is independent of the decomposition of C. For if $\underline{C}$ is the open core of C, then

$$C = D \cup K$$

immediately implies that $D \subseteq \underline{C}$ and that $\underline{C} \llcorner D$ is a null set. For any ε and any coordinate system X there is thus an open set D_0 such that

$$\underline{C} \subseteq D \cup D_0 \quad \text{and} \quad X(D_0) < \varepsilon ,$$

so that for $\omega = f dX$ and $|f| \leqslant c$ in C we get

$$\left| \int_{\underline{C}} \omega - \int_D \omega \right| \leqslant cX(D_0) < c\varepsilon .$$

The above definition of $\int_C$ could therefore be replaced by 258

$$\int_C \omega = \int_{\underline{C}} \omega .$$

We are clearly forced to adopt these definitions if we want the integral to retain the basic properties it has had thus far (linearity, monotony, and additivity). These definitions do preserve those properties. In fact, the integral is again *completely additive* for almost open sets. The union of a definite sequence of open domains is another open domain, and the union of a definite sequence of null sets is itself a null set. Consequently, the union $\mathbf{U}(D_* \cup K_*)$ of a definite sequence $D_* \cup K_*$ of almost open sets is also almost open, and if the members of D_* are pairwise disjoint, we have

$$\int_{\mathbf{U}(D_* \cup K_*)} \omega = \int_{\mathbf{U}D_* \cup \mathbf{U}K_*} \omega = \int_{\mathbf{U}D_*} \omega = \sum \int_{D_*} \omega = \sum \int_{D_* \cup K_*} \omega .$$

In the almost open sets we now have a fitting generalization of simple figures. By the approximation theorem the simple figures are almost open, since the boundary is a null set. Compared with simple figures (regarded as point sets, with or without boundary), however, the almost open sets have the decisive advantage that the intersection of two (or finitely many) almost open sets is itself almost open:

$$D \cup K \,\dot{\cap}\, D' \cup K' = D \cap D' \,\dot{\cup}\, D \cap K' \,\dot{\cup}\, D' \cap K \,\dot{\cup}\, K \cap K' .$$

On the right-hand side $D \cap D'$ is open, and the rest is a null set, as it is contained in the null set $K \cup K'$. In terms of lattice theory, we have proved that the almost open sets constitute a σ-*lattice* (finite intersections, definitely denumerable unions). For present purposes, however, we shall only make use of the fact that the almost open sets form a *lattice* (finite intersections and unions). Similarly, we make use of the additivity but not of the complete additivity of the integral for decompositions within a simple figure.

Note that almost open sets do not form a *Boolean lattice*. The complement of an open set (which will of course be a closed set) need not be almost open. If each rational number in $[0|1]$ is surrounded by an open interval U_i such that $\sum_i x(U_i) = 1/2$, then $[0|1] \llcorner \mathbf{U}U_*$ is closed, but contains no interior points and is not a null set.

Only now can we define 2-dimensional integrals for composite figures.

259 Let $\mathfrak{R}$ be an oriented (composite) 2-dimensional figure. The task of defining $\int_{\mathfrak{R}} \omega$ only makes sense for a differential form ω which is defined for all simple subsegments of $\mathfrak{R}$. Independently of the decomposition of $\mathfrak{R}$ into simple segments, this situation can obtain by virtue of $\mathfrak{R}$'s being contained in a simple N-dimensional figure for $N > 2$ (e.g., the sides of a rectangular parallelepiped might all be contained in a 3-dimensional cube), where a differential form ω is given for the N-dimensional figure.

In this situation we now represent the set of points of $\mathfrak{R}$ as a finite union of pairwise disjoint almost open sets C_i each of which lies within a simple 2-dimensional figure. Let ω be a differential form defined on the entire N-dimensional figure (or at any rate in an open N-dimensional set containing $\mathfrak{R}$).

For each of the almost open sets C_i,

$$\int_{C_i} \omega$$

is defined, for C_i lies in a simple 2-dimensional figure.

If we now settle upon

$$\int_{\mathfrak{R}} \omega = \sum_i \int_{C_i} \omega, \tag{22.25}$$

this definition will be independent of the choice of decomposition. For if the set of points of $\mathfrak{R}$ is also representable as a finite union of pairwise disjoint almost open sets C_j' (again, each of which lies within a simple 2-dimensional figure), then we shall have

$$\sum_j \int_{C_j'} \omega = \sum_i \int_{C_i} \omega. \tag{22.26}$$

To prove it we have only to form the almost open sets $C_j' \cap C_i$ and to note that

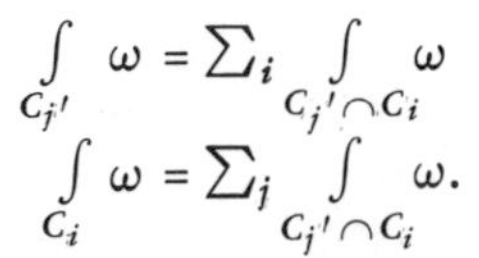

$$\int_{C_j'} \omega = \sum_i \int_{C_j' \cap C_i} \omega$$
$$\int_{C_i} \omega = \sum_j \int_{C_j' \cap C_i} \omega.$$

This simple proof shows that complete additivity would also make it possible to treat figures that only admit of infinite decompositions into definite sequences C_* of almost open sets (with each C_i inside a simple figure). $\sum \int_{C_*} \omega$ then makes sense as soon as this sum becomes absolutely convergent. In this introduction we confine ourselves to finite sums so as to avoid the complication of a convergence condition. In terms of analytic topology, this means that we restrict ourselves to compact manifolds (with the appropriate differentiability requirements). (For a constructive treatment of the Lebesgue integral, cf. Lorenzen 1955.) *260*

The basic properties of the 2-dimensional integral result directly from what we have just done. The integral is

(1) linear: $\int (c_1\omega_1 + c_2\omega_2) = c_1 \int \omega_1 + c_2 \int \omega_2$

(2) monotonic: $f_1 \leqslant f_2$, X positively-sensed $\rightarrow \int f_1 dX \leqslant \int f_2 dX$

(3) additive: $\int_{\sum_i \mathfrak{R}_i} \omega = \sum_i \int_{\mathfrak{R}_i} \omega.$

But the *transformation formula* for the integral,

$$\int_{\mathfrak{R}} f dg_1 dg_2 = \int_{\mathfrak{R}} \sum_{i,j}^{<} f \frac{(g_1, g_2)}{(x^i, x^j)} dx^i dx^j, \tag{22.27}$$

likewise holds trivially on the basis of our derivation, since the integrand is the same differential form on both sides.

By Definition 22.1 *Stokes's formula*,

$$\int_{\mathfrak{R}_i} d\Omega = \oint_{\mathfrak{R}_i} \Omega,$$

holds for every simple figure. Thus it likewise follows for composite figures $\mathfrak{R} = \sum_i \mathfrak{R}_i$ with boundary $\mathfrak{R} = \sum_i$ boundary $\mathfrak{R}_i$ that

$$\int_{\mathfrak{R}} d\Omega = \sum_i \int_{\mathfrak{R}_i} d\Omega = \sum_i \oint_{\mathfrak{R}_i} \Omega = \oint_{\mathfrak{R}} \Omega. \tag{22.28}$$

To compute a 2-dimensional integral it is not always necessary to carry out an explicit decomposition into simple segments. Instead, the boundary integral can often be computed directly. This is already of interest in the case where the entire 2-dimensional figure lies within a simple 2-dimensional figure.

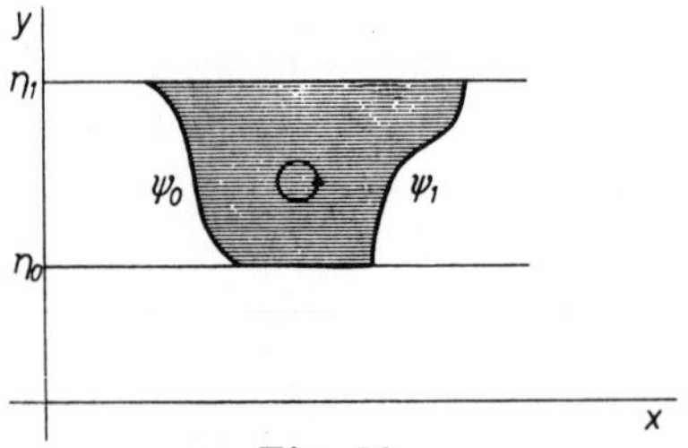

For example, suppose we are given with respect to a coordinate system x, y of positive sense a so-called *normal domain* $\mathfrak{N}$:

$\mathfrak{N}$ is to contain the points P such that

(1) $\eta_0 \leqslant y(P) \leqslant \eta_1$

(2) $\psi_0(y(P)) \leqslant x(P) \leqslant \psi_1(y(P))$

Fig. 28

261 where ψ_0 and ψ_1 are twice continuously differentiable functions. We first of all assume that $\psi_0 < \psi_1$ in $[\eta_0 | \eta_1]$. The normal domain is then clearly a simple figure. A coordinate system $\bar{x}, \bar{y}$ which maps onto the interval $[0, 0 | 1, 1]$ is created by

$$\bar{x} = \frac{x - \psi_0(y)}{\psi_1(y) - \psi_0(y)}$$

$$\bar{y} = \frac{y - \eta_0}{\eta_1 - \eta_0}.$$

But even without using this coordinate system we can employ Stokes's formula,

$$\int_{\mathfrak{N}} f dx dy = \oint_{\mathfrak{N}} F dy \quad \text{where} \quad \frac{\partial F}{\partial x} = f,$$

to compute the integral. Since the boundary integral vanishes along the curves $y = \eta_0$ and $y = \eta_1$ (because $dy = 0$), we obtain

$$\int_{\mathfrak{R}} f dx dy = \int_{x=\psi_1(y)} F dy - \int_{x=\psi_0(y)} F dy.$$

Setting $f = \phi(x, y)$, we get

$$\frac{\partial F}{\partial x} = \phi(x, y),$$

and hence, let us say,

$$F = \overset{x}{\underset{0}{\mathcal{J}_\xi}} \phi(\xi, y).$$

On the curve segments $x = \psi_0(y)$ and $x = \psi_1(y)$, y is a parameter. Thus we obtain

$$\int_{\mathfrak{R}} f dx dy = \overset{\eta_1}{\underset{\eta_0}{\mathcal{J}_\eta}} \overset{\psi_1(\eta)}{\underset{0}{\mathcal{J}_\xi}} \phi(\xi, \eta) - \overset{\eta_1}{\underset{\eta_0}{\mathcal{J}_\eta}} \overset{\psi_0(\eta)}{\underset{0}{\mathcal{J}_\xi}} \phi(\xi, \eta)$$

and hence altogether

$$\int_{\mathfrak{R}} \phi(x, y) dx dy = \overset{\eta_1}{\underset{\eta_0}{\mathcal{J}_\eta}} \overset{\psi_1(\eta)}{\underset{\psi_0(\eta)}{\mathcal{J}_\xi}} \phi(\xi, \eta). \tag{22.29}$$

Formula (22.9) is contained in this as a special case for intervals.

We can see immediately that formula (22.29) remains valid when only $\psi_0 \leq \psi_1$ rather than $\psi_0 < \psi_1$ is satisfied. Thus, e.g., this formula can be used to calculate integrals over the surface of a circle.

On the basis of the approximation theorem, formula (22.29)—and thus by the same token Stokes's formula—can also be extended to normal domains in which ψ_0, ψ_1 are just simply continuously differentiable. However, we shall not carry that out here. *262*

Instead, we turn now to integration over 3-dimensional figures. Essentially, all we have to do is retrace the line of thought that led us from 1- to 2-dimensional integrals. In fact, all the steps we took can be executed for an arbitrary number of dimensions. However, simply because the notation is more manageable, we will sketch the necessary course for the case $N = 3$.

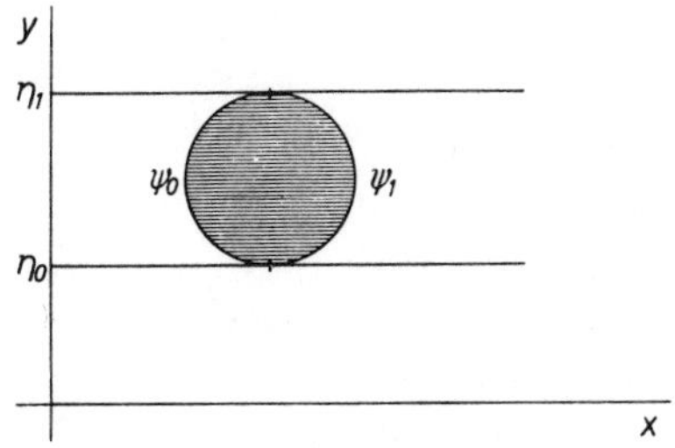

Fig. 29

Let $\mathfrak{R}$, then, be a simple 3-dimensional figure with a coordinate system $X = x^1, x^2, x^3$. Every differential form ω on $\mathfrak{R}$ can be represented as

$$\omega = f dx^1 dx^2 dx^3 .$$

If f is continuously differentiable, there will be a differential form Ω such that

$$d\Omega = \omega .$$

We have only to choose $\frac{\partial F}{\partial x^1} = f$ such that $\Omega = F dx^2 dx^3$. Ω is uniquely determined up to a differential form Ω_0 such that $d\Omega_0 = 0$. Accordingly, if we form the boundary integral $\oint_{\mathfrak{R}} \Omega$ (an integral over a composite 2-dimensional figure!), it will be determined by ω and $\mathfrak{R}$ alone: we shall have

$$\oint_{\mathfrak{R}} \Omega_0 = 0 ,$$

for by Poincaré's theorem $\Omega_0 = d\Omega_1$ (where Ω_1 is a first-degree differential form), and by Stokes's theorem

$$\oint_{\mathfrak{R}} d\Omega_1 = \int_{\text{boundary } \mathfrak{R}} d\Omega_1$$

is equal to the boundary integral $\oint_{\text{boundary } \mathfrak{R}} \Omega_1$ over the boundary of the boundary of $\mathfrak{R}$. Since for simple figures the boundary of the boundary is clearly always 0, it follows that

$$\oint_{\mathfrak{R}} \Omega_0 = 0 .$$

Thus we can define $\int_{\mathfrak{R}} \omega$ as

$$\int_{\mathfrak{R}} \omega = \oint_{\mathfrak{R}} \Omega \quad \text{for every } \Omega \text{ such that} \quad d\Omega = \omega . \qquad (22.30)$$

263 An elementary calculation gives us the connection between this and the numerical integral, as in (22.8):

$$\int_{\mathfrak{R}} \frac{\partial^3 f}{\partial x^1 \partial x^2 \partial x^3} dx^1 dx^2 dx^3 = \mathcal{J}^3 \mathcal{D}^3 \phi \Big|_{\Xi_0}^{\Xi_1} = \Delta^3 \phi \Big|_{\Xi_0}^{\Xi_1}$$

where $f = \phi(x^1, x^2, x^3)$ and $X(P_0) = \Xi_0$, $X(P_1) = \Xi_1$ for the end points P_0, P_1 of the X-interval $\mathfrak{R}$.

If a simple 3-dimensional figure is contained in an N-dimensional figure ($N > 3$), then for every Nth-degree differential form ω, $\int_{\mathfrak{R}} \omega$ will be defined by the fact that ω uniquely induces a third-degree differential form on $\mathfrak{R}$.

Since alternating differentiation is invariant with respect to parameters, Stokes's theorem

$$\int_{\mathfrak{R}} d\Omega = \oint_{\mathfrak{R}} \Omega$$

follows for arbitrary second-degree differential forms Ω in N-dimensional figures.

The integral thus far defined only for simple 3-dimensional figures is

(1) linear: $\int(c_1\omega_1 + c_2\omega_2) = c_1\int\omega_1 + c_2\int\omega_2$

(2) monotonic: $f_1 \leqslant f_2 \rightarrow \int f_1\,dX \leqslant \int f_2\,dX$ for X of positive sense

(3) additive: $\int\limits_{\sum_i \mathfrak{R}_i} \omega = \sum_i \int\limits_{\mathfrak{R}_i} \omega.$

$\sum_i \mathfrak{R}_i$ here means a combinatory decomposition of a simple figure into simple segments such that, even for the (canonically oriented) boundaries, the boundary of the sum is the sum of the boundaries:

$$\text{boundary } \sum \mathfrak{R}_i = \sum \text{ boundary } \mathfrak{R}_i.$$

In order to be able to extend the integral to (finitely) composite figures, we again define the integral for almost open sets within a simple figure (always with the orientation of the simple figure). First we define the X-measure for open domains and then the integral over open domains represented as unions of up to infinitely many X-intervals. Its independence from the coordinate system is established by the approximation theorem, which was already proved for any number of dimensions. Having introduced null sets, we ultimately obtain the completely additive integral

$$\int\limits_C \omega$$

for almost open domains C within a simple figure. Thereupon the integral can be extended to composite figures $\mathfrak{R}$. $\int\limits_{\mathfrak{R}} \omega$ is then dependent on $\mathfrak{R}$ and the differential form ω alone (i.e., the transformation formula holds). It is furthermore linear, monotonic, and additive, and Stokes's formula 264

$$\int d = \oint$$

follows for composite figures too.

From this sketch of the path that leads from the 2-dimensional to the 3-dimensional integral, we see that literally the same line of development will bring us to the 4-dimensional, . . . , and generally to the N-dimensional integral.

To compute the N-dimensional integral it is again advisable to decompose the figure into normal domains. If the $(N - 1)$-dimensional normal domains are already defined, then an N-dimensional *normal domain* (with respect to a coordinate system x^1, . . . ,

x^N) is understood to be a set of points P with the following properties:

(1) The coordinate values $x^2(P), \ldots, x^N(P)$ form an $(N-1)$-dimensional normal domain in number space.

(2) For twice continuously differentiable functions ψ_0, ψ_1,

$$\psi_0(x^2, \ldots, x^N) \leq x^1 \leq \psi_1(x^2, \ldots, x^N).$$

If P is a point with the coordinate values $x^1(P), \ldots, x^N(P)$, then let P be mapped onto the point $\overline{P}$ in such a way that

$$x^1(\overline{P}) = 0, x^2(\overline{P}) = x^2(P), \ldots, x^N(\overline{P}) = x^N(P).$$

The set of points $\overline{P}$ is called the x^1-projection D_1 of the domain D. Condition (1) then stipulates that the projection D_1 is to be an $(N-1)$-dimensional normal domain (with respect to $x^2, \ldots, x^N$).

For the function

$$f = \phi(x^1, \ldots, x^N)$$

and

$$F = \mathcal{I}_{\xi \, 0}^{\;x^1} \phi(\xi, x^2, \ldots, x^N),$$

Stokes's formula yields

$$\int_D f dx^1 \ldots dx^N =$$

$$\int_{x^1=\psi_1(x^2, \ldots, x^N)} F dx^2 \ldots dx^N - \int_{x^1=\psi_0(x^2, \ldots, x^N)} F dx^2 \ldots dx^N,$$

265 since the integral over the remaining boundary segments of D vanishes in view of $dx^2 \ldots dx^N = 0$. For on the boundary segments of the x^1-projection D_1 there are functional dependencies among $x^2, \ldots, x^N$, such as, say,

$$x^2 = \psi(x^3, \ldots, x^N).$$

Thus we get the *reduction formula*

$$\int_D f dx^1 \ldots dx^N = \int_{D_1} \mathcal{I}_{\xi \;\; \psi_0(x^2, \ldots, x^N)}^{\;\psi_1(x^2, \ldots, x^N)} \phi(\xi, x^2, \ldots, x^N) dx^2 \ldots dx^N. \tag{22.31}$$

Iterated application of the reduction formula yields a representation of the N-dimensional integral as an N-fold integral. With that we have finally reduced all instances of integration to the inverse of differentiation.

§23. Metric Differential Geometry

The beginnings of the infinitesimal calculus lie in Greek mathematics. The problems to be solved were primarily the determination of geometric quantities (volumes, surface areas, lengths). The area of an equilateral triangle (with a side of 1) is already the irrational number $\frac{1}{4}\sqrt{3}$. But though in a plane any triangle, and hence any polygon, can be transformed by a finite decomposition into a rectangle of equal area, in space the volume of a pyramid can only be determined by a sequence of approximating decompositions (as M. Dehn proved for the first time in 1900; cf. Hilbert 1968).

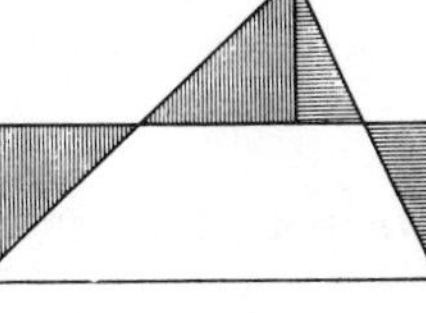
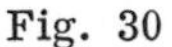
Fig. 30

To compute the length of curved line segments or the area of surface segments with curvilinear boundaries, infinitesimal methods are already necessary for plane figures. Eudoxos laid the foundations for this (as handed down in Book V of Euclid's *Elements*): he had already determined the volume of the pyramid and the circumference and the area of the circle. Archimedes then added numerous applications: volume and area of the sphere, squaring of the parabola, finding the center of gravity, and many more.

The development of calculation with decimal numbers and the exploitation of the inverse relation between differentiation and integration (Newton, Leibniz) made a systematic treatment of the problems of integration possible. The means we now have at our disposal make routine tasks out of Archimedes' brilliant calculations of volume and area.

In what follows, therefore, we shall assemble only the most important definitions for the application of differentials and integrals to Euclidean geometry. 266

Euclidean space is characterized—analytically—by the designation of a class of "*Cartesian*" coordinate systems $X = x^1, \ldots, x^N$ which are transformed into one another by the equations

$$\bar{x}^i = c^i_j x^j + d^i$$

under the condition of *orthogonality*,

$$\textstyle\sum_k c^k_i c^k_j = \delta_{ij} \tag{23.1}$$

(i.e., the transposed matrix is the inverse matrix). δ_{ij} is a so-called *Kronecker symbol*:

$$\delta_{ij} = \begin{cases} 1 & \text{if} \quad i = j \\ 0 & \text{if} \quad i \neq j. \end{cases}$$

The X-image of Euclidean space is the entire N-dimensional number space.

With the help of Cartesian coordinates we can single out from the various curves the *line segments*, which are represented by

$$x^i = a^i t + b^i,$$

where t is a mapping onto an interval. For any point P on the segment, the *vector*

$$a^i = \frac{dx^i}{dt}$$

has the same Cartesian coordinates. The vector $\mathfrak{a}$ with Cartesian coordinates a^i belongs at any arbitrary point P_0 to the line segment leading from P_0 to the point P_1 with coordinates

$$x^i(P_1) = x^i(P_0) + a^i.$$

Conversely, the vector $\mathfrak{a}$ is uniquely determined by any point pair P_0, P_1 such that

$$x^i(P_1) - x^i(P_0) = a^i.$$

Accordingly we write

$$\mathfrak{a} = \overrightarrow{P_0 P_1}.$$

Only for *affine coordinate systems* does the vector $\mathfrak{a} = \overrightarrow{P_0 P_1}$ take as coordinate values the difference of the coordinate values of
267 the points P_0, P_1. These coordinate systems arise from Cartesian ones by

$$\overline{x}^i = c^i_j x^i + d^i,$$

where c^i_j is any invertible matrix.

Because the condition (23.1) is required for the transformation of Cartesian coordinates into one another, for two vectors $\mathfrak{a}$, $\mathfrak{b}$ with Cartesian coordinate values a^i, b^i it is possible in Euclidean geometry to define an "*inner composition*":

$$\mathfrak{a} \cdot \mathfrak{b} = \sum_i a^i b^i. \tag{23.2}$$

Usually one speaks here of the *inner "product."* Since, however, the composition is not associative, we avoid the words "product" and "multiplication."

(23.2) defines the number $\mathfrak{a} \cdot \mathfrak{b}$ invariantly, for it will hold of any other Cartesian coordinate system $\overline{x}^1, \ldots, \overline{x}^N$ (leaving out summation signs) that

$$\overline{a}^i \overline{b}^i = c^i_k a^k c^i_l b^l = \delta_{kl} a^k b^l = a^i b^i.$$

The term $\mathfrak{a} \cdot \mathfrak{b}$ represents as a function of $\mathfrak{a}, \mathfrak{b}$ a *bilinear* or 2-place linear vector function, i.e., a second-degree tensor. This tensor is known as the *fundamental tensor*.

In particular we set

$$|\mathfrak{a}| = \sqrt{\mathfrak{a} \cdot \mathfrak{a}},$$

where $|\mathfrak{a}|$ is called the *absolute value* or the *length* of $\mathfrak{a}$.

For any (not necessarily affine) coordinate system with basis vectors $\mathfrak{x}_1, \ldots, \mathfrak{x}_N$ (cf. §19) there are for every point P_0 representations

$$\mathfrak{a} = a^i \mathfrak{x}_i$$

$$\mathfrak{b} = b^j \mathfrak{x}_j,$$

from which it follows that

$$\mathfrak{a} \cdot \mathfrak{b} = a^i b^j \mathfrak{x}_i \cdot \mathfrak{x}_j.$$

We let

$$G_{ij} = \mathfrak{x}_i \cdot \mathfrak{x}_j \tag{23.3}$$

so that

$$\mathfrak{a} \cdot \mathfrak{b} = G_{ij} a^i b^j.$$

The numbers G_{ij} are called the coordinate values of the fundamental tensor G. Obviously

$$G_{ij} = G_{ji},$$

i.e., G is a *symmetrical tensor* of the second degree.

G is already characterized by the fact that for Cartesian coordinates, 268

$$G_{ij} = \delta_{ij}.$$

To compute the coordinate values of the fundamental tensor for arbitrary coordinate systems, we start with a Cartesian coordinate system $\mathring{x}^1, \ldots, \mathring{x}^N$ and we obtain (since $\mathring{G}_{ij} = \delta_{ij}$)

$$G_{kl} = \mathring{G}_{ij} \frac{\partial \mathring{x}^i}{\partial x^k} \frac{\partial \mathring{x}^j}{\partial x^l} = \sum_j \frac{\partial \mathring{x}^j}{\partial x^k} \frac{\partial \mathring{x}^j}{\partial x^l}. \tag{23.4}$$

For convenience in calculations involving the inner composition, we introduce further coordinate values a_j by the definition

$$a_j = G_{ij} a^i \tag{23.5}$$

in addition to the *contravariant coordinate values* a^i which we already had. For the transformation of these new coordinate values we have

$$\overline{a}_i = \overline{G}_{ji}\,a^j = G_{kl}\,\frac{\partial x^k}{\partial \overline{x}^l}\,\frac{\partial x^l}{\partial \overline{x}^i}\,a^h\,\frac{\partial \overline{x}^j}{\partial x^h}$$

$$= G_{kl}\,a^h\,\delta_{hk}\,\frac{\partial x^l}{\partial \overline{x}^i}$$

$$= G_{kl}\,a^k\,\frac{\partial x^l}{\partial \overline{x}^i}$$

$$= a_l\,\frac{\partial x^l}{\partial \overline{x}^i}.$$

Thus these coordinate values transform covariantly, whence they are called *covariant coordinate values*.

Using both kinds of coordinate values, we get—for an arbitrary coordinate system—the simple formula

$$\mathfrak{a} \cdot \mathfrak{b} = a_i b^i. \tag{23.6}$$

(23.4) immediately gives us the determinant g of the matrix G_{ij}:

$$g = \det_{ij} G_{ij} = \left(\frac{d(\overset{\circ}{x}{}^1, \ldots, \overset{\circ}{x}{}^N)}{d(x^1, \ldots, x^N)}\right)^2,$$

so that in any case $g > 0$. g is not, however, an invariant. Since $g \neq 0$, G_{ij} has an inverse matrix G^{kl} defined by

$$G^{ki}G_{ij} = \delta_{kj}.$$

269 Corresponding to $a_j = G_{ij}a^i$ we then also have

$$a^i = G^{ij}a_j.$$

Multiplication by G^{ij}s or G_{ij}s can, as they say, *raise* or *lower* the indices.

In Euclidean space it is advisable to drop the distinction between vectors and tensors of the first degree and between multivectors and alternating tensors of higher degree. We shall now refer to vectors and multivectors as "*tensors*," keeping the word "*vector*" as an alternative name for first-degree tensors. We denote the coordinate values of a tensor simply by attached indices; e.g., the coordinate values of a second-degree tensor A will be A_{ij} or A^{ij}. This can only create misunderstandings when additional right-hand superscripts or subscripts are used, as, e.g., for coordinates x^i. Thus we should have to write $\overset{i}{x}$ or ${}^{(i)}x$ rather than x^i, but since misunderstandings are unlikely, we shall stay with x^i.

In addition we use the letters $\mathfrak{a}, \mathfrak{b}, \ldots$ for vectors, whereas tensors will normally be denoted by Latin capitals $A, B, \ldots$

If we use Cartesian coordinates exclusively, the distinction between co- and contravariant coordinate values becomes superfluous. For then of course we have, e.g.,

$$a_i = a^i$$
$$G^{ij} = G_{ij} = \delta_{ij}.$$

With non-Cartesian coordinates, however, this distinction (and thus the placement of indices as subscripts or superscripts) retains its importance even in Euclidean space.

The possibility of singling out a fundamental tensor that is invariant with respect to coordinates (and which then defines a length for every vector) also holds for any surface segment S located in a Euclidean space. If $u^1, \ldots, u^p$ is a parameter system of a surface segment in the N-dimensional Euclidean space E^N, then the parameters t of the curves in S are of course also curve parameters in E^N. Each tangential vector of S at a point P is thus also a vector in E^N. Accordingly, for two tangent vectors of S (at the same point), their inner composition is already defined by virtue of their being situated in E^N.

If $\mathfrak{a}$, $\mathfrak{b}$ are tangent vectors with parameter values

$$a^k, b^l \quad (k, l = 1, \ldots, p)$$

in the parameter system $u^1, \ldots, u^p$, then

$$a^i = a^k \frac{\partial x^i}{\partial u^k} \qquad (i, j = 1, \ldots, N)$$
$$b^j = b^l \frac{\partial x^j}{\partial u^l}$$

270

are the coordinate values in a coordinate system $x^1, \ldots, x^N$. It follows that

$$\mathfrak{a} \cdot \mathfrak{b} = G_{ij} a^i b^j = G_{ij} \frac{\partial x^i}{\partial u^k} \frac{\partial x^j}{\partial u^l} a^k b^l;$$

hence

$$\mathfrak{a} \cdot \mathfrak{b} = \Gamma_{kl} a^k b^l,$$

where

$$\Gamma_{kl} = G_{ij} \frac{\partial x^i}{\partial u^k} \frac{\partial x^j}{\partial u^l}. \tag{23.7}$$

Γ is the *fundamental tensor of the surface segment S*.

By the general multiplication theorem for determinants, we get as the determinant γ of Γ_{kl} for Cartesian coordinates

$$\gamma = \sum^{<}_{j_1 \ldots j_p} \left(\frac{\partial(x^{j_1}, \ldots, x^{j_p})}{\partial(u^1, \ldots, u^p)}\right)^2 .$$

For the differential geometry of surface segments in E^N (as far as we shall consider it here), all we need to know is the fundamental tensor Γ. Any figure in which a symmetrical second-degree tensor with a positive determinant is defined (the positivity of the determinant is invariant with respect to coordinates) is called a *metric figure* (or, to distinguish it from the metric spaces of topology, a *locally metric* or *Riemannian space*).

We therefore dispense with Cartesian coordinate systems from here on and consider an N-dimensional figure $\mathfrak{R}$ with a fundamental tensor G having a positive determinant g.

Furthermore, let the figure $\mathfrak{R}$ be oriented. We have established that g is not an invariant: transformation to another coordinate system gives

$$\bar{g} = \left(\frac{d(x^1, \ldots, x^N)}{d(\bar{x}^1, \ldots, \bar{x}^N)}\right)^2 g,$$

and hence for the positive square roots

$$\sqrt{\bar{g}} = \left|\frac{d(x^1, \ldots, x^N)}{d(\bar{x}^1, \ldots, \bar{x}^N)}\right| \sqrt{g} .$$

271 Together with

$$d\bar{x}^1 \ldots d\bar{x}^N = \frac{d(\bar{x}^1, \ldots, \bar{x}^N)}{d(x^1, \ldots, x^N)} dx^1 \ldots dx^N$$

this yields the equation

$$\sqrt{\bar{g}}\, d\bar{x}^1 \ldots d\bar{x}^N = \sqrt{g}\, dx^1 \ldots dx^N$$

where the coordinate systems $\bar{x}^1, \ldots, \bar{x}^N$ and $x^1, \ldots, x^N$ have the same sense.

Consequently, for all coordinate systems of positive sense,

$$\sqrt{g}\, dx^1 \ldots dx^N$$

yields the same differential form, one which is singled out in the oriented metric figure as invariant with respect to coordinates. We shall call this differential form the *content differential form*, denoted by Θ.

If we set

$$\operatorname{sgn} X = +1 \quad \text{or} \quad \operatorname{sgn} X = -1$$

according as X has positive or negative sense respectively, then we get the general result that

$$\Theta = \operatorname{sgn} X \sqrt{g}\, dx^1 \ldots dx^N. \tag{23.8}$$

The "*content*" of the figure $\mathfrak{R}$ is defined by

$$C(\mathfrak{R}) = \int_{\mathfrak{R}} \Theta. \tag{23.9}$$

The content, unlike the content differential form, is independent of the orientation. If we choose the other orientation of $\mathfrak{R}$, the content differential form will be $-\Theta$.

In Euclidean space with a Cartesian coordinate system $x^1, \ldots, x^N$ of positive sense, (23.9) means nothing else than

$$C(\mathfrak{R}) = \int_{\mathfrak{R}} dx^1 \ldots dx^N.$$

The content is thus the X-measure, and we have already seen that this coincides with the *elementary content*, the product of the edges, in the case of simple figures (rectangles, rectangular parallelepipeds, etc.).

Content is defined here invariantly with respect to coordinates (or, as they say in geometry, invariantly with respect to displacement). It is furthermore additive, indeed completely so. It is equal to 0 only for null sets; otherwise it is always positive. It thus satisfies all requirements we could reasonably make of content. It goes without saying that one can extend the notion of content beyond the class of almost open sets to still other sets, e.g., to all closed sets—for that, however, the reader is referred to textbooks on measure theory.

As one example, let us calculate the area of the circle $\mathfrak{R}$. In a Cartesian coordinate system x, y the boundary or circumference $\mathfrak{K}$ is given by a number $r > 0$ such that *272*

$$\begin{aligned} x &= r \cos t \\ y &= r \sin t \end{aligned} \qquad 0 \leq t \leq 2\pi.$$

We get

$$\begin{aligned} \int_{\mathfrak{R}} dxdy &= \oint_{\mathfrak{R}} xdy \\ &= \int_{\mathfrak{K}} r \cos t \cdot r \cos t\, dt \\ &= r^2 \int_{\mathfrak{K}} \cos^2 t\, dt \\ &= r^2 \mathcal{J}_0^{2\pi} \cos I \\ &= r^2 \Delta_0^{2\pi} \left(\frac{I}{2} + \frac{1}{2} \sin I \cos I \right), \end{aligned}$$

since

$$\left(\frac{I}{2} + \frac{1}{2}\sin I \cos I\right)' = \frac{1}{2} + \frac{1}{2}\cos^2 I - \frac{1}{2}\sin^2 I = \cos^2 I.$$

Thus

$$C(\mathfrak{R}) = \pi r^2.$$

If an oriented surface segment $\mathfrak{S}$ is given in a metric figure, the situation is not changed at all. Where Γ is the fundamental tensor of $\mathfrak{S}$, we let

$$C(\mathfrak{S}) = \int_{\mathfrak{S}} \sqrt{\gamma}\, du^1 \ldots du^p \tag{23.10}$$

for every parameter system $u^1, \ldots, u^p$ of positive sense. For a Cartesian coordinate system $x^1, \ldots, x^N$ of the figure, this means that

$$C(\mathfrak{S}) = \int_{\mathfrak{S}} \sqrt{\sum\nolimits^{<}_{i_1 \ldots i_p} \left(\frac{\partial(x^{i_1}, \ldots, x^{i_p})}{\partial(u^1, \ldots, u^p)}\right)^2}\, du^1 \ldots du^p.$$

Cartesian coordinate systems of opposite sense yield the same result here.

For the *length* of curve segments $\mathfrak{K}$ we get in particular

$$C(\mathfrak{K}) = \int_{\mathfrak{K}} \sqrt{\sum\nolimits_i \frac{dx^i}{dt}\frac{dx^i}{dt}}\, dt.$$

273 The circumference of the circle $\mathfrak{K}$ is accordingly

$$\begin{aligned}\int_{\mathfrak{K}} \sqrt{\left(\frac{dx}{dt}\right)^2 + \left(\frac{dy}{dt}\right)^2}\, dt &= r\int_{\mathfrak{K}} \sqrt{\sin^2 t + \cos^2 t}\, dt \\ &= r\int_{\mathfrak{K}} dt \\ &= r\underset{\mathfrak{K}}{\Delta} t = r\overset{2\pi}{\underset{0}{\Delta}} I = 2\pi r.\end{aligned}$$

This trivial calculation simply shows that the actual work of determining the circumference is already accomplished with the definition of π $\left(\text{in the determination of } \lim_{\delta \to 0} \frac{\operatorname{Sin} \delta}{\delta}\right)$.

Besides the determination of content, the fundamental tensor G of a figure $\mathfrak{R}$ also makes possible the introduction of *orthogonality*. Two vectors $\mathfrak{a}$, $\mathfrak{b}$ in the tangential space are said to be *orthogonal*, $\mathfrak{a} \perp \mathfrak{b}$, when for the coordinates a^i, b^i,

$$G_{ij} a^i b^j = 0.$$

This can be abbreviated to

$$a_i b^i = 0,$$

or in symbolic notation,

$$\mathfrak{a} \cdot \mathfrak{b} = 0.$$

For $N - 1$ linearly independent vectors $\overset{1}{a}_i, \ldots, \overset{N-1}{a}_i$ (at a common point P_0), the condition

$$\overset{j}{a}_i b^i = 0 \qquad (j = 1, \ldots, N - 1)$$

uniquely determines a vector b^i up to a factor. By the determinant-development theorem the coordinate values b^i are proportional to certain $(N - 1)$-rowed subdeterminants of the matrix

$$\begin{pmatrix} \overset{1}{a}_1 & \ldots & \overset{N-1}{a}_1 \\ \vdots & & \vdots \\ \overset{1}{a}_N & \ldots & \overset{N-1}{a}_N \end{pmatrix}.$$

Geometrically, a solution b^i can be singled out as follows. The content differential form

$$\Theta = \operatorname{sgn} X \sqrt{g}\, dx^1 \ldots dx^N$$

is an Nth-degree alternating tensor. We denote its coordinate values by $\varepsilon_{i_1 \ldots i_N}$. Thus for coordinate systems $x^1, \ldots, x^N$ of positive sense we have 274

$$\varepsilon_{i_1 \ldots i_N} = \operatorname{sgn}(i_1, \ldots, i_N) \sqrt{g} \qquad (23.11)$$

where $\operatorname{sgn}(i_1, \ldots, i_N)$ again means the number +1 or -1 according as $i_1, \ldots, i_N$ is an even or an odd permutation of $1, \ldots, N$. For the contravariant coordinate values we shall have

$$\varepsilon^{i_1 \ldots i_N} = \operatorname{sgn}(i_1, \ldots, i_N) \frac{1}{\sqrt{g}}.$$

We call the tensor $\varepsilon_{i_1 \ldots i_N}$ the *content tensor* or the ε-tensor for short. For Cartesian coordinate systems all coordinate values are either +1 or -1. If the Cartesian coordinate system has positive sense, then e.g.

$$\varepsilon_{1 \ldots N} = +1,$$

whereas if its sense is negative, then

$$\varepsilon_{1 \ldots N} = -1.$$

For each vector system $\overset{1}{\mathfrak{a}}, \ldots, \overset{N-1}{\mathfrak{a}}$ we now form the vector

$$b^i = \varepsilon^{i_1 \ldots i_{N-1} i} \overset{1}{a}_{i_1} \ldots \overset{N-1}{a}_{i_{N-1}}.$$

Since $a_i b^i$ is accordingly the determinant of the matrix with the columns $\overset{1}{a}_i, \ldots, \overset{N-1}{a}_i$ and a_i up to the factor $\frac{1}{\sqrt{g}}$, b^i is orthogonal to $\overset{1}{a}_i, \ldots, \overset{N-1}{a}_i$ and $\overset{1}{\mathfrak{a}}, \overset{2}{\mathfrak{a}}, \ldots, \overset{N-1}{\mathfrak{a}}$, $\mathfrak{b}$ has positive sense.

b^i is frequently referred to as the "*outer product*" of $\overset{1}{\mathfrak{a}}, \ldots, \overset{N-1}{\mathfrak{a}}$. For $N = 3$ in particular we have two factors. However (besides the fact that this "product" is not associative for the case $N = 3$ either), it is more to the point to break the operation leading from the $\overset{1}{a}_i, \ldots, \overset{N-1}{a}_i$s to b^i down into two steps. For if we take the alternating product of the vectors $\overset{1}{\mathfrak{a}}, \ldots, \overset{N-1}{\mathfrak{a}}$ [i.e., by §21, an $(N - 1)$-vector the coordinate values of which are the coefficients produced by alternating multiplication of

$$\overset{1}{\mathfrak{a}} = \sum \overset{1}{a}{}^i \mathfrak{x}_i$$
$$\vdots$$
$$\overset{N-1}{\mathfrak{a}} = \sum \overset{N-1}{a}{}^i \mathfrak{x}_i \,],$$

we easily see that b^i is already uniquely determined by the alternating product $[\overset{1}{\mathfrak{a}} \ldots \overset{N-1}{\mathfrak{a}}]$:

$$b^i = \frac{1}{(N-1)!} \varepsilon^{i_1 \ldots i_{N-1} i} [\overset{1}{\mathfrak{a}} \ldots \overset{N-1}{\mathfrak{a}}]_{i_1 \ldots i_{N-1}}.$$

275 What we do, then, is first to take the alternating product $[\overset{1}{\mathfrak{a}} \ldots \overset{N-1}{\mathfrak{a}}]$ of the vectors $\overset{1}{\mathfrak{a}}, \ldots, \overset{N-1}{\mathfrak{a}}$, and then to form a new vector for this alternating tensor (multivector) with the aid of its coordinate values $A_{i_1 \ldots i_{N-1}}$:

$$b^i = \frac{1}{(N-1)!} \varepsilon^{i_1 \ldots i_{N-1} i} A_{i_1 \ldots i_{N-1}}.$$

We generalize this last procedure by correlating each alternating tensor A of degree p (with coordinate values $A_{j_1 \ldots j_p}$) with an alternating tensor $^*A$ of degree q such that $p + q = N$ with coordinate values

$$^*A^{i_1 \ldots i_q} = \frac{1}{p!} \varepsilon^{j_1 \ldots j_p i_1 \ldots i_q} A_{j_1 \ldots j_p}. \qquad (23.12)$$

The tensor $^*A$ is called the *orthocomplement* of A. In place of the "outer product" of $N-1$ vectors $\overset{1}{\mathfrak{a}}, \ldots, \overset{N-1}{\mathfrak{a}}$, then, we have the orthocomplement of the alternating product:

$$^*[\overset{1}{\mathfrak{a}} \ldots \overset{N-1}{\mathfrak{a}}].$$

If one wishes, one can generalize the usual *outer composition*

$$\mathfrak{a} \times \mathfrak{b} = {}^*[\mathfrak{a}\mathfrak{b}]$$

for $N = 3$ to an arbitrary number of dimensions by

$$A \times B = {}^*[AB]$$

for alternating tensors A, B, but evidently little is gained thereby.

It is, on the other hand, important that the *inner composition* can be generalized from vectors to alternating tensors:

$$A \cdot B = {}^*[A\, {}^*B]. \tag{23.13}$$

In case A and B are of like degree, say p, $^*B$ is a tensor of degree q $(p+q=N)$, so that $[A\, {}^*B]$ is a tensor of degree N and $A \cdot B$ is an invariant.

This invariant indeed generalizes the inner composition of vectors, since we have for tensors of like degree

$$A \cdot B = \sum\nolimits^{<}_{i_1 \ldots i_p} A_{i_1 \ldots i_p} B^{i_1 \ldots i_p}. \tag{23.14}$$

Proof. To begin with we have, for a coordinate system of positive sense, let us say, 276

$$^*[A\, {}^*B] = \frac{1}{N!}\, \varepsilon_{k_1 \ldots k_N} [A\, {}^*B]^{k_1 \ldots k_N} = \sqrt{g}\, [A\, {}^*B]^{1 \ldots N}.$$

To compute the coordinate values of $[A\, {}^*B]$, we use

$$A = \sum\nolimits^{<} A^{i_1 \ldots i_p} [\mathfrak{x}_{i_1} \ldots \mathfrak{x}_{i_p}] = \frac{1}{p!} \sum A^{i_1 \ldots i_p} [\mathfrak{x}_{i_1} \ldots \mathfrak{x}_{i_p}]$$

and correspondingly

$$^*B = \frac{1}{q!} \sum {}^*B^{i_{p+1} \ldots i_N} [\mathfrak{x}_{i_{p+1}} \ldots \mathfrak{x}_{i_N}].$$

From these we get

$$^*[A\, {}^*B] =$$

$$\sqrt{g}\, \frac{1}{p!\,q!} \sum\nolimits_{i_1 \ldots i_N} \operatorname{sgn}(i_1, \ldots, i_N) A^{i_1 \ldots i_p}\, {}^*B^{i_{p+1} \ldots i_N}.$$

By the definition of the orthocomplement it further follows that

$$*[A*B] = \frac{1}{p!q!}\sum_{i_1 \ldots i_N} \frac{1}{p!}\sum_{j_1 \ldots j_p} \operatorname{sgn}(i_1, \ldots, i_N)$$

$$\operatorname{sgn}(j_1, \ldots, j_p, i_{p+1}, \ldots, i_N)\, A^{i_1 \ldots i_p} B_{j_1 \ldots j_p}.$$

Only for the $p!$ permutations $j_1 \ldots j_p$ of $i_1 \ldots i_p$ do we have here in each case $q!$ equal summands such that

$$\operatorname{sgn}(j_1, \ldots, j_p, i_{p+1}, \ldots, i) \neq 0.$$

Consequently we get

$$*[A*B] =$$

$$\frac{1}{p!}\sum_{i_1 \ldots i_p} \operatorname{sgn}(i_1, \ldots, i_N)\operatorname{sgn}(i_1, \ldots, i_N) A^{i_1 \ldots i_p} B_{i_1 \ldots i_p}$$

$$= \frac{1}{p!}\sum_{i_1 \ldots i_p} A^{i_1 \ldots i_p} B_{i_1 \ldots i_p}$$

$$= \sum^{<} A^{i_1 \ldots i_p} B_{i_1 \ldots i_p}.$$

In particular, we set

$$|A| = \sqrt{A \cdot A} = \sqrt{*[A*A]}\,.$$

$|A|$ is the *absolute value* or the *measure* of A.

As long as we confine ourselves to alternating tensors, and in particular to vectors, the basic algebraic formations are—as Grassmann already correctly noted in 1862—the alternating product and the orthocomplement, and not the so-called inner or outer "product."

The orthocomplement is *involutory* up to a sign:

$$**A = (-1)^{pq}A \tag{23.15}$$

where A is an alternating tensor of degree p and $p + q = N$.

277 *Proof.* For $j_1 < \ldots < j_p$ we have to show that

$$**A^{j_1 \ldots j_p} = (-1)^{pq} A^{j_1 \ldots j_p}.$$

Let X have positive sense. Then

$$**A^{j_1 \ldots j_p} = \frac{1}{q!}\sum \varepsilon^{i_1 \ldots i_q j_1 \ldots j_p} *A_{i_1 \ldots i_q}$$

$$= \frac{1}{\sqrt{g}} \operatorname{sgn}(i_1, \ldots, i_q, j_1, \ldots, j_p) *A_{i_1 \ldots i_q},$$

where there is only one system $i_1 < \ldots < i_q$ such that

$$\operatorname{sgn}(i_1, \ldots, i_q, j_1, \ldots, j_p) \neq 0.$$

It thus follows further that

$$**A^{j_1 \cdots j_p}$$

$$= \frac{1}{\sqrt{g}} \operatorname{sgn}(i_1, \ldots, i_q, j_1, \ldots, j_p) \frac{1}{p!} \sum \varepsilon_{k_1 \cdots k_p i_1 \cdots i_q} A^{k_1 \cdots k_p}$$

$$= \operatorname{sgn}(i_1, \ldots, i_q, j_1, \ldots, j_p) \operatorname{sgn}(k_1, \ldots, k_p, i_1, \ldots, i_q) A^{k_1 \cdots k_p}.$$

Again there is just one system $k_1 < \ldots < k_p$ such that

$$\operatorname{sgn}(k_1, \ldots, k_p, i_1, \ldots, i_q) \neq 0,$$

viz., $j_1 < \ldots < j_p$. Thus we get

$$**A^{j_1 \cdots j_p} =$$

$$\operatorname{sgn}(i_1, \ldots, i_q, j_1, \ldots, j_p) \operatorname{sgn}(j_1, \ldots, j_p, i_1, \ldots, i_q) A^{j_1 \cdots j_p}.$$

The only other thing we have to take into account is that

$$\operatorname{sgn}(i_1, \ldots, i_q, j_1, \ldots, j_p) = (-1)^{pq} \operatorname{sgn}(j_1, \ldots, j_p, i_1, \ldots, i_q).$$

If N is odd (e.g. $= 3$), p or q will always be even, in which case

$$**A = A.$$

If N is even, then

$$**A = (-1)^p A = (-1)^q A.$$

Since by (23.14)

$$A \cdot B = B \cdot A$$

holds trivially for tensors A, B of like degree, and

$$[AB] = (-1)^{pq} [BA]$$

holds generally for a tensor A of degree p and a tensor B of degree q (as it does for differential forms), it follows for tensors of like degree that

$$*A \cdot *B = A \cdot B. \tag{23.16}$$

Proof: $$*A \cdot *B = *[*A **B] = (-1)^{pq} *[*AB]$$ 278

$$= *[B*A] = B \cdot A.$$

In particular, $|{*A}| = |A|$: Every tensor has the same measure as its orthocomplement.

In addition to the algebraic operations, we now consider differentiation d. In metric spaces we also have a "*codifferentiation*" operation $\delta = {*d*}$ at our disposal in addition to d. Whereas for an alternating tensor A of degree p the tensor dA has the degree $p + 1$, δA has the degree $p - 1$, i.e., $N - (N - p + 1)$. As with d we clearly have

$$\delta\delta = 0,$$

and Poincaré's theorem (§21) likewise extends from d to δ.

The usual *differential operators*

$$\text{grad, rot, div}$$

of 3-dimensional Euclidean *vector analysis* are easily generalizable by means of the operators $*$, d, and δ to metric spaces in such a way as to be invariant with respect to coordinates, without the necessity of resorting to Cartesian coordinates.

Where f is an invariant,

$$df = \sum \frac{\partial f}{\partial x^i}\, dx^i$$

is a vector (first-degree tensor) with coordinate values

$$(df)_i = \frac{\partial f}{\partial x^i}.$$

This vector is called the *gradient* of f, denoted in vector analysis by grad f.

For a vector A, dA is a second-degree alternating tensor with the coordinate values

$$(dA)_{ij} = \frac{\partial A_i}{\partial x^j} - \frac{\partial A_j}{\partial x^i}.$$

Only for $N = 3$ is the orthocomplement $*dA$ itself a vector. It is called the *rotation* of A, rot A. Its coordinate values are

$$(\text{rot } A)^i = \frac{1}{2}\, \varepsilon^{jki} \left(\frac{\partial A_j}{\partial x^k} - \frac{\partial A_k}{\partial x^j}\right).$$

279 For a coordinate system of positive sense, then, we have

$$(\text{rot} A)_1 = \frac{\partial A_2}{\partial x^3} - \frac{\partial A_3}{\partial x^2}$$

$$(\text{rot } A)_2 = \frac{\partial A_3}{\partial x^1} - \frac{\partial A_1}{\partial x^3}$$

$$(\text{rot}\,A)_3 = \frac{\partial A_1}{\partial x^2} - \frac{\partial A_2}{\partial x^1}.$$

The application of the operator δ to a vector A leads to the invariant div A, the *divergence* of A. For an arbitrary coordinate system X, we have

$$\delta A = \frac{1}{\sqrt{g}} \sum \frac{\partial(\sqrt{g}\,A^i)}{\partial x^i}.$$

Proof. If the sense of a coordinate system X is positive, then

$$*A_{i_1 \ldots i_{N-1}} = \varepsilon_{i i_1 \ldots i_{N-1}} A^i = \sqrt{g}\ \text{sgn}(i, i_1, \ldots, i_{N-1}) A^i.$$

$*A$ can be represented as a differential form by

$$*A = \sum^{<}_{i_1 \ldots i_{N-1}} *A_{i_1 \ldots i_{N-1}} dx^{i_1} \ldots dx^{i_{N-1}}.$$

This consequently yields

$$d{*A} =$$

$$\sum^{<}_{i_1 \ldots i_{N-1}} \sum_i \frac{\partial \sqrt{g}\,A^i}{\partial x^i}\ \text{sgn}(i, i_1, \ldots, i_{N-1})\, dx^i dx^{i_1} \ldots dx^{i_{N-1}}$$

$$= \sum_i \frac{\partial \sqrt{g}\,A^i}{\partial x^i}\, dx^1 \ldots dx^N,$$

whence

$$(d{*A})_{1 \ldots N} = \sum \frac{\partial \sqrt{g}\,A^i}{\partial x^i}$$

and

$$\delta A = \frac{1}{N!}\, \varepsilon^{i_1 \ldots i_N} (d{*A})_{i_1 \ldots i_N} = \frac{1}{\sqrt{g}} \sum \frac{\partial \sqrt{g}\,A^i}{\partial x^i}.$$

The same formula results when the sense of X is negative. For affine coordinate systems, and in particular Cartesian ones, we have

$$\text{div}\,A = \sum \frac{\partial A^i}{\partial x^i}.$$

In metric differential geometry, the fundamental formula *280*

$$\int d\Omega = \oint \Omega$$

gives rise to theorems that are important for applications to geometry and physics.

To this end let $\mathfrak{S}$ be an oriented p-dimensional surface segment in an N-dimensional metric figure. Every point of $\mathfrak{S}$ will have a positively sensed p-leg of tangential vectors $\mathfrak{t}_1, \ldots, \mathfrak{t}_p$. Up to a factor, these determine the pth-degree tensor $[\mathfrak{t}_1 \ldots \mathfrak{t}_p]$. If we normalize this factor in such a way that its absolute value is 1, we obtain the *unit tangent tensor*

$$T^p = \frac{[\mathfrak{t}_1 \ldots \mathfrak{t}_p]}{|[\mathfrak{t}_1 \ldots \mathfrak{t}_p]|}.$$

It will hold of a parameter system $t^1, \ldots, t^p$ that

$$[\mathfrak{t}_1 \ldots \mathfrak{t}_p]^{j_1 \ldots j_p} = \frac{d(x^{j_1}, \ldots, x^{j_p})}{d(t^1, \ldots, t^p)}.$$

By (23.7) and (23.14) it follows that

$$|[\mathfrak{t}_1 \ldots \mathfrak{t}_p]| = \sqrt{\gamma}$$

where γ is the determinant of the fundamental tensor Γ of $\mathfrak{S}$. Thus the coordinate values of the tensor T^p are

$$T^{j_1 \ldots j_p} = \frac{1}{\sqrt{\gamma}} \frac{d(x^{j_1}, \ldots, x^{j_p})}{d(t^1, \ldots, t^p)}.$$

Multiplying this by the content differential form

$$\Theta^p = \sqrt{\gamma}\, dt^1 \ldots dt^p$$

yields

$$T^{j_1 \ldots j_p} \Theta^p = dx^{j_1} \ldots dx^{j_p}.$$

Accordingly, for a differential form

$$\omega = \sum\nolimits^{<} A_{j_1 \ldots j_p}\, dx^{j_1} \ldots dx^{j_p},$$

we can also represent the differential form as restricted to $\mathfrak{S}$,

$$\omega_{\mathfrak{S}} = \sum\nolimits^{<} A_{j_1 \ldots j_p} \frac{d(x^{j_1}, \ldots, x^{j_p})}{d(t^1, \ldots, t^p)}\, dt^1 \ldots dt^p,$$

281 by

$$\omega_{\mathfrak{S}} = \sum\nolimits^{<} A_{j_1 \ldots j_p} T^{j_1 \ldots j_p} \Theta^p$$
$$= (A \cdot T^p)\Theta^p.$$

In this way, for metric spaces the fundamental formula $\oint = \int d$ turns into the formula

$$\oint (A \cdot T^p)\Theta^p = \int (dA \cdot T^{p+1})\Theta^{p+1}. \tag{23.17}$$

A and T^p here are tensors of like degree, just as are dA and T^{p+1}.

Besides the unit tangent tensor, we shall also introduce the *unit normal tensor*:

$$S^q = {}^*T^p \quad \text{where} \quad p + q = N.$$

In accordance with this definition,

$$S^q = \frac{1}{\sqrt{\gamma}}{}^*[\mathfrak{t}_1 \ldots \mathfrak{t}_p].$$

For $p = N - 1$, S^1 is a vector such that $\mathfrak{t}_1, \ldots, \mathfrak{t}_{N-1}, S^1$ has positive sense. Therefore the system

$$(-1)^{N-1}S^1, \mathfrak{t}_1, \ldots, \mathfrak{t}_{N-1}$$

likewise has positive sense; i.e., by the canonical orientation convention, $(-1)^{N-1}S^1$ is the *unit normal* pointing toward the exterior of the $(N - 1)$-dimensional surface segment.

In view of (23.16), the fundamental formula (23.17) now becomes

$$\oint (A \cdot T^p)\Theta^p = \int ({}^*dA \cdot S^{q-1})\Theta^{p+1} \tag{23.18}$$

and in particular, by (23.15), with ${}^*A$ substituted for A,

$$(-1)^{pq}\oint(A \cdot S^q)\Theta^p = \int(\delta A \cdot S^{q-1})\Theta^{p+1}. \tag{23.19}$$

These formulas contain as special cases where $N = 3$ and p or q = 1 for vectors A

$$\oint(A \cdot T^1)\Theta^1 = \int(\operatorname{rot} A \cdot S^1)\Theta^2$$
$$\oint(A \cdot S^1)\Theta^2 = \int(\operatorname{div} A \cdot S^0)\Theta^3.$$

Here we have the original *Stokes* and *Gauss formulas*. In the traditional but unsystematic notation one uses

$$d\mathfrak{t} \quad \text{for} \quad T^1\Theta^1$$
$$d\mathfrak{f} \quad \text{for} \quad S^1\Theta^2$$
$$dV \quad \text{for} \quad \Theta^3.$$

Since $S^0 = 1$, we get the usual formulations

282

$$\oint\mathfrak{a} \cdot d\mathfrak{t} = \int\operatorname{rot} \mathfrak{a} \cdot d\mathfrak{f} \quad \text{(Stokes)}$$
$$\oint\mathfrak{a} \cdot d\mathfrak{f} = \int\operatorname{div} \mathfrak{a} \cdot dV \quad \text{(Gauss)}.$$

The application of these formulas to hydrodynamics explains the names "*rotation*" and "*divergence*." If $\mathfrak{a}$ is the velocity vector of an incompressible stationary fluid flow, then for a curve $\mathfrak{K}$,

$$\int_{\mathfrak{R}} \mathfrak{a} \cdot d\mathfrak{t}$$

is a measure of the "*circulation*" of the flow along $\mathfrak{R}$. By Stokes's theorem, rot $\mathfrak{a}$ is the "*local circulation*" of the flow $\mathfrak{a}$.

For a surface segment $\mathfrak{S}$, on the other hand,

$$\int_{\mathfrak{S}} \mathfrak{a} \cdot d\mathfrak{f}$$

is a measure of the "*flux*" of the flow over $\mathfrak{S}$. By Gauss's theorem, div $\mathfrak{a}$ is the "*local flux*" of the flow $\mathfrak{a}$.

In affine space—if we limit ourselves to affine coordinate systems—we can even introduce integrals over *generalized differential forms*, which take tensors rather than point functions as coefficients. Where $\mathfrak{S}$ is a p-dimensional surface segment in an affine N-dimensional space (and $u^1, \ldots, u^p$ is a parameter system of $\mathfrak{S}$), we define for every tensor A of degree r a tensor B (of degree r) by

$$B_{i_1 \ldots i_r} = \int_{\mathfrak{S}} A_{i_1 \ldots i_r} du^1 \ldots du^p.$$

Because the transformation from one affine coordinate system to another is a linear transformation with constant coefficients, and because the integral is a linear operator, this definition of B is invariant with respect to coordinates. We may therefore write

$$B = \int_{\mathfrak{S}} A du^1 \ldots du^p.$$

Consequently, it is only in Euclidean spaces and with affine coordinate systems that formula (23.17) can be extended to tensors A of degree $r < p$ if (23.14) is generalized to

$$(A \cdot B)^{i_{r+1} \ldots i_p} = \frac{1}{r!} A_{i_1 \ldots i_r} B^{i_1 \ldots i_r i_{r+1} \ldots i_p}. \quad (23.20)$$

283 A generalization of the proof of (23.14) then gives in place of (23.13)

$$A \cdot {}^*B = {}^*[BA]. \quad (23.21)$$

Using the outer composition here for once, we get from (23.17), because $A \cdot B = {}^{***}B \times A$, the result that

$$\oint ({}^{**}S^q \times A)\Theta^p = \int ({}^{**}S^{q-1} \times dA)\Theta^{p+1},$$

and hence

$$\oint (S^q \times A)\Theta^p = (-1)^{N-1} \int (S^{q-1} \times dA)\Theta^{p+1}. \quad (23.22)$$

As a special case for $N = 3, q = 1$, we get for vectors A

$$\oint (S^1 \times A)\Theta^2 = \int (S^0 \times dA)\Theta^3;$$

i.e., in traditional notation, since $S^0 = 1$ and $1 \times dA = *[1\ dA] = *dA$,

$$\oint (d\mathfrak{f} \times \mathfrak{a}) = \int \text{rot}\ \mathfrak{a}\ dV.$$

These few examples suffice to show that the differential calculus with its operations $*$ and d empowers us to handle the infinitesimal summations (integrations) which are important for geometry and physics. Only in the higher development of differential geometry, as begun by Gauss and Riemann, does it become necessary to differentiate arbitrary tensors rather than just the alternating ones. For that, all we can do here is to refer the reader to the literature on tensor analysis and Riemannian geometry.

BIBLIOGRAPHY

Textbooks

Behnke, H., and F. Bachmann, eds. *Grundzüge der Mathematik,* Vol. 3, *Analysis.* Göttingen: Vandenhoeck & Ruprecht, 1962.

Dieudonné, J. *Foundations of Modern Analysis.* New York: Academic Press, 1960.

Duschek, A., and A. Hochrainer. *Grundzüge der Tensorrechnung in analytischer Darstellung.* Vienna: Springer, part 1, 5th ed., 1968; part 2, 2nd ed., 1961; part 3, 1955.

Erwe, F. *Differential and Integral Calculus.* New York: Hafner; Edinburgh-London: Oliver & Boyd, 1967.

Haupt, O., G. Aumann, and C. Y. Pauc. *Differential- und Integralrechnung.* Berlin: De Gruyter, vol. 1, 2nd ed., 1948; vol. 2, 2nd ed., 1950; vol. 3, 2nd ed., 1955.

Maak, W. *An Introduction to Modern Calculus.* New York: Holt, Rinehart & Winston, 1963.

Mangoldt, H. von. *Einführung in die höhere Mathematik für Studierende und zum Selbststudium,* ed. and rev. K. Knopp. Stuttgart: Hirzel, vol. 1, 12th ed., 1963; vol. 2, 12th ed., 1964; vol. 3, 12th ed., 1963.

On the History of Infinitesimal Mathematics

Hofmann, J. E. *Classical Mathematics: A Concise History of the Classical Era in Mathematics.* New York: Philosophical Library; London: Vision, 1959.

———. *The History of Mathematics.* New York: Philosophical Library, 1957.

Lorenzen, P. *Die Entstehung der exakten Wissenschaften.* Berlin-Göttingen-Heidelberg: Springer, 1960.

Toeplitz, O. *The Calculus: A Genetic Approach.* Chicago: University of Chicago Press, 1963.

Sources Cited

Brouwer, L. E. J. *Wiskunde, Waarheid, Werkelijkheid.* Groningen: Noordhoff, 1919.

Dingler, H. *Aufbau der exakten Fundamentalwissenschaften.* Munich: Eidos-Verlag, 1964.

Grassmann, H. *Die Ausdehnungslehre.* Berlin: Enslin, 1862.

Hilbert, D. *The Foundations of Geometry.* LaSalle, Ill.: Open Court, 1902.

Landau, E. *Foundations of Analysis*. 3rd ed. New York: Chelsea, 1966.
Lorenzen, P. *Einführung in die operative Logik und Mathematik*. Berlin-Göttingen-Heidelberg: Springer, 1955.
———. *Metamathematik*. Mannheim: Bibliographisches Institut, 1962.
Quine, W. V. *Set Theory and Its Logic*. Cambridge: Harvard University Press, 1963.
Ritt, J. F. *Differential Algebra*. 2nd ed. New York: Dover, 1966.
Schneider, T. *Einführung in die transzendenten Zahlen*. Berlin-Göttingen-Heidelberg: Springer, 1957.
Weyl, H. *Das Kontinuum*. Leipzig: De Gruyter, 1918.

INDEX OF SYMBOLS

Chapter 1

Arithmetic

Symbol	Page
$\vert$	6
$=$	8
$\ddagger$	9
$+$	13
$\cdot$	13
$-$	14, 19
$//$	16
$/$	16
$<$	18
$>$	18
$\leqslant$	19
$\geqslant$	19
$\vert\ \vert$	20
1, 2, 3, ...	14
0	19
μ	27
!	34

Logic

Symbol	Page
$\Rightarrow$	6
$\leftrightharpoons$	9
$\rightarrow$	7
$\leftrightarrow$	9
$\wedge$	8
$\vee$	14 f.
$\neg$	9
Λ	7
$\boldsymbol{\Lambda}$	10
V	14 f.
$\mathbf{V}$	15
⅄	26

Set theory

Symbol	Page
$=$	22, 24 f.
ι	27
ϵ	24, 26
$\notin$	25
$\{\ \}$	25
˥	21 f., 28 f.
Γ	28
χ	29
$\cap$	25
$\cup$	25
$\llcorner$	25
$\sqcup$	25
$\subset$	26
$\subseteq$	26
$\boldsymbol{\cap}$	25
$\boldsymbol{\cup}$	25

Chapters II-III

Symbol	Page
*	43 f.
†	43 f.
$\ddagger$	43
∞	52
[$\vert$]	54
($\vert$)	55
glb	67
lub	67
lim	50, 57, 75 f., 98
$\rightsquigarrow$	50, 75 f.
I	84
$\surd$	87
$\overset{\rightsquigarrow}{=}$	99
$\sum$	100, 106
i	92
$\vert\ \vert$	92
. . .	101
. . . .	101
. . . . .	101
$'$	116
$\mathscr{D}$	116
Δ	117
e	126
π	129
$\binom{}{}$	137
$\mathscr{T}$	143

Chapter IV

Symbol	Pages
$(\mid)$	156
$[\mid]$	156
$\overline{\ }$	158
$\underline{\ }$	158
lim	158, 163
$\rightsquigarrow$	158, 163
$+$	159
$-$	159
$\mid\ \mid$	159, 177
$\Vert\ \Vert$	159
$\mid\ \mid_+$	160
I	166
$'$	167, 176
$\mathscr{D}$	167 f., 176 f., 184 f.
$[\ \ ]$	175
$\{ \mid \}$	175
$\longrightarrow$	179
Δ	183 f.
$\mathscr{T}$	186 f., 191 ff.

Chapter V

Symbol	Pages
$\Vert\ \Vert$	201
$\check{\ }$	205
$\rightsquigarrow$	212
$'$	212, 223
d	213, 218, 221, 224, 289
Δ	214
∂	216, 227
$\sum^{<}$	228
sgn	226
$[\ \ ]$	203, 206, 221, 224, 228 f.
$\cdot$	237, 274, 289
$\int$	243
$\stackrel{+}{\sum}$	247
$\sum$	251
$\oint$	248
$\mid\ \mid$	275, 284
$\perp$	280
$*$	283
$\times$	283
δ	286
grad	286
rot	286
div	287

INDEX OF SUBJECTS

Abstraction, 6, 15 ff.
Addition, 12 f.
    theorem, 94
Additivity, complete, 252, 256
Almost open, 265
Alternating
    differentiation, 235 ff.
    multiplication, 233 f.
Analysis, 1 ff.
    tensor, 291
    vector, 286
Analytic notation, 223
Antecedent, 7
Applicable, 14, 18 f., 56, 72
Approximate solution, 43
Approximation theorem, 260
Archimedean property, 60
Area, surface, 149
Argument, 21
    domain, 71
Arithmetic, laws of, 14
Assertion, 7, 10
Associativity, 14
Axiomatic
    method, 2
    set theory, 37

Borel's theorem, 255
Bound, 67
    greatest lower, 67
    least upper, 67
    -theorem, 116, 184;
        strengthened, 118, 188
Boundary, 209
Bounded, 47, 67, 158
    variation, 113

Chain rule, 122, 169, 176 f.
Characteristic, 29
Choice, principle of, 76 f., 161
Chord, 90
    semi-, 94
Circle, 246
Class, 40
    left, 68
Closure, 158
Codifferentiation, 286
Coefficients
    binomial, 137
    sequence of, 100
    Taylor, 132
Column, unit, 208
Commutativity, 14
Compact, 158
Comparativity, 11
Complement, 25
    ortho-, 283
Complete, 63, 160
    additivity, 252, 256
Complex
    associated, 107
    infinite, 107
    summable, 107
Composition, 29, 82
    inner, 274, 283
    outer, 283
Concave, 128
Concentrated, 46
Configuration, 10
Connected, simply, 198
Connective, 14
Consequent, 7
Constant, 6
    partially, 183
Construction, 6, 8
    language, 40 f.
    rule of, 6, 13
Content, 279
Continuity
    rational, 76
    uniform, 73, 164, 175

Continuous, 72
  at, 75, 164, 174
  jump-, 110, 189
  left-, 75
  right-, 75
Continuously differentiable, 120, 168
Contravariant, 223, 275
Convergence
  absolute, 104
  criterion, Cauchy, 62, 160
  interval, 103
  uniform, 99, 176
Convergent, 50, 98
  improperly, 52
Convex, 124
Coordinate
  system, 93, 201; affine, 274; cartesian, 201, 273
  transformation, 202 f.
  value, 93 f.
Coordinates, 200, 203
  polar, 178
Core, open, 158
Covariant, 222, 275
Covering, 161
  theorem, 162
Cube, 158 f.
Curve, 245
  boundary, 248
  closed, 247
  segment, 205; simple, 203

Decomposition, 246, 252
Decreasing, 48, 79
Definite, 10, 39
Definiteness, rational, 74
Definition, 9
  inductive, 30 ff.
Definitional schema, 30
Density, 63
Denumerable, 36
Dependency, 72
Dependent, 181
  linearly, 182
Derivative, 116, 212
  determinant, 177
  higher-order, 120
  mixed, 184
  partial, 167, 227
Description, 27
  pseudo-, 27
  term, 27
Determinant
  derivative, 177
  transformation, 206
Diagonal argument, Cantor's, 35
Diagonal procedure, Cauchy's, 62
Difference, 25
  mixed, 184
  quotient, 115; mixed-, 184
Differentiability
  double continuous, 168
  free, 120
  n-fold continuous, 168
Differentiable, 115, 205
  continuously, 120, 168; n times, 168; twice, 168; up to the nth order, 168
  left-, 121
  mixed-, continuously, 185
  partially, 167
  right-, 121
  total, 167, 176
Differential, 213, 218, 223
  form, 215, 219; content, 278; generalized, 290; total, 220, 240
  geometry, 203
  operator, 286
  quotient, 115, 213, 223; partial 227
Differentiation
  alternating, 235 ff.
  co-, 286
  member-by-member, 135
Disjoint, 26
Distributivity, 14
Divergence, 287
Division, 14
  of a line, 200 f.
Domain, 156, 158
  argument, 71
  closed, 158
  connected, 157
  normal, 268
  open, 157, 253
Dots, 101

Enumeration, 35 f.
Equal, 17

Equality, 8, 17
angular, 92
Equivalence relation, 17
Everywhere, ω-, 119
Existential quantifier, 15
Expression, indeterminate, 52

Field, 42
Archimedean, 60
Figure, 6, 251
simple, 206
Finally, 46
Formula, 24
Gauss, 250, 289
induction, 33
Leibniz, 193 f.
reduction, 272
Stokes, 250, 268, 270 ff., 289
Taylor, 145
transformation, 192, 244, 250, 267
Fraction, 16
Function, 21, 72
analytic, 101
arc, 97
basic elementary, 83
circle, 89
constant, 83
elementary, 84
exponential, 87
fractional-rational, 85
general linear, 84
identity, 84
integral, 143
integral-rational, 84 f.
inverse of, 81
limit, 98
linear, 84
logarithm, 89
many-place, 28, 155
number, 200, 205
partially constant, 183
point, 200, 205
power, 87
primitive, 143, 188
principal, 97
projection, 166
rational, 85
as restricted to, 74, 231
sequence, 98
sine, 90
step, 109, 188
sub-, 74, 230
subsidiary, 97
tangential, 115
trigonometric, 89, 97
vector, 222
Functional, 213
Fundamental theorem, 148, 191

Geometry
differential, 203
Riemannian, 291
Global, 80
Gradient, 286
Graph, 28

Identical, 8, 17
Identity, 8 f.
laws of, 11
principle of, 10
Increasing, 46, 79
properly, 47
Indefinite, 10, 35
Induction
analytic, 69
arithmetical, 8
formula, 33
Inequality, 9
Infinity point, 86
Integer, 20
Integrable, 143
Integral, 143
boundary, 248
curve, 244
geometric, 244, 248
mixed, 186
numerical, 244, 248
remainder, 145
Integration, 186
member-by-member, 154
partial, 151
by substitution, 152
Intermediate-value theorem, 79
Intersection, 25
Interval, 54 f., 156
closed, 44, 156
convergence, 103
open, 55, 156
sum, 253
Intuitionism, 33, 77
Invariant, 16, 233

Inverse
of a function, 81
mapping, 177
Invertible, 80
locally, 178
Involutory, 284
Isomorphic, 65
Isotony, 144

Jump, 110

Kronecker symbol, 273

Language, 35 f.
construction, 40 f.
Lattice, 266
σ-, 266
Left
class, 68
-differentiable, 121
-hereditary, 69
-normalized, 111
Length, 159, 275, 280
Limit, 50
iterated, 163
left-hand, 75
partial, 163
rational, 76
right-hand, 75
total, 163
Line segment, 274
Linear, 173
Local, 77
Logarithm, 89
natural, 126
Logic
classical, 14 f.
constructive, 15, 34, 77
Lower, 276

Majorant, 102
Manifold, 210, 267
Mapping, 172 f.
affine, 202
from, 173
into, 173
inverse, 177
locally affine, 202
of, 173
onto, 173
topological, 203

Marker, place, 44, 84
Matrix, 173 f.
derivative-, 176
orthogonal, 201
Mean-value theorem, 118
Measure, 254, 284
Member, 24, 43
Metric, 278
Monotonic, 48, 79, 255
Monotony, 19
theorem, 144
Multiplication, 13
alternating, 233 f.
Multiplicity, 107

Neighborhood, 74, 157
deleted, 75
one-sided, 75
Neighboring, 198
Nonidentity, 9
Norm, 159
topologically equivalent, 159
Normal, unit, 289; tensor, 289
tensor, 289
Normal domain, 268
Notation
analytic, 223
symbolic, 223
*n*tuple, 26
Number, 6, 10 f.
complex, 92
Euler, 126
natural, 6
negative, 18 ff.
positive rational, 17
real, 56
whole, 20
Numeral, 6

Object, 9
abstract, 17
Open, almost, 265
Operator
differential, 286
ϵ-, 24
ι-, 21
Orientation, 204, 206
coherent, 210
convention, 210; canonical, 210

Orthocomplement, 282 f.
Orthogonality, 273, 280
Outward, 210

Parameter, 194, 206
Partial sum, 100
Particle, logical, 7
Partition, 187 f., 201
Path-independence, 247
Peano axioms, 9
Period, 97
Place marker, 44, 84
Poincaré's theorem, 240
Point, 75, 93, 200, 203
  boundary, 158, 204, 209
  end, 204
  exterior, 158
  image, 172
  infinity, 86
  initial, 204
  interior, 158
  limit, 160
  original, 172
  unit, 173
Polynomial, 85
  quotient, 85
  Taylor, 132
Product
  inner, 274
  outer, 282
Proof, indirect, 50
Pseudodescription, 27

Quantifier, 7
  existential, 15
  universal, 14
Quotient
  difference, 115
  differential, 115, 213, 223
  mixed-difference, 184
  partial differential, 227
  polynomial, 85

Raise, 276
Range of values, 9
Reduction, 85
  formula, 272
Refinement, 246, 252
Reflexivity, 11
Region
  closed, 158
  open, 157
Relation, 26
  equivalence, 17
  ordering, 18, 59
Remainder, 131 f.
  estimate: Cauchy's, 133;
    Lagrange's, 146
Removable, 110
Representable, 133
Representation, reduced, 85
Riemannian
  geometry, 291
  space, 278
Rolle's theorem, 118
Rotation, 286

Saltus, 110
Segment
  curve, 205
  line, 274
  simple curve, 203
  surface, 211
Sense, sameness of, 204, 206
Sequence, 43
  of coefficients, 100
  double, 62
  function, 98
  natural, 47
  null, 51
  sub-, 47
Series
  double, 104
  Leibniz, 142
  power, 100
  Taylor, 133
Set, 24
  associated, 63
  empty, 26
  finite, 25
  Jordan null, 264
  Lebesgue null, 264
  power, 39
  proper sub-, 26
  sub-, 26
  theory, axiomatic, 37
Sign, 10
Simple, 198
Simply connected, 198

Simultaneously, 231
Smooth, 207, 211 f.
Solution, approximate, 43
Space
  Euclidean, 273
  number, 156
  Riemannian, 278
  tangent, 206 ff.
Statement, 6, 24
  arithmetical, 10
  equality, 8
  form, 24
  identity, 8
  variable, 10
Subfunction, 74, 230
Subjunction, 7
Subsequence, 47
Subset, 26
  proper, 26
Subtraction, 14
Sufficiently large, 45
Sum, 100, 107
  Boolean, 25
  combinatory, 251, 264, 271
  interval, 253
  partial, 100
Summation convention, 221
Superadditive, 88
Surface
  area, 149
  segment, 211
Symbolic notation, 223
Symmetry, 12
System, 25
  member, 26

Tangent, 115
Tensor, 222, 276
  analysis, 291
  content, 281
  ε-, 281
  fundamental, 275, 277
  unit normal, 289
  unit tangent, 288
Term, 13
  constant, 13
  description, 27
  elementary, 84
  ι-, 27
  μ-, 27
  rational, 21
Topological, 158
Topologically equivalent, 159
Topology, 203
Total differential form, 220, 240
Totality, 19
Transitivity, 17, 19
Transformation
  determinant, 206
  formula, 192, 244, 250, 267
  law, 215
Transposed, 219
True, 6
Types, theory of, 3

Union, 25, 264
Universal quantifier, 7, 14

Valid, 6
Value
  absolute, 92, 275, 284
  coordinate, 93
  of a function, 21
  of a variable, 9
Variable, 6
  free, 23
  range of, 9
  statement, 10
Variation, bounded, 113
Vector, 93, 207, 274, 276
  analysis, 286
  basis, 208
  multi-, 234

INDEX OF NAMES

Archimedes, 44, 50, 90, 150, 273
Aristarchos, 128
Aristotle, 201

Bernoulli, 22, 145
Bolzano, 121
Boole, 25, 256, 266
Borel, 255 f.
Bourbaki, 2
Brouwer, 33

Cantor, 35 f.
Cartan, E., 5, 200
Cauchy, 46, 62, 99, 133, 137, 160

Dehn, 273
Dieudonné, 4

Euclid, 44, 50, 201, 204, 273
Eudoxos, 44, 50, 201, 273
Euler, 126

Fibonacci, 55

Gauss, 250, 289 ff.
Gentzen, 33
Grassmann, 284
Gregory, 142

Hilbert, 33, 37

Jordan, 264 f.

Kant, 54
Klein, 202
Kronecker, 273

Lagrange, 146
Laplace, 233
Lebesgue, 264, 267
Leibniz, 5, 22, 142, 150, 194 ff., 199 f., 203, 214, 244, 273
Liouville, 150

Möbius, 210

Newton, 150, 273

Peano, 2 f., 9, 21, 24, 27
Poincaré, 5, 240 f., 248, 270, 286
Ptolemy, 128

Riemann, 5, 278, 291

Stokes, 1, 5, 193, 250, 268 ff., 289

Taylor, 1, 100, 131 ff., 145 f., 154

Wallis, 52
Weyl, vii, 37